Molecular Engineering

Other McGraw-Hill Chemical Engineering Books of Interest

AUSTIN ET AL. • *Shreve's Chemical Process Industries*

CHOPEY, HICKS • *Handbook of Chemical Engineering Calculations*

COOK, DUMONT • *Process Drying Practice*

DEAN • *Lange's Handbook of Chemistry*

DILLON • *Corrosion Control in the Chemical Process Industries*

FREEMAN • *Hazardous Waste Minimization*

FREEMAN • *Standard Handbook of Hazardous Waste Treatment and Disposal*

GRANT, GRANT • *Grant & Hackh's Chemical Dictionary*

KISTER • *Distillation Operation*

MILLER • *Flow Measurement Engineering Handbook*

PERRY, GREEN • *Perry's Chemical Engineers' Handbook*

REID ET AL. • *Properties of Gases and Liquids*

REIST • *Introduction to Aerosol Science*

RYANS, ROPER • *Process Vacuum System, Design, and Operation*

SANDLER, LUCKIEWICZ • *Practical Process Engineering*

SATTERFIELD • *Heterogeneous Catalysis in Industrial Practice*

SCHWEITZER • *Handbook of Separation Techniques for Chemical Engineers*

SHINSKEY • *Process Control Systems*

SHUGAR, DEAN • *The Chemist's Ready Reference Handbook*

SHUGAR, BALLINGER • *The Chemical Technician's Ready Reference Handbook*

SMITH, VAN LAAN • *Piping and Pipe Support Systems*

STOCK • *AI in Process Control*

TATTERSON • *Fluid Mixing and Gas Dispersion in Agitated Tanks*

YOKELL • *A Working Guide to Shell-and-Tube Exchangers*

Molecular Engineering

Henry A. McGee, Jr.

Chemical Engineering Department
Virginia Polytechnic Institute and State University
Blacksburg, Virginia

McGraw-Hill, Inc.

New York St. Louis San Francisco Auckland Bogotá
Caracas Hamburg Lisbon London Madrid
Mexico Milan Montreal New Delhi Paris
San Juan São Paulo Singapore
Sydney Tokyo Toronto

Library of Congress Cataloging-in-Publication Data

McGee, Henry A.
Molecular engineering / Henry A. McGee, Jr.
p. cm.
Includes bibliographical references.
ISBN 0-07-044977-5
1. Chemical engineering. 2. Chemical reactions. 3. Molecular theory. I. Title
TP149.M38 1991
660—dc20 90-22630
CIP

1 2 3 4 5 6 7 8 9 0 DOC/DOC 9 7 6 5 4 3 2 1

ISBN 0-07-044977-5

The sponsoring editor for this book was Gail F. Nalven, the editing supervisor was David E. Fogarty, and the production supervisor was Pamela A. Pelton. This book was set in Century Schoolbook. It was composed by McGraw-Hill's Professional Book Group composition unit.

Printed and bound by R. R. Donnelley & Sons Company.

To Dr. Waldemar T. Ziegler ("Wal"), Regents Professor of Chemical Engineering (retired), Georgia Institute of Technology, who introduced the author and so many more of his students to the power and beauty of the molecular perspective in engineering

Contents

Preface

Scientific understanding and engineering design have increasingly become parallel rather than sequential activities. Applied science seems inevitably multidisciplinary, and practitioners of each discipline must enjoy real scientific communication if we are to ensure optimum progress. Engineers typically have well-developed continuum points of view, but their molecular perspective is usually weak.

The language of all of the physical sciences is the language of molecules, and if the applied scientist or engineer is to communicate with physicists or with chemists, the engineer must speak the language of molecular orbitals, energy states, vibrational modes, and all the rest. Molecular theory forms the common language of both pure and applied science. Even modern techniques of analysis such as infrared, nuclear magnetic resonance (NMR), x-ray diffraction, or coherent anti-Stokes Raman spectroscopy (CARS) can be understood as analytical schemes only in terms of the molecular behavior of matter. The engineer then is unable to judge the suitability of some instrumental scheme without such insights.

The structural and dynamic properties of a molecule are described by quantum mechanics. But the behavior of 10^{23} of these molecules is described by statistical mechanics, which is itself divided into two parts—equilibrium statistical mechanics or statistical thermodynamics and nonequilibrium statistical mechanics or kinetic theory. The macroscopic world presents a complex array of phenomena and processes. All of these may be divided into either equilibrium processes or rate processes, that is, into either thermodynamic or transport-kinetic processes and properties. Macroscopic descriptions of thermodynamic and transport-kinetic processes inevitably involve phenomenological coefficients such as heat capacity, viscosity, or specific rate coefficient. Statistical thermodynamics is the bridge from the properties of one molecule to the value of these phenomenological coefficients, that is, the thermodynamic properties of 10^{23} of those molecules. Similarly, kinetic theory is the bridge from the description of individual molecular collisions to the transport and kinetic coefficients of 10^{23} molecules.

We understand complex macroscopic pheonomena as mere artifacts of innumerable molecular events. If we can understand the energetics, the structures, and the interactions of molecules, we will be able to understand the bulk behavior of matter. More important from an engineering perspective, this understanding allows the engineer to manipulate molecular behavior to yield some desired macroscopic property or process. This is the definition of molecular engineering.

This book is concerned with statistical thermodynamics and kinetic theory with an assumed starting point of basic chemistry, physics, and mathematics through some aspects of ordinary and partial differential equations. This book explores in detail the molecular understanding and the computation of phenomenological coefficients, and each idea and technique is illustrated with many examples. The goal is to build a sufficient scaffold to see over into more esoteric areas and to develop a manipulative competency with useful calculational schemes. The goal is to be able to converse with intelligence and understanding with the chemical physicists or the physical chemists about new developments in their more esoteric areas and to then be able to make an independent judgment as to the value of the new insight or the new theory in one's own particular applied problem area. With each topic, the reader is taken to the point of greatly diminishing returns regarding applicability to real problems. As quickly as possible after presenting some inevitably abstract idea, I have tried to present applications of physical meaning or intuitive feel. All of this is designed to help make the abstract idea real.

Molecular partition functions are used throughout for ease of comprehension rather than the more general, but conceptually troublesome, ensemble approach. Molecular theory is here demystified.

The following pages contain most of what the applications-oriented person needs to know about statistical thermodynamics, kinetic theory, and chemical reaction dynamics. There is nothing herein that must be unlearned or modified if one should later elect to become expert in this subject matter. The reader is, however, completely dependent on my judgment as to the position of many dividing lines that are drawn between the "useful" and the "esoteric." Here one must make the leap of faith and then see if, in one's subsequent practice, he or she will have found my judgments to be accurate or wanting. The theoretical arguments in this book are terribly impressive, and they are extraordinarily valuable to the applied scientist, but they are also frequently incomplete and inexact. If it were not so, there would be computer codes that would calculate any property of any substance under any experimental condition. This happy state of affairs does not yet exist. We must learn herein the boundaries of what existence there is.

We here seek the development of attitudes and perspectives in thermodynamics, transport, and rate phenomena that would be essential to persons concerned with the invention and design of perhaps a laser-based process for isotope separation. Or one might be concerned with fundamental problems in combustion leading to greater fuel efficiencies and less pollution. Or one might be concerned with ion implantation for the development of new alloys or new and unusually doped materials of interest in electronics. Or one's application may require some property of matter that may be unknown. The applied scientist must work with too many substances, both pure and in an infinite array of mixtures, for experimental measurement to always be a viable route to a needed property. So theories and predictive correlations are needed, and the best of them are based on molecular arguments. From whatever the perspective, a molecular view is essential and a purely traditional or classical perspective unacceptably slows invention, hinders creativity, and frustrates original design.

This book will have served its purpose if its attitudes can be internalized. That is, long after we have forgotten just exactly how this or that particular argument or calculation goes, we will nonetheless instinctively think about any applied problem that we face in terms of what the molecules must be doing. That is the real, bottom-line goal.

Henry A. McGee, Jr.

Molecular Engineering

Chapter 1

Introduction

The goals of this discussion are to develop all of the current molecular theory that is applicable to the calculation of values for each of the thermodynamic, transport, and kinetic properties. We wish to understand these theoretical notions, but to limit our scope to only those ideas of practical utility. We will herein take our discussion of principles to some point that we will arbitrarily define as no longer practical in order that we may provide a vantage point to gaze over the ridge into those more uncertain or tentative areas that may yet someday become a part of our practical vocabulary. The applied scientist who understands the following arguments can well converse with the theorist and thereby reach a wise decision as to the value or relevance of any new idea to the real-world problems with which the applied scientist must deal. From a utilitarian perspective, the following discussion is a comprehensive one.

All of the relationships of classical thermodynamics are absolutely true whether molecules exist or not. But molecules do exist, and intuitively we believe that if we could but understand molecular structure, molecular energetics, and intermolecular interactions, we should somehow be able to calculate numerical values for thermodynamic properties. Statistical thermodynamics is the science that addresses this problem, and some strikingly successful theory has been developed. Our goal then is the calculation of numerical values of heat capacity, entropy, enthalpy, etc. at all temperatures and pressures for all substances. The input data will be bond angles, bond distances, intramolecular vibration frequencies, atomic masses, and the like. Many of these molecular parameters can be measured with very high accuracy.

Extensive compilations of the thermodynamic properties of matter have been developed by using statistical thermodynamical formalism, and once we understand this formalism, we can then use existing and tabulated data with confidence, as well as calculate the properties of heretofore unstudied species in which we might have a particular interest.

There will be an additional more philosophical or attitudinal result from our study of molecular engineering that may be more valuable than the computational skill itself. Classical thermodynamics offers no explanation of why matter has the properties that it does, but we will now see that the macroscopic properties of matter have an underlying form and structure that are based on the molecular properties of matter. We will see that the various macroscopic properties are really just artifacts of the molecular character of the species. For example, we can use data on thermal conductivity to deduce the character of the intermolecular interaction from which we can then deduce the equation of state. Or vice versa. Thus the transport properties and the thermodynamic properties are linked through the molecular properties. Or we can make optical measurements of vibration frequencies and from those frequencies calculate the calorimetric quantities of heat capacity or entropy. Thus optical and calorimetric properties are linked through the molecular properties. Or we can measure the velocity of sound in a solid and from that calculate the heat capacity of that solid. Thus acoustical and calorimetric properties are linked through the molecular properties. And so on. These are powerful insights, and they are essential if one would hope to comprehend the physical world.

Our development will be divided into two parts. The introductory arguments of Chap. 2 are easy to follow, and they result in the introduction of the Boltzmann distribution function,

$$n_i = \frac{n e^{-\epsilon_i/kT}}{\sum_i e^{-\epsilon_i/kT}} \tag{1.1}$$

and the association (note, not derivation) of the macroscopic concept of entropy with microscopic statistics,

$$S = k \ln W \tag{1.2}$$

The meaning of the terms in these relationships will be made abundantly clear subsequently.

We will see too that molecules will distribute themselves among a set of energy levels in this exponential manner through the action of no force other than that of pure chance. In systems at thermodynamic equilibrium, the odds are overwhelming that the distribution will be exponential. But other unnatural distributions can be produced for short times, and indeed, the very important laser and maser phenomena depend totally on the production of a so-called population inversion, that is, nonexponential distributions.

Although these arguments are all simple, one may also with impunity accept the two relationships of Eqs. (1.1) and (1.2) as valid, omit Chap. 2 altogether, and move immediately to the following chapters where these two results are utilized in the calculation of the thermodynamic, transport, and kinetic behavior of matter.

The second part of this book is concerned with nonequilibrium properties and behavior. Much of engineering concern with nonequilibrium behavior relates to the translational motions of molecules. Translation is classical rather than quantum in nature, so the Maxwell-Boltzmann distribution, Eq. (1.1), is recast in classical form and then used to develop what is called the kinetic theory of gases. These ideas allow a good conceptual understanding of the transport properties in all situations as well as useful schemes for calculating actual values of the transport coefficients for gases at up to modest pressures.

Kinetic theory also allows insight into chemical reaction kinetics. Chemical reaction does not occur in one step as we typically write an overall stoichiometric change. Rather, reaction occurs by a complex array of usually bimolecular encounters which together constitute the reaction mechanism. The rate of each of these steps of the mechanism depends on its particular reactant collision frequency, the relative energy involved in the collision, the energy states of each colliding species, and the relative geometry of the colliding species at the moment of impact. Reaction occurs only in collisions that occur with above some minimum threshold energy and even then only in collisions that occur with certain geometric orientations. Finally, the macroscopic or observed rate of the overall stoichiometric change is some sort of a complex average of these many different microscopic events.

Whether concern is with physical or chemical rate processes, we imagine all macroscopic or empirical behavior to be artifacts of innumerable microscopic events. We must understand these individual molecular events, and then we must properly average these events to deduce the macroscopic behavior that is always sought in engineering practice.

Chapter

2

Statistical Background

Energy is quantized, and the science of quantum mechanics gives us an array of energy levels for each type of molecular motion including, for example, the translational motion of the molecules of a gas confined to a volume V. Given these sets of energy levels from quantum mechanics, statistical thermodynamics is that branch of science that is concerned with determining the distribution of any number, say Avogadro's number, of molecules within that set of energy levels, and from that distribution to then calculate, say, the enthalpy of the system in joules per mole as well as values for all of the other thermodynamic properties. We will see that the molecules are irresistibly driven to an exponential distribution with most molecules in the lowest energy level. And this is by the action of no force other than that of simple chance.

Consider, for example, an array of levels spaced e units of energy apart, and let us assume that we have three molecules which are to share $3e$ units of energy. How will the molecules be distributed? The arrangement of Fig. 2.1*a* can occur in only one way, that of Fig. 2.1*b* can occur in three ways by simple permuting of the distinguishable molecules,† and that of Fig. 2.1*c* is similarly seen to occur in six ways. Let us also assume that there is free interchange of energy among the molecules by collisions. Then if we observed this system over a period of time, we would expect to see these 10 distributions with equal frequency, and hence 60 percent of the time, the distributions will be as depicted in Fig. 2.1*c*.

If the same three molecules share twice as much energy, 6 units, there will be 7 different arrangements rather than 3, and these can occur in a total of 28 different distributions rather than 10. The occu-

†The question of distinguishability is an important one. For now, let us take the molecules to be distinguishable, and we will see later the implications of relaxing this idea.

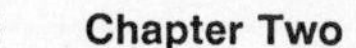

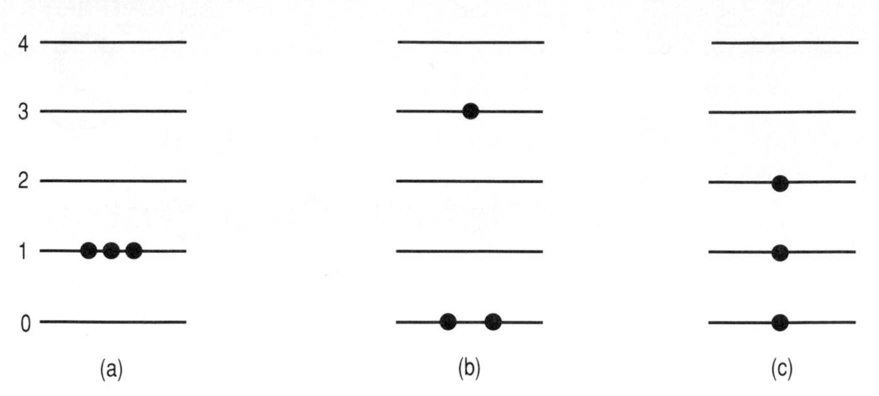

Figure 2.1 Distribution of three molecules containing $3e$ units of energy.

pation of the energy levels will always decrease with increasing energy of the level because one molecule can have a relatively high energy only if two molecules have a relatively low energy.

To enumerate the possible distributions for a group containing more than three or four molecules having more than three or four units of energy is too long a task. Let us merely assert that the distribution that one would find in nature would surely be that one that could occur in the greatest number of ways. Thus, if a set of 20 molecules shares 20 units of energy in a way that can occur 380 ways, we can safely imagine this system moving to another distribution that can occur 29,070 ways, and then to another distribution that can occur 10^5 ways, and so on. If each arrangement is *a priori* equally likely—i.e., let us take each of the 10 possibilities of Fig. 2.1 as equally likely—then the distribution of Fig. 2.1*c* will be the most likely. Then clearly the chance that our larger system will be in a distribution that can occur 380 ways will be slim compared to an arrangement that can occur 10^5 ways. When extended to very large groups of molecules, say 10^{23} molecules, there are so many distributions that have essentially the same shape that all others may be neglected, as we shall see momentarily.

The determination of the number of ways, w, that a particular distribution can arise is the familiar combinatorial problem of how many ways n distinguishable objects can be placed in p piles. If we would arrange n objects (or molecules) in a line, we have n choices for the first, $(n - 1)$ choices for the second, and so on. Thus we can produce $(n)(n - 1)(n - 2)\cdots(3)(2)(1)$ or $n!$ arrangements. But if the first n_0 of these are in one pile, then the permuting of each of these in the original selection will yield the same distribution. So the original listing must be reduced by a factor of $n_0!$ And this may be repeated for each of the piles. Thus the total ways that a particular distribution of n objects into p piles can arise is given by

$$w = \frac{n!}{n_0!n_1!n_2!\cdots} \tag{2.1}$$

where there are n_0 objects in the zeroth pile, n_1 in the first pile, n_2 in the second pile, and so on over all of the piles. Of course $\Sigma n_i = n$.

An example may be instructive. If we have 6 molecules arranged as shown in Fig. 2.2*a*, the number of different ways this can occur is 6! However, in Fig. 2.2*b*, there are 6 choices for the molecules to go into level 3, 5 choices to go into level 2, 4 to go into level 1, but all 3 remaining molecules must go into level 0. Thus $w = 6!/1!1!1!3!$ or $6 \times 5 \times 4$ or 120. In Fig. 2.2*c* the result is $w = 6!/1!2!3!$ or 60. We have then derived a simple way to calculate the number of ways that any distribution of molecules among a set of energy levels may occur. And we have previously asserted that the distribution that will actually occur is that for which w will be a maximum. In the above case, the array of 6 molecules will be in the arrangement in Fig. 2.2*a* by a fraction 720/(720 + 120 + 60) or 80 percent of the time, or with a probability of 0.8.

Recall that quantum mechanics gives us a set of energy levels and that it is this discontinuity of energy, rather than the previously accepted smooth continuity, that is the essence of quantum mechanics. Quantum mechanics says nothing about the occupation of these levels

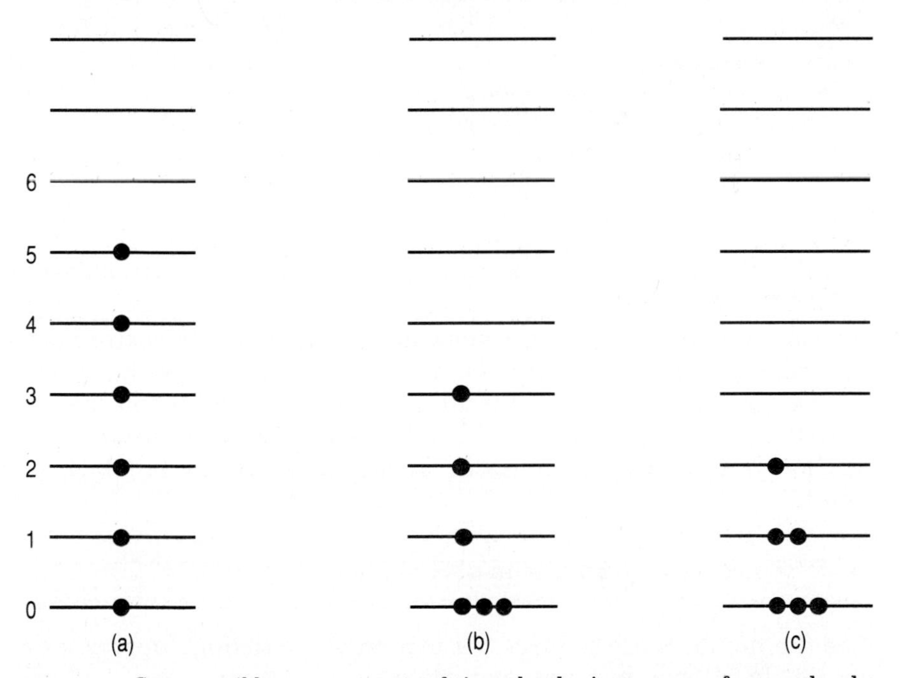

Figure 2.2 Some possible arrangements of six molecules in an array of energy levels.

by real molecules. This is rather a question of statistics, and this is precisely the question that we are here examining.

Fortunately we can readily derive a general expression for that distribution that will maximize w by use of the common mathematical technique of lagrangian undetermined multipliers. The langrangian technique, for example, is commonly used to determine the equilibrium composition of a phase in which several chemical reactions are occurring simultaneously. Equilibrium is at those concentrations of each species which together minimize the Gibbs free energy G of the phase. So we merely write an expression for G as a function of the concentrations and minimize the function with respect to each of the concentrations but under the constraints of fixed total moles of each atom.

Example Problem 2.1 As a simple example of the lagrangian technique, determine those values of x and y that maximize the area of a rectangle subject to the constraint of a constant perimeter.

$$A = xy$$

$$2x + 2y = p = \text{constant}$$

Then the lagrangian sum function becomes

$$S = xy + \lambda(2x + 2y - p)$$

Upon differentiation,

$$\frac{\partial s}{\partial x} = y + 2\lambda = 0$$

$$\frac{\partial s}{\partial y} = x + 2\lambda = 0$$

or $x = y$, since λ is a constant. The rectangle of maximum area, but with fixed perimeter, is then a square.

The lagrangian technique is a general one for finding the extremum of a function subject to constraints, and in the case of w, we have two constraints—that of a constant total number of molecules and that of constant total energy. We seek then that set of occupation numbers, i.e., the n_i's that will maximize w, or just as easily $\ln w$, which is given by

$$\ln w = \ln n! - \sum_i \ln n_i!$$

Since the numbers are so large, we can employ Stirling's approximation, i.e., $\ln n! \cong n \ln n - n$, and write

$$\ln w = n \ln n - \sum n_i \ln n_i$$

The constraint relations are

$$\sum n_i = n$$

and

$$\sum n_i \epsilon_i = U$$

where n and U are constants; that is, we are concerned with a fixed number of molecules, which together share a fixed energy. From this problem statement, all textbooks on statistical thermodynamics show the straightforward lagrangian scheme which ends with w(max)† being given by an exponential distribution,‡

$$n_i = \frac{ne^{-\mu\epsilon_i}}{\sum e^{-\mu\epsilon_i}} \tag{2.2}$$

where μ is a positive constant.

Consider for a moment equally spaced energy levels where the exponential distribution is also a geometric progression, for clearly then

$$\frac{n_1}{n_2} = \frac{n_2}{n_3}$$

The fact that numbers in a geometric progression have the largest possible value of W is made readily acceptable by considering any three such numbers, say 3000, 300, 30, describing the occupation of three equally spaced energy levels. We will shift 5 molecules from the middle level to the upper level and similarly 5 molecules to the lower level, which will keep the total energy constant; that is, the new distribution is 3005, 290, 35, which is only slightly changed. But yet the value of W is now down by a factor of almost 2:

$$\frac{W(\text{after})}{W(\text{before})} = \frac{3000!\ 300!\ 30!}{3005!\ 290!\ 35!}$$

$$= 0.533$$

But such occupation numbers are much too small to be of any significance. So let us consider a situation more reasonable from a molecular perspective—one in which the distribution is 10^{20}, 10^{19}, and 10^{18}, that is, still in a geometric progression—and let us again move molecules in equal numbers from the upper and lower levels and into the

†We will consistently use the upper case W when referring to the exponential distribution.

‡This development is reproduced in App. A.

middle level so as to maintain the energy constant. How many molecules have been moved when W has fallen to one-thousandth of its maximum value? Let us again move molecules up and down in equally spaced energy levels such as to maintain a constant total energy. Thus, after the shift of p molecules, the ratio of W before and after the shift is again

$$\frac{W(\text{after})}{W(\text{before})} = \frac{n_0!n_1!n_2!}{(n_0 - p)!\,(n_1 + 2_p)!\,(n_2 - p)!}$$

Using these realistic values for n_0, n_1, and n_2, and taking the ratio of W(after)/W(before) to be 0.001, analysis reveals that p is about $\sqrt{n_2}$ or about 10^9. Or the distribution that reduces W by a factor of 1000 is different from the geometric progression by about 1 part in 10^8, a totally insignificant amount. With still larger occupation numbers, even smaller perturbations from the geometric distribution will produce corresponding decreases in W.

None of these arguments depend on our having equally spaced energy levels. If we move one molecule from level j to level k, the ratio of W after the change to W before the change is just

$$\frac{W(\text{after})}{W(\text{before})} = \frac{n_j!\,n_k!}{(n_j - 1)!\,(n_k + 1)!} \cong \frac{n_j}{n_k} \tag{2.3}$$

where n_j and n_k are very large compared to unity. Similarly, if we move q molecules, provided only that q is of the order of $\sqrt{n_j}$ (where n_j is less than n_k) or less,

$$\frac{W(\text{after})}{W(\text{before})} = \left(\frac{n_j}{n_k}\right)^q \tag{2.4}$$

Now imagine three levels unevenly spaced as in Fig. 2.3, so that we must move p molecules down a units of energy and q molecules up b units to maintain the condition of constant energy

$$ap = bq$$

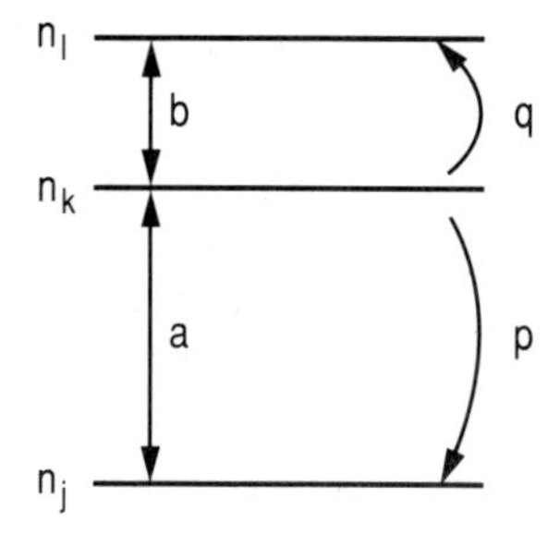

Figure 2.3 Shifting molecules among three unevenly spaced energy levels.

According to the relationship just derived for a geometric progression, the ratio of the values of W before and after the shift indicated in Fig. 2.3 would be

$$\frac{W(\text{after})}{W(\text{before})} = \frac{n_k^{p+q}}{n_j^p n_l^q}$$

provided only that the number of molecules moved, $p + q$, is small relative to the square root of the occupation numbers themselves, n_j, n_k, and n_l.

If we assume the occupation numbers of the three levels to be in a geometric progression with the lowest level assigned $\epsilon_0 = 0$, then, as we have seen, the occupation numbers will be given by

$$n_i = \frac{ne^{-\mu\epsilon_i}}{f}$$

where for ease of writing we have replaced $\Sigma e^{-\mu\epsilon_i}$ by f, and W(after)/W(before) will be equal to unity, that is, W will have its maximum value if

$$n_k^{p+q} = n_j^p n_l^q$$

$$\text{or} \quad \left(\frac{n}{f}\right)^{p+q} e^{-\mu\epsilon_k(p+q)} = \left(\frac{n}{f}\right)^p e^{-\mu\epsilon_j p}\left(\frac{n}{f}\right)^q e^{-\mu\epsilon_l q}$$

which demands only that

$$\epsilon_k(p + q) = \epsilon_j p + \epsilon_l q$$

Now from the given spacing of the energy levels, $\epsilon_k = \epsilon_j + a$ and $\epsilon_l = \epsilon_j + a + b$, or

$$(\epsilon_j + a)(p + q) = \epsilon_j p + (\epsilon_j + a + b)q$$

which reduces to $ap = bq$. But this is just the constraint of constant energy under which the molecules were moved in the first place. We see then that the geometric distribution that maximized W will equally well describe systems whatever the energy level spacing may be.

With the zero of energy defined at the zeroth energy level, i.e., $\epsilon_0 \equiv 0$, the denominator of the Boltzmann equation becomes

$$f = 1 + e^{-\mu\epsilon_1} + e^{-\mu\epsilon_2} + \cdots$$

This infinite series is called the *partition function* or the *sum over states*, and it is not possible to overemphasize its importance in relating microscopic characteristics to macroscopic properties. As we shall see, it will be possible to express the totality of classical thermodynamics in terms of this partition function.

To summarize, we have shown that (1) the most probable distribution of molecules in a set of energy levels is the exponential or Boltzmann distribution, (2) distributions even insignificantly off-exponential are so very unlikely as to be totally neglectable, and (3) the mechanism forcing this distribution is simple chance. The argument assumes that the molecules are distinguishable and that there is no limit to the number of molecules that can occupy a given energy level. These are the assumptions of so-called Maxwell-Boltzmann statistics. The molecules of a gas are indistinguishable, and we must examine this point. The molecules of a solid are also indistinguishable, but their lattice sites are labelable, and, as we shall see, Maxwell-Boltzmann statistics is applicable to solids. The limitation on the number of molecules that may occupy a single level involves the Pauli exclusion principle and leads to what is called Fermi-Dirac statistics, which need not concern us here.

Thermodynamics and Statistics

Imagine an amount of heat added to a total of n molecules occupying a set of energy levels. The energy can be absorbed by one molecule moving from ϵ_j to ϵ_k or by q molecules moving from ϵ_m to ϵ_n, where obviously $q(\epsilon_n - \epsilon_m) = (\epsilon_k - \epsilon_j)$. According to Eq. (2.4), the change in W for the movement of the single molecule is n_j/n_k, whereas for the movement of q molecules, the change in W would be $(n_m/n_n)^q$. If the distribution is an exponential one, $n_j/n_k = e^{-\mu(\epsilon_j - \epsilon_k)}$ and $(n_m/n_n)^q = e^{-\mu(\epsilon_m - \epsilon_n)q}$ are identical since $(\epsilon_j - \epsilon_k) = q(\epsilon_m - \epsilon_n)$. Thus for an exponential distribution, a change in W will depend only on the amount of energy added or removed and not at all on the number of molecules that are shifted or the particular energy levels between which they are shifted. Thus, moving one molecule from level i to level j will change W:

$$\ln W(\text{after}) - \ln W(\text{before}) = -\mu(\epsilon_i - \epsilon_j)$$

and moving however many molecules one at a time from any level i to any level j will result in a total change in W of

$$\ln W(\text{after}) - \ln W(\text{before}) = -\mu\sum(\epsilon_i - \epsilon_j) \tag{2.5}$$

where the sum is over all molecules that are moved between whatever energy levels. If the levels ϵ_j are higher in energy than the levels ϵ_i, we say an amount of heat δQ, equal to the sum of all these shifts, has been added to the system, or since δQ is just $\Sigma(\epsilon_i - \epsilon_j)$, we can rewrite Eq. (2.5) as

$$d \ln W = \mu \, \delta Q \tag{2.6}$$

Recall that d means differential of a state function (independent of path), while δ means a line differential (depends on path). In some cases, the addition or removal of energy from the system can change the spacing of the energy levels. Equation (2.6) continues to hold for such systems as well as for situations of fixed energy levels, and the equation describes the change in W for any system undergoing positive, negative, or zero heat additions. We note statistically that all systems are driven to that configuration that can occur in the maximum number of ways, i.e., to the configuration of maximum W. On the other hand, we also recall thermodynamically that all systems are driven to that state of maximum entropy. Furthermore, entropy is an extensive property; that is, its values for two systems are additive, whereas W for two systems is multiplicative. This multiplicative character is intuitively clear, for W is just the number of different ways that a set of occupation numbers can occur, and for every one such arrangement in system A, there will be W arrangements of system B. These observations suggest a simple proportional association between S and $\ln W$, which we write with Boltzmann† as

$$S = k \ln W \tag{2.7}$$

where k is a constant. Note that this relationship has not and cannot be derived. It is rather a sort of premise, and we will accept it more as we see the agreement with experiment that it affords. This relationship is enormously important, for it forms the bridge that connects statistical arguments to thermodynamics.

The meaning of the lagrangian constant μ may now be derived from this association of entropy and W, which in differential form may be written

$$dS = k \, d \ln W$$

and when substituted into Eq. (2.6) yields

$$dS = k\mu \, \delta Q$$

but since entropy is defined as $dS \equiv dQ/T$, we see that μ must be just $1/kT$. The set of occupation numbers that maximizes w is then

†This powerful insight is attributable to Boltzmann (1844–1906), and this relationship is inscribed on his tombstone in Vienna.

$$n_i = \frac{ne^{-\epsilon_i/kT}}{\sum e^{-\epsilon_i/kT}} \tag{2.8}$$

This is one of the most important relationships in all of science, for in it we see the fundamental origin of the exponential dependence of all macroscopically observed measures of the intensity of some process. For example, the vapor pressure depends on exp $(\Delta H_{vap}/RT)$, where ΔH_{vap} is the heat of vaporization; the rate of chemical reaction depends on exp $(-E_{act}/RT)$, where E_{act} is called an *activation energy*; the position of a chemical equilibrium depends on exp $(-\Delta G/RT)$, where ΔG is the change in free energy for the reaction; and so on throughout physical chemistry.

Suppose two separate systems, each at its own temperature and at its maximum value of W, are brought into thermal contact with each other. We might inquire as to the values of W in each system when equilibrium has been attained. For each system,

$$d \ln W_A = \frac{\delta Q_A}{kT_A}$$

and

$$d \ln W_B = \frac{\delta Q_B}{kT_B}$$

but $\delta Q_A = -\delta Q_B$ and at equilibrium $T_A = T_B$. Then $\delta \ln W_A + \delta \ln W_B = 0$. Therefore, at thermal equilibrium, we can assert that $\ln W_A W_B$ will be at its maximum value. We expect this result, and we indeed could have arrived at the same conclusion by realizing that W_A is independent of W_B, and equilibrium will surely correspond, as always before, to that condition of the combined system that can occur in the most number of ways. And this implies that the product $W_A W_B$ must be at its maximum value. We shall return to this point later when we discuss physical and chemical equilibrium, for, here too, not unlike thermal equilibrium, the potential for mass transfer and for chemical reaction will reach equilibrium when ΠW_i has its maximum value.

The Boltzmann expression for entropy in terms of W, Eq. (2.7), will be more useful if we can express W in terms of the partition function. Thus, since

$$W = \frac{n!}{\pi n_i!}$$

then, with Stirling's approximation and a geometric distribution, the above equation may be rewritten,

$$\ln W = n \ln n - n - \sum(n_i \ln n_i - n_i)$$

$$= n \ln n - \frac{n}{f}\sum e^{-\epsilon_i/kT} \ln\left[\left(\frac{n}{f}\right)(e^{-\epsilon_1/kT})\right]$$

$$= n \ln n - \frac{n}{f}\sum e^{-\epsilon_i/kT}\left(\ln n - \ln f - \frac{\epsilon_i}{kT}\right)$$

$$= n \ln f + \frac{1}{kT}\left(\frac{n}{f}\right)\sum \epsilon_i e^{-\epsilon_i/kT} \tag{2.9}$$

Now since the thermodynamic internal energy U is just

$$U = \sum n_i \epsilon_i$$

$$= \frac{n}{f}\sum \epsilon_i e^{-\epsilon_i/kT} \tag{2.10}$$

by combining Eqs. (2.9) and (2.10) we can write

$$\ln W = n \ln f + \frac{U}{kT}$$

or

$$S = nk \ln f + \frac{U}{T} \tag{2.11}$$

The internal energy is

$$U = \sum n_i \epsilon_i = \frac{n}{f}\sum \epsilon_i e^{-\epsilon_i/kT}$$

but this can be related to the partition function by first forming the derivative

$$\left(\frac{\delta f}{\delta T}\right)_{V,n} = \sum \frac{\epsilon_i}{kT^2} e^{-\epsilon_i/kT}$$

and noting that

$$\frac{1}{f}\left(\frac{\delta f}{\delta T}\right)_V = \left(\frac{\delta \ln f}{\delta T}\right)_V = \frac{1}{f}\sum \frac{\epsilon_i}{kT^2} e^{-\epsilon_i/kT}$$

But the last part of this equation is almost the internal energy. In fact, comparing it with Eq. (2.10), we see that

$$U = nkT^2\left(\frac{\delta \ln f}{\delta T}\right)_V \tag{2.12}$$

These are partial derivatives with respect to temperature for a constant total number of molecules, n, and a constant volume V. Volume must be held constant, for, as we will see subsequently, some energy levels are functions of volume.

The Helmholtz free energy is $A = U - TS$ or

$$A = -nkT \ln f \tag{2.13}$$

and, since $dA = -p\,dV - S\,dT$, the pressure p, and thus the equation of state, is given by

$$p = -\left(\frac{\delta A}{\delta V}\right)_T = nkT\left(\frac{\delta \ln f}{\delta V}\right)_T \tag{2.14}$$

The fundamental relationships for all of the properties are summarized in Table 2.1, and we see that we have very neatly expressed all of the properties of all species that obey Maxwell-Boltzmann statistics in terms of the logarithm of the partition function and its derivatives. We have not at all solved the problem of our desire to calculate macroscopic properties in terms of molecular parameters, but this is certainly a very neat and compact statement of what we must do. Note that the partition function exists only for the Maxwell-Boltzmann exponential distribution. This is the only distribution for which temperature has meaning, and thermodynamics, as a science, then applies only to such systems. In the language of classical thermodynamics, we say the same things, but in other words, by demanding that the system always be in a state of thermal, mechanical, and chemical equilibrium.

This is the end of the exercise unless we can both evaluate f for a real substance and manipulate it through the formal relations of Table 2.1. We shall see subsequently that both of these demands are troublesome. For example, even if we knew exactly how two molecules interact with each other, the formalism has proven to be too complex to yet allow an accurate development of the equation of state, i.e., an evaluation of p from Table 2.1. But even so, we do not know how to write an accurate expression for the intermolecular forces either. But

TABLE 2.1 Summary of All Thermodynamic Properties of Localized Molecules in Terms of the Partition Function

$U = nkT^2\left(\frac{\partial \ln f}{\partial T}\right)_V$	$p = nkT\left(\frac{\partial \ln f}{\partial V}\right)_T$
$C_V = \left(\frac{\partial U}{\partial T}\right)_V$	$H = nkT^2\left(\frac{\partial \ln f}{\partial T}\right)_V + nkTV\left(\frac{\partial \ln f}{\partial V}\right)_T$
$S = nk \ln f + nkT\left(\frac{\partial \ln f}{\partial T}\right)_V$	$C_p = \left(\frac{\partial H}{\partial T}\right)_p$
$A = -nkT \ln f$	$G = -nkT \ln f + nkTV\left(\frac{\partial \ln f}{\partial V}\right)_T$

all is not lost, for reasonable approximations have been made that allow many useful insights and calculations, as we shall see.

Energy Scale

The concept of an absolute energy has no meaning. We have arbitrarily assigned $\epsilon_0 = 0$ and counted all other energies relative to this datum. Suppose we assign an energy of α to the zeroth level; the partition function then becomes

$$F = e^{-\alpha/kT} + e^{-(\alpha+\epsilon_1)/kT} + e^{-(\alpha+\epsilon_2)/kT} + \cdots$$

which we may obviously express in terms of the original partition function as

$$F = fe^{-\alpha/kT}$$

The original internal energy of, let us say, U_0 is

$$U_0 = nkT^2\left(\frac{\partial \ln f}{\partial T}\right)_V$$

which becomes, with this arbitrary assignment of $\epsilon_0 = \alpha$,

$$U = nkT^2\left(\frac{\partial \ln fe^{-\alpha/kT}}{\partial T}\right)_V \tag{2.15}$$

$$\text{or} \quad U = \alpha n + nkT^2\left(\frac{\partial \ln f}{\partial T}\right)_V$$

The new internal energy based on the new arbitrarily assigned zero of energy, represented by U_α, can be written

$$U_\alpha = U_0 + \alpha n$$

Therefore, changing the zero of energy for the molecular energy levels by an arbitrary constant of α is simply reflected in an additive constant of $n\alpha$ to the macroscopic internal energy. Clearly the question is of no importance.

Nonlocalized Molecules

Calculations of U, H, C_V, and C_p using the relations of Table 2.1 will all reveal good agreement with experiment, as we shall soon see. The entropy, however, does not agree with experiment, and consequently the Gibbs and Helmholtz free energies, which depend on entropy, are also erroneous. Why?

All of the preceding arguments have been based on the expression

$$w = \frac{n!}{\pi n_i!}$$

as the number of distinguishable ways that a set of occupation numbers can arise when there is no restriction on the number of molecules in each energy level and when the fixed sites occupied by the molecule are distinguishable. That is, the copper atoms all look alike; they are indistinguishable. But we can, in principle, label each site in the three-dimensional solid lattice. A distinguishability is thus imposed on identical atoms, and the above expression for w is then obtained. But in a gas, the atoms occupy no one site, but rather each may roam the entire volume. Now the statistical question is "How many ways can n indistinguishable molecules be placed in p distinguishable energy levels with no restriction on the number of molecules per level?"

Quantum mechanics reveals that the number of energy levels is larger the larger the volume in which the molecule is free to move. In fact, for a gas there are many, many more translational energy levels than there are molecules to go into these levels, and each level will then have only one molecule or no molecules in it. Thus we have a dense set of levels, sparsely populated by the molecules. Another problem has to do with the individual distinguishability of molecules. If we are told which quantum states are occupied, that distribution can occur in only one way, for no distinguishable rearrangement is possible; i.e., the molecules all look exactly alike. So how may we count identifiable arrangements so that we can determine that one that is statistically preferred? This is the problem of Bose-Einstein statistics that we will now briefly examine.

Let us divide the closely spaced levels into narrow bands so that we can with little loss of accuracy ascribe every molecule in any energy level within this interval as having energy ϵ_i. Let us say that there are p_i levels in the interval and an average of n_i molecules occupy those p_i levels, where $n_i \ll p_i$. One cubic centimeter of helium at standard temperature and pressure (STP) contains about 3×10^{19} atoms, and we may certainly arbitrarily divide the energy levels that may possibly be occupied at some temperature into, say, 10^{16} unequal intervals, such that each interval will contain about 3000 atoms. This is a number sufficiently large relative to unity to permit us to use Stirling's approximation in subsequent developments. Quantum mechanics suggests that there may be 10^5 energy levels in each interval, and let us then place 3000 helium atoms into these 10^5 energy levels all of which have substantially the same energy. How many distinguishable ways may they be arranged? Let us focus on only one such interval, and assert that an arrangement within that interval will be characterized by the number of molecules in each of the numbered levels, since the molecules themselves are not distinguishable. Pictorially, consider a listing of numbered lines; there will be 10^5 such lines in our example, and indistinguishable asterisks as follows:

|**|*|*****| | |*|*| ...

1 6 3 10 11 2 20 80

If we always start with the line numbered one, there will be $(n + p - 1)!$ ways of arranging the n asterisks and the remaining $(p - 1)$ lines. Let us say that each numbered line with the asterisks immediately following it represents a particular numbered energy level and the molecules in that level. Thus, in the above example, there are 2 molecules in the first level, 1 in the sixth, 5 in the third, and so on. But in this arrangement, all $n!$ possible reorderings of the n asterisks represent in fact the same arrangement, for the asterisks are indistinguishable one from another. Similarly, all $(p - 1)!$ reorderings of the numbered lines and their following asterisks describe the same arrangement of asterisks in the same numbered levels. That is, the $(p - 1)!$ permutations merely reorder the listing without changing the number of molecules in level 6, or 10, or whatever. Therefore, the general expression for the number of identifiable arrangements of n indistinguishable molecules in p numbered energy levels is

$$\frac{(n + p - 1)!}{n!\,(p - 1)!} \tag{2.16}$$

The arguments leading to Eq. (2.16) parallel earlier arguments that led to the expression for w for localized molecules, Eq. (2.1). Let us see how this relationship might work with two molecules occupying four energy levels rather than 3000 occupying 10^5 levels. By simple construction, the 10 arrangements that appear in Fig. 2.4 are possible,

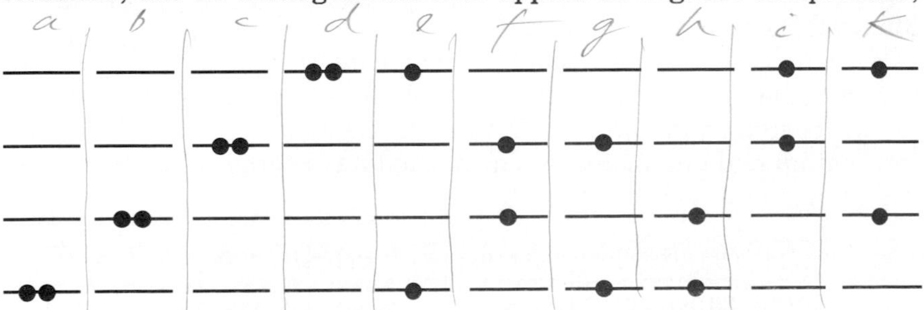

Figure 2.4 Identifiable arrangements of two molecules in four energy levels.

but the magic formula, Eq. (2.16), also says that there will be 10 such arrangements, $(2 + 4 - 1)!/(2)!(4 - 1)!$. If p is very large, as it always is for gases, we can with little loss of accuracy omit the -1 of Eq. (2.16) and write

$$w = \frac{(n + p)!}{n!p!}$$

as the number of arrangements of the n molecules in the p energy levels. Returning to the entire system, we recognize that the arrangement within any one interval is independent of that in all other intervals of energy levels, and we can then write the following expression for w for the whole population of 10^{23} molecules.

$$w = \frac{(n_0 + p_0)!}{n_0!p_0!} \times \frac{(n_1 + p_1)!}{n_1!p_1!} \times \frac{(n_2 + p_2)!}{n_2!p_2!} \cdots \tag{2.17}$$

or

$$w = \prod_i \frac{(n_i + p_i)!}{n_i!p_i!}$$

where the subscripts refer to the numbered intervals of energy, and in our example there are 10^{16} of these. This entire argument regarding the counting of indistinguishable molecules is pictorially evident in Fig. 2.5.

Equation (2.17) for nonlocalized molecules is the equivalent of

$$w = \frac{n!}{\pi n_i!}$$

for localized particles. In all statistical arguments with gases, Eq. (2.17) must be used for w. This way of counting is called Bose-Einstein statistics, and with this relationship we will be led to either similar or somewhat different conclusions to those we drew from the earlier Maxwell-Boltzmann statement.

We can again develop valuable insight into the nature of the occupation of the intervals of energy by molecules by considering three intervals spaced equally apart. If we shift one molecule from the center interval up and one molecule down, the total energy is constant, and the change in w will be given by

$$\frac{w(\text{after})}{w(\text{before})} = \left[\frac{n_0 + p_0}{n_0}\right]\left[\frac{n_1 + p_1}{n_1}\right]^{-2}\left[\frac{n_2 + p_2}{n_2}\right]$$

where we have neglected small integers relative to the size of n and p.

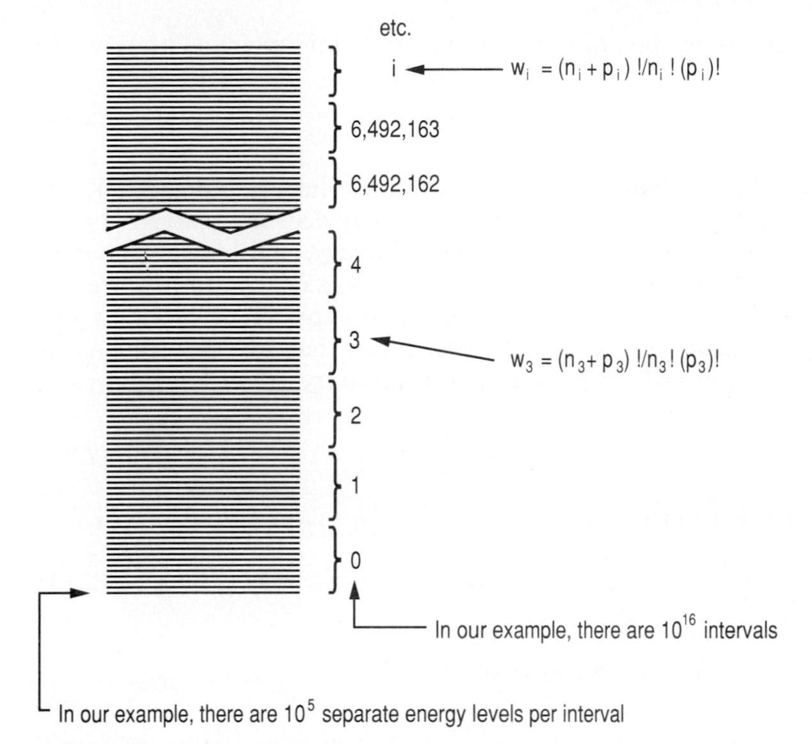

Figure 2.5 Schematic of Bose-Einstein statistics.

If w is at its maximum value, the ratio w(after)/w(before) must be equal to one, and the three quantities $[(n + p)/n]_i$ must be in a geometric progression. This is the same requirement that we earlier deduced for localized molecules.

We can develop the distribution that gives a maximum value to w just as before by using the technique of lagrangian undetermined multipliers,† and we find the maximum W occurs when

$$\frac{n_i}{p_i} = \frac{1}{Ae^{\mu\epsilon_i} - 1} \tag{2.18}$$

where A and μ are the constant lagrangian multipliers and where again we have an exponential decrease in occupation with increasing energy. The intervals of lowest energy are the most densely populated. Quantum mechanics reveals that the number of translational energy levels for molecules in the gas phase depends on the volume of the container. For any actual number density of a gas, that is, molecules per cubic centimeter, there will be many more

†See App. A.

levels than molecules, and thus $n_i/p_i \ll 1$, and we can safely write

$$\frac{n_i}{p_i} = Be^{-\mu\epsilon_i} \tag{2.19}$$

Equation (2.19) then is the distribution that maximizes W for nonlocalized molecules. We note that it really does not matter what n_i and p_i actually are. The division was arbitrary, and we now have an expression for the density of molecules in states near ϵ_i. In fact, we can with impunity redefine n_i as the average occupation of energy states near ϵ_i, wherein we will well recognize that in this new conceptualization n_i will have values less than unity. Then, since $\Sigma n_i = n$,

$$n_i = \frac{ne^{-\mu\epsilon_i}}{f} \tag{2.20}$$

where, just as before,

$$f \equiv \sum e^{-\mu\epsilon_i} \tag{2.21}$$

and, similarly as before, $\mu = 1/kT$.

Equation (2.17) may be rewritten as

$$\ln w = \sum n_i \ln\left(\frac{n_i + p_i}{n_i}\right) + \sum p_i \ln\left(\frac{n_i + p_i}{p_i}\right)$$

$$= \sum n_i \ln\left(1 + \frac{p_i}{n_i}\right) + \sum p_i \ln\left(\frac{n_i}{p_i} + 1\right)$$

but n_i/p_i is very small compared to unity, and the second summation becomes $\Sigma p_i(n_i/p_i)$ or Σn_i or n. Then, using the most probable distribution of Eq. (2.20), one obtains

$$\ln W = \sum n_i \ln\left[\frac{f}{ne^{-\epsilon_i/kT}}\right] + n$$

$$= n \ln\left(\frac{f}{n}\right) + \sum \frac{n_i\epsilon_i}{kT} + n \tag{2.22}$$

and, recalling the obvious fact that the thermodynamic internal energy U can only be $\Sigma n_i\epsilon_i$,

$$\ln W = n\left[\ln\frac{f}{n} + 1\right] + \frac{U}{kT} \tag{2.23}$$

Since the partition function f for nonlocalized molecules is the same as that for localized molecules as shown by Eq. (2.21), the internal energy is given, just as before [see the development of Eq. (2.12)], by

$$U = nkT^2\left(\frac{\partial \ln f}{\partial T}\right)_V$$

Since $S = k \ln W$, Eq. (2.23) allows one to immediately write

$$S = nk\left(\ln \frac{f}{n} + 1\right) + nkT\left(\frac{\partial \ln f}{\partial T}\right)_V$$

Equation (2.23) for nonlocalized molecules may be compared with the earlier and similar expression for localized molecules,

$$\ln W = n \ln f + \frac{U}{kT} \tag{2.24}$$

We can write for the Helmholtz free energy

$$A \equiv U - TS = -\,nkT\left(\ln \frac{f}{n} + 1\right) \tag{2.25}$$

for nonlocalized molecules, which we might compare with the earlier expression

$$A = -\,nkT \ln f \tag{2.26}$$

for localized molecules. Table 2.2 summarizes the expressions for the thermodynamic properties for nonlocalized molecules (compare Table 2.1).

Near 0 K the molecules will crowd into the lowest levels and the approximation of $p_i/n_i \gg 1$ will no longer be reasonable, and we are led to still a third counting scheme that is called Fermi-Dirac statistics. But in the real world, ideal gases do not exist at such temperatures

TABLE 2.2 Summary of All Thermodynamic Properties of Nonlocalized Molecules in Terms of the Partition Function

$$U = nkT^2\left(\frac{\partial \ln f}{\partial T}\right)_V$$

$$C_V = \left(\frac{\partial U}{\partial T}\right)_V$$

$$S = nk\left(\ln \frac{f}{n} + 1\right) + nkT\left(\frac{\partial \ln f}{\partial T}\right)_V$$

$$A = -\,nkT\left(\ln \frac{f}{n} + 1\right)$$

$$p = nkT\left(\frac{\partial \ln f}{\partial V}\right)_T$$

$$H = nkT^2\left(\frac{\partial \ln f}{\partial T}\right)_V + nkTV\left(\frac{\partial \ln f}{\partial V}\right)_T$$

$$C_p = \left(\frac{\partial H}{\partial T}\right)_p$$

$$G = -\,nkT\left(\ln \frac{f}{n} + 1\right) + nkTV\left(\frac{\partial \ln f}{\partial V}\right)_T$$

anyway, so this qualification on the Bose-Einstein formalism is of little significance. Fermi-Dirac statistics are applicable to elementary particles such as electrons, photons, and neutrons.

Molecular Distribution and the Laws of Thermodynamics

At 0 K, all of the molecules are in their lowest energy state, and $W = 1$. With the addition of a little thermal energy, the dispersal of the molecules among low-lying energy states begins, and W increases greatly because of the great number of choices that are now available for the molecules as ways in which to share the total energy. On a macroscopic level, we say that the entropy has increased.

There is another way, not involving thermal energy, in which W can similarly increase greatly. The quantum-mechanical expression for the translational energy levels arises from the solution of the classic textbook problem of *the particle in a box.* Here an increase in volume leads to narrower spacings between levels and the many more levels become available to the molecules even though their total shared energy has not changed. The value of W increases. If a sudden change in volume were to occur, the population would be seriously out of equilibrium. In thermodynamics we are concerned with what are called reversible expansion and compression. So we have here a molecular picture of the reversible process. That is, a reversible process occurs at such a rate that, at all stages of the process, the population has the exponential distribution appropriate to the energy shared by the molecules at that moment. For example, in the adiabatic expansion of an ideal gas, the shared energy (that is, the internal energy) remains the same, and, since the internal energy depends on temperature, the adiabatic expansion is also isothermal. But the partition function increases, as does W, and macroscopically the entropy increases.

Since all of the macroscopic properties of matter may be expressed in terms of the partition function, we may anticipate that the three laws of classical thermodynamics should also have molecular counterparts. Consider the internal energy

$$U = \sum n_i \epsilon_i$$

and then

$$dU = \sum n_i \, d\epsilon_i + \sum \epsilon_i \, dn_i$$

But in classical thermodynamics,

$$dU = \delta Q - \delta W$$

The translational energy levels of a gas depend on the size or volume

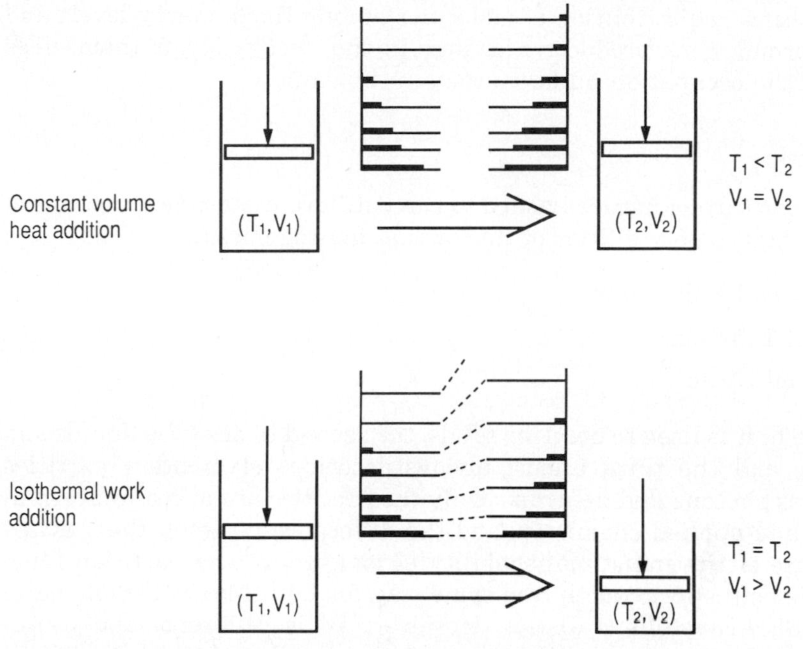

Figure 2.6 Relationship between heat and work interactions and the population of molecular energy states.

of the mass of gas. Then, as appears in Fig. 2.6, when heat is added, the molecules redistribute themselves, but when the volume is changed at constant temperature, the energy levels themselves change. Thus we make the identification of

$$\delta Q = \sum \epsilon_i \, dn_i \tag{2.27}$$

and

$$\delta W = - \sum n_i \, d\epsilon_i \tag{2.28}$$

The occupation numbers n_i do not change, for recall that

$$n_i = \frac{ne^{-\epsilon_i/kT}}{\sum e^{-\epsilon_i/kT}}$$

and when the levels are all increased by, say δ, because of a change in volume,

$$n_i = \frac{ne^{-(\epsilon_i + \delta)/kT}}{\sum e^{-(\epsilon_i + \delta)/kT}}$$

and the two expressions for the values of n_i are identical. The heat produces a redistribution of molecules among fixed energy levels and isothermal work produces a change in the energy levels themselves while the occupation numbers remain the same.

Summary

The structure of nature has led to three different statistics, or ways of counting the possibilities of this or that arrangement:

Maxwell-Boltzmann

Bose-Einstein

Fermi-Dirac

The first is used to describe solids, the second to describe liquids and gases, and the third almost always describes elementary particles such as photons and neutrons. Only the first two are of real concern in pure and applied chemistry, and the difference between them arises because of the indistinguishability of molecules. The partition function is defined in exactly the same way for both Maxwell-Boltzmann and Bose-Einstein statistics. However, W is different, and consequently the entropy and the energy functions, A and G, which depend on entropy, have different values according to the statistics that are used.

But whatever the state of matter, the partition function enables the complete calculation of the thermodynamic properties of all substances in principle, and most importantly, the formalism, with reasonable approximations, also allows the actual calculation of properties to within completely useful accuracies.

For whatever the state of matter, then, if we can write the partition function, we can evaluate all of the thermodynamic properties at all temperatures and pressures. Much of this book is devoted to just this task.

Although hopefully convincing, all of this statistical development has been somewhat cursory, but many excellent texts describe all of this in detail. Our objective here has been to make the concepts reasonable, to develop them from first and readily understandable and acceptable principles, and to remove any mystical aura that may surround the ideas. For many readers, all of this is but a sort of academic preliminary, for the main event is certainly that of predicting macroscopic properties and behavior. And to this task we must now turn.

Further Reading

1. R. W. Gurney, *Introduction to Statistical Mechanics*, McGraw Hill, New York, 1949.

Chapter

3

Thermodynamic Formalism for Gases and Plasmas

Remember that statistical mechanics is the bridge from the quantum properties of a single molecule on the one hand to the macroscopic properties of 10^{23} such molecules on the other hand. We will here merely state the familiar quantum mechanical results and explore their implications in thermodynamics.

From a utilitarian perspective, then, our problem is now one of evaluating the partition function for real systems. Let us begin this study by considering first the case of ideal gases and plasmas, then real gases and plasmas, and then solids. Because of their complexity and the resulting inaccuracies, we will mention liquids only in a conceptual way. Molecular engineering is less useful in practical sorts of calculations with the liquid phase.

In the following arguments, we will first take the total energy of a molecule as being neatly separable into energies arising from translational, rotational, vibrational, and electronic motions. This separability is usually an excellent approximation.

$$\epsilon_{\text{total}} = \epsilon_{\text{trans}} + \epsilon_{\text{rot}} + \epsilon_{\text{vib}} + \epsilon_{\text{elect}} \tag{3.1}$$

Later, we will also worry about the energy that two or more molecules share because they are near each other, that is, the intermolecular energy that causes, for example, all of the nonideality effects in real gases. Since the partition function is a sum of exponentials, it may be conveniently factored:

$$f = \sum e^{-\epsilon_{\text{total}}/kT} = \sum e^{-\epsilon_{\text{trans}}/kT} \sum e^{-\epsilon_{\text{rot}}/kT} \sum e^{-\epsilon_{\text{vib}}/kT} \sum e^{-\epsilon_{\text{elect}}/kT} \tag{3.2}$$

or $$f = f_{\text{trans}} f_{\text{rot}} f_{\text{vib}} f_{\text{elect}} \tag{3.3}$$

and the total partition function for a substance will be the product of functions wherein each has arisen from a particular mode of energy content. Now we also well recognize that this separation of energies is an excellent approximation other than for rare issues in vibration-rotation interaction. Certainly as vibrational excursions become greater, the moment of inertia becomes greater and the energy of rotation becomes greater. Thus vibration-rotation interaction exists, and we will return to this question of the separation of energy modes in this and in other situations in due course.

The thermodynamic properties of gases and plasmas when considered to behave ideally may be developed from relatively simple partition functions. Let us first consider monatomic species and then move on to molecules of increasing complexity.

Monatomic Species

Monatomic species may absorb energy by becoming electronically excited or by increased kinetic translation of the species in three-dimensional space. The translation energy levels are calculated from the familiar textbook quantum-mechanical problem of a particle in a one-dimensional box, where the first few wave functions appear as shown in Fig. 3.1. We note that the particle can never have less than the so-called zero-point energy. Although it is not fully relevant to our purposes here, we also will recall the further undergraduate textbook example of the effect of a potential barrier on these wave functions. The rather straightforward mathematical formalism reveals that a wave function can both penetrate or pass through a barrier of greater energy than the particle possesses and be partially reflected by a barrier of less energy than that of the particle. Both of these are quantum phenomena that are totally missing in the classical view of physics. The escape over a barrier is called tunneling, and this is a common physical phenomenon that occurs in the decay of radioactive isotopes and in the electronic device called the tunnel diode, to mention just two examples. The energy levels also appear in Fig. 3.1. These are called the *eigenvalues* of the wave equation, and each solution is called an *eigenfunction*. These energies are

$$\epsilon_i = \frac{h^2 n^2}{8ml^2} \qquad n = 1, 2, 3, \ldots \tag{3.4}$$

where n is the quantum number and l is the length of the line, that is, the volume of the box, in which the species moves. Then

$$f_{\text{trans}} = \sum_{n=1}^{\infty} e^{-n^2h^2/8ml^2kT} = \sum_{n=1}^{\infty} e^{-an^2}$$

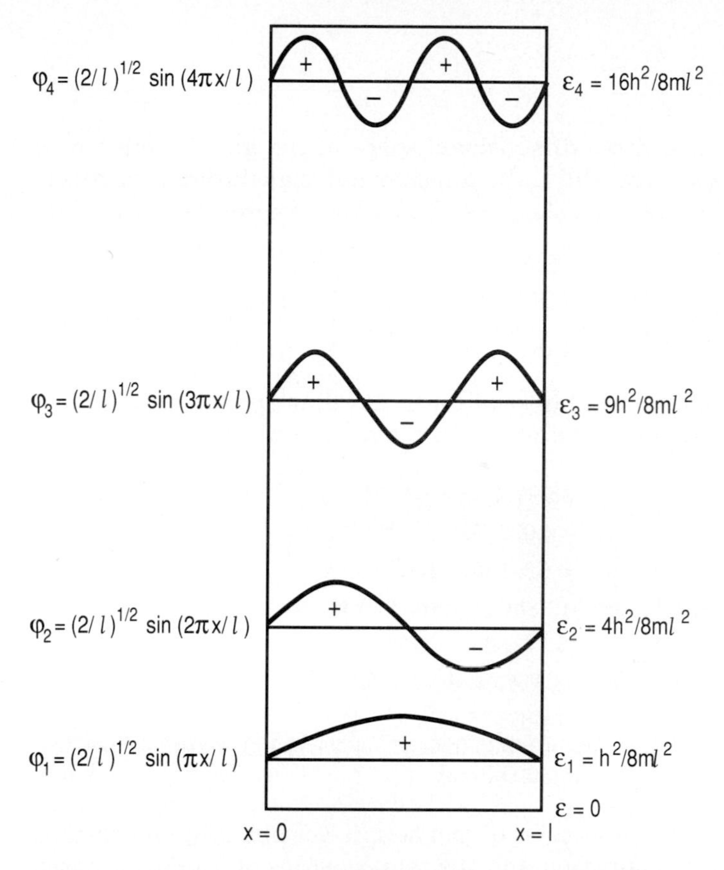

Figure 3.1 The first four wave functions and energy levels for a particle of mass m in a one-dimensional box of length l.

where a is clearly the cluster of quantities $h^2/8ml^2kT$. Now if a is sufficiently small, we can replace the summation by an integration and write

$$f = \int_0^\infty e^{-an^2}\, dn = \frac{1}{2}\sqrt{\frac{\pi}{a}}$$

or

$$f = \frac{(2\pi mkT)^{1/2}}{h} l \tag{3.5}$$

But in a gas or plasma, monatomic species are free to move in three dimensions, and a quantum state in the x direction may combine with any quantum state in either the y or z direction. That is, the molecule may have any allowed velocity (i.e., any allowed energy) of translation in three-dimensional space. The total translational energy of the atom is then

$$\epsilon = \frac{h^2}{8m}\left(\frac{n_x^2}{l^2} + \frac{n_y^2}{m^2} + \frac{n_z^2}{n^2}\right) \tag{3.6}$$

where we take the three-dimensional space of the gas to be a box of sides l, m, and n. And the three-dimensional translational partition function must be then just the cube of the above expression [Eq. (3.5)], or

$$f_{\text{trans}} = \sum exp\left[-\frac{h^2}{8m}\left(\frac{n_x^2}{l^2} + \frac{n_x^2}{m^2} + \frac{n_x^2}{n^2}\right)/kT\right] = \left(\frac{2\pi mkT}{h^2}\right)^{3/2} V \tag{3.7}$$

The several fundamental physical constants that appear in these molecular arguments and several useful conversion factors are:

Velocity of light	$c = 2.9979250 \times 10^8$ m/s
Avogadro's number	$n_A = 6.022169 \times 10^{23}$ mol^{-1}
Planck's constant	$h = 6.626186 \times 10^{-34}$ J · s
Boltzmann constant	$k = 1.380621 \times 10^{-23}$ J/K
pi	$\pi = 3.141593$
Gas constant	$R = 8.314333$ J/mol · K
Electronvolt	$R = 1.987173$ cal/mol · K 1 eV = 8,065.02 cm^{-1} = 2.417833×10^{14} s^{-1} = 23,059.6 cal/mol

The thermodynamic properties of monatomic species that one obtains using this partition function and the relationships of Table 2.2 agree well with experiment, as we shall see.

The partition function over the electronic energy levels is

$$f_{\text{elect}} = g_0 + g_1 e^{-\epsilon_1/kT} + g_2 e^{-\epsilon_2/kT} + \cdots \tag{3.8}$$

where the electronic energy levels are given by ϵ_1, ϵ_2, etc. and the multiplicity of each electronic state is given by g_0, g_1, etc. That is, the species has, for example, g_1 separate energy states, all having the same energy ϵ_1. Transitions between the electronic energy levels from which their values may be deduced appear in the visible and ultraviolet regions of the electromagnetic spectrum. The existence of the ϵ_i's and their g_i's is a fact given by nature. The spectroscopist may split and observe these multiple states by allowing the electronic transition to occur in an electric or magnetic field, that is, by studying the so-called Stark or Zeeman effects, respectively. Although these levels are experimental quantities, quantum mechanics permits an exact calculation of the electronic energy levels for the hydrogen atom, where we find

$$\epsilon_{\text{elect}} = -\left(\frac{\mu e^4}{8h^2\epsilon_0^2}\right)\frac{1}{n^2} \qquad n = 1, 2, 3, \ldots \tag{3.9}$$

where the symbols all have their usual meaning except μ, which is called the *reduced mass*, defined as $\mu = Mm/(M+m)$, where M is the mass of the nucleus and m is the mass of the electron. Clearly μ is very nearly just the mass of the electron. The small e is the charge on the electron (1.60219×10^{-19} C) and ϵ_0 is the vacuum permittivity ($8.854188 \times 10^{-12}\ \text{J}^{-1} \cdot \text{C}^2 \cdot \text{m}^{-1}$). The collection of quantities in the parentheses is called the *Rydberg constant* for the hydrogen atom, which is equal to 2.178×10^{-18} J. The integer n is called the principal quantum number because it alone determines the electronic energy of the atom. When $n = \infty$, $\epsilon_{\text{elect}} = 0$. Each of these energy levels is doubly degenerate, which means that there are two quantum states with the same energy, but that need not concern us here. These energy levels for the hydrogen atom, up to its ionization at 13.6 eV, and the energy differences between the quantum states are shown in Fig. 3.2. These energy differences are what is observed by the spectroscopist. A photon of exactly $h\nu = \epsilon_1 - \epsilon_0$, or at 121.6 nm, appears when the electron falls from $n = 2$ to $n = 1$, and it forms one of the lines in the Lyman series of the spectrum of hydrogen. Inspecting the wavelengths of the many transitions and remembering the narrow visible region of the spectrum, we would expect an electric discharge in hydrogen to be red (656.2 nm), and this is indeed the case.

One of the goals of quantum mechanics is the similar calculation of the electronic energy levels of all species. And progress is being made. Approximations of varying quality have been applied to larger atoms; for example, the alkali metals may be regarded as a sort of pseudo-hydrogen atom wherein the nucleus is shielded by the lower electrons so that the valence electron of the metal, which is responsible for its chemistry, may be imagined as moving in the effective or screened charge of the nucleus. But, in general, the electronic energy levels must still be experimentally measured by the spectroscopist rather than predicted quantum-mechanically.

The electronic partition function for all species must also be summed term-by-term, for the levels are too far apart and too irregularly spaced to allow any simple integral or series summation. This series is, in principle, infinite, but it is terminated in all real situations by the presence of the electric field of the nearby ions and electrons which change the energy of the level. This results in a general broadening of the energy levels such that those energy levels near the ionization energy have a tendency to disappear because the broadening causes these closely spaced upper levels to merge. The higher levels are also elevated because the excited electron does not move in the

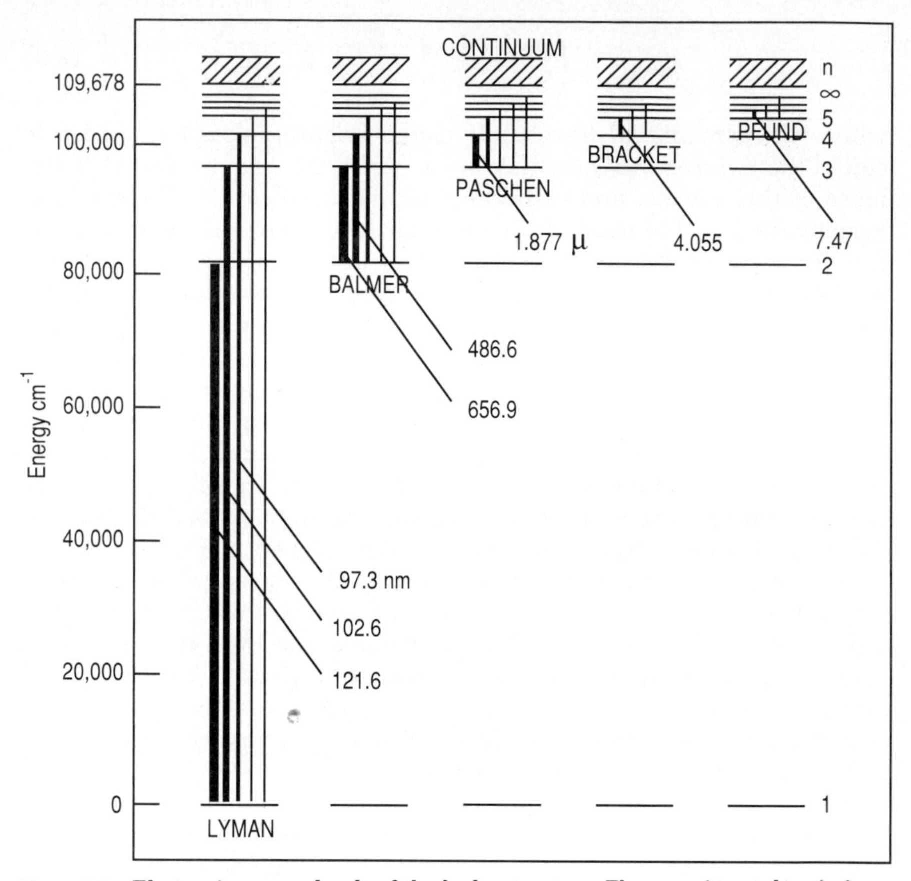

Figure 3.2 Electronic energy levels of the hydrogen atom. The experimental emissions or absorptions occur at exactly the wave lengths predicted by Eq. (3.9). The weight of the lines is an approximate indication of the intensity of that transition in the experimentally observed spectrum of hydrogen.

pure Coulomb field of its nucleus alone, but rather in a field that is shielded by the other electrons of the atom. With approximate theoretical descriptions of these phenomena, it is possible to estimate the maximum principal quantum number that will, in fact, exist. The electronic partition function is then bounded. To provide a perspective on the size of these numbers, one calculates a maximum quantum number of 10 and a decrease in the ionization potential of 0.14 eV for hydrogen at 800 K and 0.101 kPa.†

Fortunately, the difference in the energies of the electronic ground state and the first excited level is usually very large, usually about

†For further information on the thermodynamic properties of very hot gases or plasmas, see McGee and Heller,[1] Inglis and Teller,[2] and Margenau and Lewis.[3]

two-thirds of the ionization energy, as is evident for hydrogen in Fig. 3.2, where the first level is near 10 eV as compared to ionization at 13.6 eV. Consequently, the exponential terms of the electronic partition function are all effectively zero except at very high temperatures. For virtually all instances of chemical interest, then, the partition function reduces to just g_0—the ground state multiplicity. Even for the easily ionized lithium atom of ionization energy of only 5.36 eV or 517.1 kJ/mol, the first excited level, at 1.84 eV, is populated only to 1 percent at about 4500 K, as revealed by the Boltzmann equation

$$\frac{n_1}{n} = \frac{g_1 e^{-\epsilon_1/kT}}{\sum_i g_i e^{-\epsilon_i/kT}} = 0.01 \tag{3.10}$$

that is, for the numbers characteristic of lithium, this equation is satisfied at a temperature of about 4500 K.

The higher this first excited level, the higher will be the temperature at which the partition function will begin to have a value somewhat greater than g_0.

Nitric oxide is one of the very few molecules which is electronically excited at low to moderate temperatures. Even here all levels other than the first are excited only at very high temperatures, but the first excited level is only 121 cm^{-1} up. The multiplicity of both the ground and first level is two, so the electronic partition function is

$$f_{\text{elect}} = 2 + 2\exp\left(\frac{-121hc}{kT}\right) \tag{3.11}$$

and thus the first excited state is contributing 15 percent to the value of the partition function at temperatures of only 100 K. Species with at least one unpaired electron, such as CH_3, OH, H, O_2, and NO, are called *free radicals*. In general, free radicals have low-lying electronic states that may be occupied at moderate temperatures. With the exception of the very unusual O_2 and NO, free radicals do not usually exist as stable species at ordinary temperatures.

In essentially all cases, then, we can write the partition function of all gaseous monatomic species, whether neutral or charged, as

$$f_{\text{mon}} = g_0 \left(\frac{2\pi mkT}{h^2}\right)^{3/2} V \tag{3.12}$$

which is clearly just the product of the translational and electronic partition functions. The ground state degeneracy or multiplicity g_0 of many atoms is summarized in Table 3.1. The symbol in the column headed *normal state* is a spectroscopic term symbol which has evolved as a shorthand way of summarizing the electronic configuration of the species. For hydrogen, the symbol is read "doublet S one-half." A de-

TABLE 3.1 Some Values of g_0 for Monatomic Species

Periodic group	Elements	Normal state	Degeneracy
I	H, Li, Na, K, Rb, Cs; Cu, Ag, Au	$^2S_{1/2}$	2
II	Be, Mg, Ca, Sr, Ba; Zn, Cd, Hg	$^1S_0$	1
III	B, Al, Ga, In, Tl	$^2P_{1/2}$	2
IV	C, Si, Ge, Sn, Pb	$^3P_0$	3
V	N, P, As, Sb, Bi	$^4S_{3/2}$	4
VI	O, S, Se, Te, Po	$^3P_2$	3
VII	F, Cl, Br, I	$^2P_{3/2}$	2
0	He, Ne, A, Kr, Xe, Rn	$^1S_0$	1
	Electrons		2

tailed understanding of the construction of the term symbols for the electronic states of either atoms or molecules need not concern us here.

The several molar thermodynamic properties may then be readily evaluated:

$$U^0 = nkT^2\left(\frac{\partial \ln f}{\partial T}\right)_V \tag{3.13}$$

or

$$U^0 = \frac{3}{2}RT \tag{3.14}$$

and

$$C_V^0 = \left(\frac{\partial U^0}{\partial T}\right)_V = \frac{3}{2}R \tag{3.15}$$

and

$$H^0 = nkT\left[T\left(\frac{\partial \ln f}{\partial T}\right)_V + V\left(\frac{\partial \ln f}{\partial V}\right)_T\right] \tag{3.16}$$

or

$$H^0 = \frac{5}{2}RT \tag{3.17}$$

and

$$C_p^0 = \left(\frac{\partial H^0}{\partial T}\right)_p = \frac{5}{2}R \tag{3.18}$$

These results all agree well with all available experimental data. Deviations from experiment occur only at very high temperatures because of the onset of electronic excitation and at low temperatures because of the onset of nonideality effects. The entropy is, from Table 2.2,

$$S^0 = nk\left[\ln\left(\frac{f}{n}\right) + 1 + T\left(\frac{\partial \ln f}{\partial T}\right)_V\right] \tag{3.19}$$

or

$$S^0 = nk\left\{\ln\left[g_0\left(\frac{2\pi mkT}{h^2}\right)^{3/2}\frac{kT}{p}\right] + \frac{5}{2}\right\} \tag{3.20}$$

which, in practical units and in dimensionless form, becomes

$$\frac{S^0}{R} = \frac{3}{2}\ln M + \frac{5}{2}\ln T + \ln g_0 - \ln p + 3.4533 \tag{3.21}$$

where M = molecular weight
T = absolute temperature, K
p = pressure, kPa
R = 8.3143 J/mol · K (usually, although it may be in any units).

This is the so-called Sackur-Tetrode equation. We can use this relationship to calculate the absolute entropy of several monatomic species for comparison with experimental values from low-temperature measurements of heat capacity and heats of transition. The calorimetric absolute entropy is given by

$$S^0(T, p) = \int_0^{T_{\mathrm{mp}}} \frac{C_p(\text{solid})}{T}\,dT + \frac{\Delta H_f}{T_{\mathrm{mp}}} + \int_{T_{\mathrm{mp}}}^{T_{\mathrm{bp}}} \frac{C_p(\text{liquid})}{T}\,dT + \frac{\Delta H_v}{T_{\mathrm{bp}}} + \int_{T_{\mathrm{bp}}}^{T} \frac{C_p(\text{gas})}{T}\,dT - [S(T, p) - S^0(T, p)] \tag{3.22}$$

where all of the indicated data—heat capacities as a function of temperature for all phases, heat of fusion and melting point, heat of vaporization and boiling point—must be calorimetrically determined, and we even need an equation of state to evaluate the nonideality term ($S - S^0$). Some experimental comparisons of Eqs. (3.21) and (3.22) appear in Table 3.2.

The Sackur-Tetrode equation clearly fails in the limit of $T = 0$, for it predicts that the entropy becomes $-\infty$. The reason for this failure goes back to our initial assumptions regarding statistics wherein we took the number of quantum states or energy levels to be very large compared to the number of molecules. At very low temperatures this is no longer justified, for the molecules pile up in the lowest levels.

TABLE 3.2 Calculated versus Experimental Entropies for Several Monatomic Species at 298.15 K

Species	S (calculated)	S (experimental)
He	30.11	30.4
Na	36.70	37.2
Hg	41.78	42.2
A	36.98	36.85

SOURCE: From E. A. Moelwyn-Hughes, *Physical Chemistry*, Pergamon, London, 1957.

The temperature at which our statistical assumptions begin to fail depends on the molecular weight and the pressure, and for helium at about 100 kPa, this temperature is about 0.5 K. So, for all practical purposes, we can forget about this problem with Eq. (3.21).

It is interesting that statistical-mechanical formalism permits one to calculate the thermodynamic properties of a two-dimensional gas. Physically this may sometimes serve as a reasonable model for species adsorbed at an interface but free to move around laterally on that interface. The two-dimensional partition function is clearly

$$f = g_0\left(\frac{2\pi mkT}{h^2}\right)A$$

and then $$U^0 - U_0^0 = n_0kT = RT$$

$$C_V^0 = n_0k = R$$

and $$\frac{S^0}{R} = \ln g_0 + \ln M + \ln T + \ln\left(\frac{A}{n_0}\right) + 42.332$$

where A/n_0 is the area available in square meters per molecule. For argon at 298.15 K and 101.325 kPa, the atoms are about 3.4 nm apart, and we might imagine a corresponding area per molecule on a surface to be $(3.4)^2$ nm^2/molecule. Such a two-dimensional argon has an S^0 of 105.7 J/mol · K. This may be compared with an S^0 of 154.7 J/mol · K, or about half again as much as the two-dimensional value, for regular three-dimensional argon gas.

Just for completeness, we also show here the other thermodynamic properties.

$$A = -nkT\left[\ln\left(\frac{f}{n}\right) + 1\right] = -nkT\ln\left[g_0\left(\frac{2\pi mkT}{h^2}\right)^{3/2}\frac{kT}{P}\right] - nkT \tag{3.23}$$

Recall that the argument of the logarithm is dimensionless, as is clear if one consistently uses International System (SI) units of m in kg, k in J/K, h in J · s, and p in N/m^2. In more practical units, Eq. (3.23) becomes

$$\frac{A^0(T,p)}{RT} = -\frac{3}{2}\ln M - \frac{5}{2}\ln T - \ln g_0 + \ln p - 1.9533 \tag{3.24}$$

where M is the molecular weight in grams, T is the temperature in kelvins, and p is the pressure in kilopascals.

Finally, the Gibbs free energy is given by

$$G = -nkT\left[\ln\left(\frac{f}{n}\right) + 1 - V\left(\frac{\partial \ln f}{\partial V}\right)_T\right] \tag{3.25}$$

which, in practical units, is

$$\frac{G^0(T,p)}{RT} = -\ln g_0 - \frac{3}{2}\ln M - \frac{5}{2}\ln T + \ln p - 0.9533 \tag{3.26}$$

This last function is particularly useful, and data compilations present tables of this function as $(G^0 - H_0^0)/RT$, where the Gibbs free energy is tabulated relative to the enthalpy of the species at 0 K, which is normally set equal to zero. Thus the numerical values of G^0/RT and $(G^0 - H_0^0)/RT$ are identical.

The relationships developed in this section can be readily used to calculate all of the properties of monatomic species and some typical values are collected in Table 3.3. Note that we have not listed the dimensionless heat capacity and enthalpy for each species, for they are all identical. Only the entropy depends on the mass of the atom.

The Electronic Partition Function

Equation (3.10) revealed that temperatures near 4500 K are necessary to populate the first excited state of lithium to only 1 percent. Although subsequent chapters will deal with chemical equilibrium in some detail, it is nonetheless instructive to consider here the degree of ionization of lithium vapor at 4500 K and 1 kPa (about 7.5 torr). Neglecting, for the moment, the electronic partition functions for both Li and Li^+, for unionized lithium, Eq. (3.26) becomes

$$\begin{aligned}\frac{G^0 - H_0^0}{RT} &= -\ln 2 - \frac{3}{2}\ln 6.94 - \frac{5}{2}\ln 4500 + \ln 1 - 0.9533\\ &= -25.582\end{aligned}$$

for Li^+,

$$\begin{aligned}\frac{G^0 - H_0^0}{RT} &= -\ln 1 - \frac{3}{2}\ln 6.94 - \frac{5}{2}\ln 4500 + \ln 1 - 0.9533\\ &= -24.889\end{aligned}$$

and for the free electrons,

$$\begin{aligned}\frac{G^0 - H_0^0}{RT} &= -\ln 2 - \frac{3}{2}\ln 5.49 \times 10^{-4} - \frac{5}{2}\ln 4500 + \ln 1 - 0.9533\\ &= -11.415\end{aligned}$$

Therefore, for the reaction

TABLE 3.3 Thermodynamic Properties of Some Monatomic Species

	H						
T	C^0p/R	$(H^0 - H^0_0)/RT$	S^0/R	Ar, S^0/R	Li, S^0/R	Hg, S^0/R	C, S^0/R
100	2.500	2.500	11.054	15.880	13.952	13.294	—
200	2.500	2.500	12.789	17.613	15.684	20.026	—
300	2.500	2.500	13.800	18.627	16.694	21.036	19.015
400	2.500	2.500	14.519	19.346	17.416	21.758	19.737
600	2.500	2.500	15.533	20.360	18.426	22.768	20.747
800	2.500	2.500	16.252	21.079	—	—	21.469
1000	2.500	2.500	16.810	21.637	19.701	—	22.034

SOURCE: Data from Landolt-Börnstein, *Kalorische Zustandsgrossen*, Springer-Verlag, Berlin, 1961.

$$\text{Li} = \text{Li}^+ + e$$

the free energy change in the ideal gas phase is

$$\frac{\Delta G^0}{RT} = \sum_i \frac{G^0 - H_0^0}{RT} + \frac{\Delta H_0^0}{RT}$$

$$= (-24.889) + (-11.415) - (-25.582) + \frac{517{,}100}{(8.314)(4500)}$$

$$= 3.099$$

where 517,100 J/mol is the heat of reaction, also called, in this case, the *ionization potential*. This ΔG^0 corresponds to

$$K = \exp -(3.099) = 0.045$$

which corresponds to an extent of ionization of about 20 percent. The 1 percent occupation of the first excited level of Li means its electronic partition function will be essentially the ground-state multiplicity of 2. Ignoring the electronic partition function for Li^+ was an excellent approximation, for the first excited level of Li^+, which is isoelectronic with He, has a high energy relative to kT at 4500 K. The Boltzmann distribution demands that occupation of excited levels depends on temperature only. But pressure too has a large effect on this dissociation of one particle into two. Dissociation produces a large increase in entropy which offsets the ΔH^0 of 517,100 J/mol to produce a ΔG^0 of only 115,900 J/mol and a K of 0.045. It is this entropy effect that produces significant higher-energy ionization even though there is insignificant lower-energy excitation. One percent of lithium atoms are in their first excited state near 4500 K, while 1 percent of these same atoms are already completely dissociated at near 3400 K.

The calculation of properties when the electronic partition function is significant may be illustrated by considering atomic bromine at 2000 K. Here the ground state is a $^2\text{P}_{3/2}$ (see Table 3.1) with $g_0 = 2J + 1$, or $2(\frac{3}{2}) + 1$, or 4. The first excited state is a $^2\text{P}_{1/2}$ with $g_1 = 2J + 1 = 2(\frac{1}{2}) + 1 = 2$. This level is experimentally observed to lie 3685 cm^{-1} above the ground state. Interestingly, the corresponding first excited state of atomic fluorine lies 404 cm^{-1} above the ground state, and hence its effect upon the thermodynamic properties will occur at lower temperatures than in the case of bromine. The next levels in bromine are much higher at about 65,000 cm^{-1}, and they will influence the thermodynamic properties only at much higher temperatures than will the first excited level at 3685 cm^{-1}. The electronic partition function for atomic bromine is

$$f_{el} = 4 + 2 \exp\left[\frac{-3685(3 \times 10^{10})(6.626 \times 10^{-34})}{(1.3806 \times 10^{-23})T}\right]$$

$$+ \text{increasingly negligible terms}$$

or, to a good approximation,

$$f_{el} = 4 + 2 \exp\left(\frac{-5248}{T}\right)$$

The internal energy is

$$U^0 - U_0^0 = nkT^2\left[\frac{\partial \ln f_{tr}}{\partial T} + \frac{\partial \ln f_{elect}}{\partial T}\right] = \frac{3}{2}nkT + nkT^2\left[\frac{\partial \ln f_{elect}}{\partial T}\right]$$

where

$$\frac{\partial \ln f_{elect}}{\partial T} = \frac{5248}{T^2\,[2 \exp(5248/T) + 1]}$$

At 2000 K, $U^0 - U_0^0$ = 25,305 J/mol, of which 365 J/mol or 1.4 percent was due to electronic excitation. C_V^0 = 12.93 J/mol · K, of which 0.46 J/mol · K or 3.6 percent was due to electronic excitation. And finally, S^0(2000 K, 101.325 kPa) = 215.51 J/mol · K, of which 12.58 J/mol K or 5.8 percent was due to electronic excitation. Note that the pressure must, of course, be specified to calculate S^0, whereas it was irrelevant to both $U^0 - U_0^0$ and C_V^0.

With bromine, and with any species where there is a significant gap in the spacing of energy levels, the heat capacity will go through a maximum when electrons are being moved up to the first excited level. This movement may be completed, causing the heat capacity to drop, before movement into the next higher levels begins. With this onset of new upward motion, the heat capacity will again rise.

Atoms vary greatly in the complexity of their electronic energy levels, as is revealed, of course, by the varying complexity of their ultraviolet-visible spectra. The electronic energy levels of the transition elements are rather complex. For example, titanium has a $^3F_2$ ground state of $2J + 1 = 5$ levels, the two lowest of which are at only 170 cm^{-1} and 387 cm^{-1} above the ground state and would then be significant at low temperatures. There are some 15 levels that contribute to the properties of gaseous titanium at 2000 K.[4,5]

The required degeneracies of each electronic configuration of an atom can be deduced, if rather tediously, from the quantum-mechanical structure of each atom. The energy of each configuration is, however, much more difficult to calculate theoretically, so we rely upon experiment, and extensive tables of such data listing both energy and degeneracy are available.[6] Recall some rules from quantum mechanics:

1. No two electrons of an atom may have the same quantum numbers, n, l, m, and s, which are named the principle, azimuthal, magnetic, and spin quantum numbers, respectively.
2. l may have values of 0, 1, 2,..., $n - 1$.
3. Each electron may have $2l + 1$ values of the magnetic quantum number m_l and these are 0, ± 1,..., ± l. Each has the same energy in the absence of an external field.
4. The spin quantum number is ± ½ for each electron.

Consider atomic fluorine with two 1s electrons ($n = 0$, $l = 0$), two 2s electrons ($n = 2$, $l = 0$), and five 2p electrons ($n = 2$, $l = 1$). Without changing the principle quantum number of any electron, the electrons may be arrayed in six different ways as shown in Fig. 3.3.

Now m_l and s are measures of angular momentum which, in the lighter elements, combine for each electron to give a resultant M_l and M_s for the atom. J is the sum of M_l and M_s formed such that J takes on values of $|M_l + M_s| \cdots |M_l - M_s|$. For fluorine, M_l is − 1, 0, 1 and M_s ½, − ½, and then J is 3/2 or ½.

It is useful to understand how the electronic states of atoms as displayed in Fig. 3.3 are named. First uppercase letters are associated with absolute values of M_l as follows:

M_l	Symbol
0	S
1	P
2	D
3	F
4	G
etc.	

A left superscript to the letter of value $|2M_s + 1|$ fixes both the value of M_s as well as the number of spin configurations, i.e., $2M_s + 1$. Finally, a right subscript J fixes both the total angular momentum and the number of states, i.e., $2J + 1$, which we have also called g, that have the same energy in the absence of an external field,

$$^{|2M_s + 1|}L_J$$

In the case of fluorine, with configurations as shown in Fig. 3.3, M_l is 1, 0, or − 1, so each of these six configurations is a P state. The P state includes $M_l = 0$, just as a D state includes $M_l = 0$ and ± 1 in addition

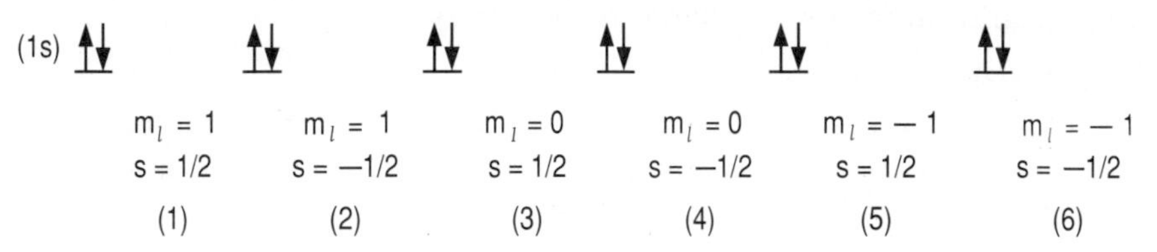

Figure 3.3 Possible configurations of the electrons of fluorine with no change of principle quantum numbers.

to ±2. M_s is ±½, so the left superscript is 2. J is then 3⁄2 or ½, and we write $^2P_{3/2}$ to represent configurations (1), (2), (5), and (6) and $^2P_{1/2}$ to represent configurations (3) and (4). It is not obvious which of these six electronic configurations has the lowest energy. However, Hund's rule states that the lowest energy will be that of largest M_s and, within that restriction, also that configuration of largest M_l. So for fluorine, the ground state is $^2P_{3/2}$ and we could write the first two terms in the electronic partition function as

$$f_{\text{elect}} = 4 + 2 \exp(-\epsilon_1/kT) + \cdots \text{smaller terms} \tag{3.27}$$

Experimentally the $^2P_{1/2}$ is only 404 cm^{-1} above the $^2P_{3/2}$ ground state, and it contributes to the thermodynamic properties at modest temperatures. The next higher energy levels demand an increase in the principle quantum number from 2 to 3, and then many more states become accessible. However, these terms in the electronic partition function are increasingly small because their energy is high and there is an exponential dependence on $-\epsilon_i/kT$.

The symbols like $^2P_{3/2}$ are called spectroscopic term symbols, and this particular example is read "doublet P three-halves."

To better explain this naming scheme for energy states, consider another example of two p electrons which would describe the lowest levels of carbon or silicon or lead. Here M_l = 2, 1, 0, − 1, − 2 and M_s = 1, 0, − 1. Rather than placing electrons in atomic orbitals as schematically shown in Fig. 3.3, it is more convenient to display the quantum-mechanically allowed configurations as in Table 3.4. The configuration and naming of states for two p electrons does not depend on the principle quantum number. So Table 3.4 describes C, Si, Pb, etc. all equally well.

TABLE 3.4 Allowed Placings of Two *p* Electrons

m_l						
+1	0	−1	M_l	M_s	Term	Possible combinations
↑↓			2	0	A ^{1}D	A M_l = 2,1,0, −1, −2
↑	↑		1	1	B ^{3}P	M_s = 0
↑	↓		1	0	A ^{1}D	
↓	↑		1	0	B ^{3}P	B M_l = 1,0, −1
↓	↓		1	−1	B ^{3}P	M_s = 1,0, −1
↑		↑	0	1	B ^{3}P	
↑		↓	0	0	A ^{1}D	C M_l = 0
↓		↑	0	0	B ^{3}P	M_s = 0
↓		↓	0	−1	B ^{3}P	
	↑↓		0	0	C ^{1}S	^{1}S,^{3}P, and ^{1}D terms
	↑	↓	−1	0	A ^{1}D	are allowed
	↓	↑	−1	0	B ^{3}P	
	↓	↓	−1	−1	B ^{3}P	
	↑	↑	−1	1	B ^{3}P	
		↑↓	−2	0	A ^{1}D	

There are a total of 15 separate configurations. All these may be described by a more limited number of term symbols, just as all six configurations of atomic fluorine were described by only two terms. Let us begin with the largest value of $M_l = 2$ which is then named a D term. Within $M_l = 2$, the allowed magnetic quantum numbers are 2, 1, 0, − 1 and − 2. For two electrons, the allowed spin quantum numbers, in principle, are 1, 0, − 1, but in this case only 0 is physically permissible at $M_l = 2$, for otherwise the two electrons would have all four quantum numbers with the same values. This restriction on M_s is applied to all configurations of the D term, even though, for $M_l = \pm 1$ and 0, the spins could also be parallel rather than opposed; i.e., M_s could be ± 1 as well as 0. These configurations of (M_l, M_s) values (2,0), (1,0), (0,0), (− 1,0), and (− 2,0) describe a ^{1}D term, i.e., $M_l = 2$ and $M_s = 0$. The resultant angular momentum quantum numbers J are (Ml+ $(M_l + M_s) \cdots (M_l - M_s)$ or only $J = 2$. Quantum mechanics reveals that there are always $2J + 1$ of these states having the same energy or five states just as we have already found. These five configurations are labeled A in Table 3.4. The complete term symbol is then $^1D_2$, read "singlet-D-two."

Continuing, with $M_l = 1$, the allowed quantum numbers are $M_l = 1$, 0, −1, and as before the allowed resultant spin quantum numbers are 1, 0, −1, depending on the spins being parallel one way, parallel the opposite way, or paired. There are then a total of nine physically allowed configurations of (M_l, M_s); that is, for each J there may be any one of three values of M_s: (1,1), (1,0), (1, −1), (0,1), (0,0), (0, −1),

(−1,1), (−1,0), (−1,1). Here M_j is $|M_l + M_s| \cdots |M_l - M_s|$ or J = 2, 1, 0. If M_s can be 1, its other allowed values of −1 and 0 are understood and each state is a triplet, that is, $M_s = 1$, $2M_s + 1 = 3$, and the left superscript on each term is 3. Each group of states has the same energy in the absence of an external field. For the J = 2 term, the allowed values of J are 2, 1, 0, −1, −2. These are indicated by a B in Table 3.4. For the J = 1 term, the allowed values of J are 1, 0, −1. These are also indicated by a B in Table 3.4. For the J = 0 term, the allowed values of J are 0, and this single state is also indicated by a B in Table 3.4. Thus these nine states may be described by three term symbols, $^3P_2$, $^3P_1$, and $^3P_0$, containing five, three, and one state, respectively.

Finally, for $M_l = 0$, M_s must be 0, for otherwise both p electrons would have all four quantum numbers alike. This corresponds to a $^1S_0$ term, i.e., $M_l = 0$, $M_s = 0$, and $J = 0$, that can occur in only one way, $2J + 1 = 1$. This configuration is indicated by a C in Table 3.4.

The electronic partition function for these two electrons becomes

$$f_{\text{elect}} = 1 + 3\exp\left(-\frac{^3P_1}{kT}\right) + 5\exp\left(-\frac{^3P_2}{kT}\right) + 5\exp\left(-\frac{^1D_2}{kT}\right) + \exp\left(-\frac{^1S_0}{kT}\right) + \text{smaller higher-energy terms due to larger azimuthal quantum numbers} \tag{3.28}$$

where Hund's rule (see earlier) has been used to write the terms in sequence of increasing energy. The sequencing of energy levels becomes more complicated for heavy atoms, and Hund's rule is not obeyed.

Data on the actual energy levels of C, Si, and Pb are summarized in Table 3.5. The electronic partition function for C is then

TABLE 3.5 Experimental Energy Levels in cm^{-1} of Three Atoms with Two *p* Outer Electrons

	C	Si	Pb
$^3P_0$	0.0 (1)	0.00	0.0
$^3P_1$	16.4 (3)	77.15	7,819.35
$^3P_2$	27.1 (5)	223.31	10,650.47
$^1D_2$	10,193.7 (5)	6,298.81	21,457.90
$^1S_0$	21,648.4 (1)	15,394.24	29,466.81
	Next at ~ 33,700	Next at ~ 39,700	Next at ~ 35,000

SOURCE: From Moore.[6]

$$f_{\text{elect}}(\text{C}) = 1 + 3\exp\left[\frac{-16.4(3\times10^{10})(6.626\times10^{-34})}{(1.381\times10^{-23})\text{T}}\right] + 5\exp\left[\frac{-27.1(3\times10^{10})(6.626\times10^{-34})}{(1.381\times10^{-23})\text{T}}\right] + \cdots$$

Advanced Propulsion Concepts

From the thermodynamic description of the nozzle expansion of a hot gas, it is clear that the specific impulse that is produced will vary directly with the specific enthalpy of the flame gas. One requires then the highest possible joules per kilogram for maximum I_{sp}. Stated another way, I_{sp} varies as $(T/\overline{\text{MW}})^{1/2}$, so one requires highest temperatures and lowest average molecular weight. The H_2/O_2 propellant combination is operated fuel-rich, which somewhat reduces T, but reduces $\overline{\text{MW}}$ more, to yield a net improvement of I_{sp}.

One might imagine that atomic hydrogen, or metallic H analogous to metallic Li or Na, would "burn" to H_2, releasing 104 kcal/mol (52 kcal/g) and producing an $\overline{\text{MW}}$ of about 2 to thereby be an excellent monopropellant. But how would one make pure solid H? It is possible to make a beam of H and to separate the two spin states in what is called a Stern-Gerlach experiment. Here the beam is passed through an inhomogenous magnetic field to separate the spin states. The experiment was first done with silver atoms because they were easy to produce, but it has also been done with H atoms as well. Now two spin-aligned atoms cannot combine to form H_2. So the propellant designer might imagine freezing out the spin-aligned H at perhaps liquid helium temperatures to make solid H(+½) or solid H(−½). These solids might then be catalytically burned to H_2 in the rocket engine.

This experiment was conducted at the National Bureau of Standards, but it failed, for the separated species are too easily converted one to another and only H_2 is condensed. But it is nonetheless an interesting idea in propulsion chemistry.

Linear Species

All polyatomic species can absorb energy in translational and electronic excitation just as did monatomic species, but, in addition, energy can be absorbed by the rotation of the molecule as a whole and by its internal vibrations. In all that follows, we will frequently refer to the "degrees of freedom" of a species. By this we will mean the total number of independent terms necessary to describe the total kinetic energy of a molecule, and this is clearly just $3n$, where n is the number of atoms in the molecule. That is, the entire motion of the molecule and hence its entire kinetic energy (neglecting potential energy effects arising from vibration or from external force fields) is given if we fix the velocity of each atom of the molecule in each of the three dimensions in which it moves. Now it will be more convenient to separate the translational motion of the center of mass of the molecule from the coordinated motion of the atoms within the molecule, but however we choose to describe their motion, we see that it will nevertheless be necessary to specify $3n$ independent variables to fix this total kinetic energy. These $3n$ variables are the degrees of freedom of the species. Of the six modes of kinetic energy content of all diatomic species, it is convenient to ascribe three to translation of the center of mass in three dimensions, two to rotation, and one to vibration, while for linear acetylene three are in translation, two in rotation, and seven are in vibration. We will model the molecule as a rigid rotator and a harmonic oscillator in a plane, and we will also take the several energy modes to be separable. We recognize that this separability is an approximation, for wider vibrational excursions will surely produce a greater moment of inertia and thus affect the rotational energy. These sorts of interactions in reality are small, but we will return to this issue of a better model subsequently. The quantum mechanical problem of the rigid rotator in a plane is another undergraduate textbook problem that is readily solved to give energy levels of

$$\epsilon_{\text{rot}} = \frac{J(J+1)h^2}{8\pi^2 I} \qquad J = 0, 1, 2, 3, \ldots \tag{3.29}$$

$$= J(J+1)Bh$$

where $B = h/8\pi^2 I$ is the *rotational constant* and I is the moment of inertia about the center of mass. Then ϵ_{rot} are the energies for the rotation of the rigid rotator about its center of mass. With h in $\text{J} \cdot \text{s}$ and I in $\text{kg} \cdot \text{m}^2$, ϵ_{rot} will be given by Eq. (3.29) in units of joules. J is a quantum number, and each energy level exhibits a degeneracy of $(2J + 1)$. All linear molecules have equal moments of inertia, and hence equal energy levels about two of its axes, but a zero moment about its third axis. That is, the atoms are considered to be point masses along the line of centers. We also note that the quantum of energy that is absorbed by a rotator in moving from J to $(J + 1)$ is just $(J + 1)h^2/4\pi^2 I$, and we see that the greater the rotational quantum number, the further apart will be the absorption or emission lines in the spectrum.

The harmonic vibrational motion is still another undergraduate textbook problem in quantum mechanics which readily leads to nondegenerate energy levels of

$$\epsilon_{vib} = (n + \tfrac{1}{2})h\nu \qquad n = 0, 1, 2, 3, \ldots \tag{3.30}$$

where n is a quantum number and ν is the fixed harmonic frequency in s^{-1}. We note that the energy levels are all equally spaced and that transitions will be of energy $h\nu$.

The relative sense of the electronic, vibrational, and rotational energy levels are schematically evident in Fig. 3.4. The arrow indicates a transition from electronic state B, vibrational quantum number 1, and rotational quantum number of 10 to electronic state A, vibrational quantum number of 0, and rotational quantum number of 11. And the

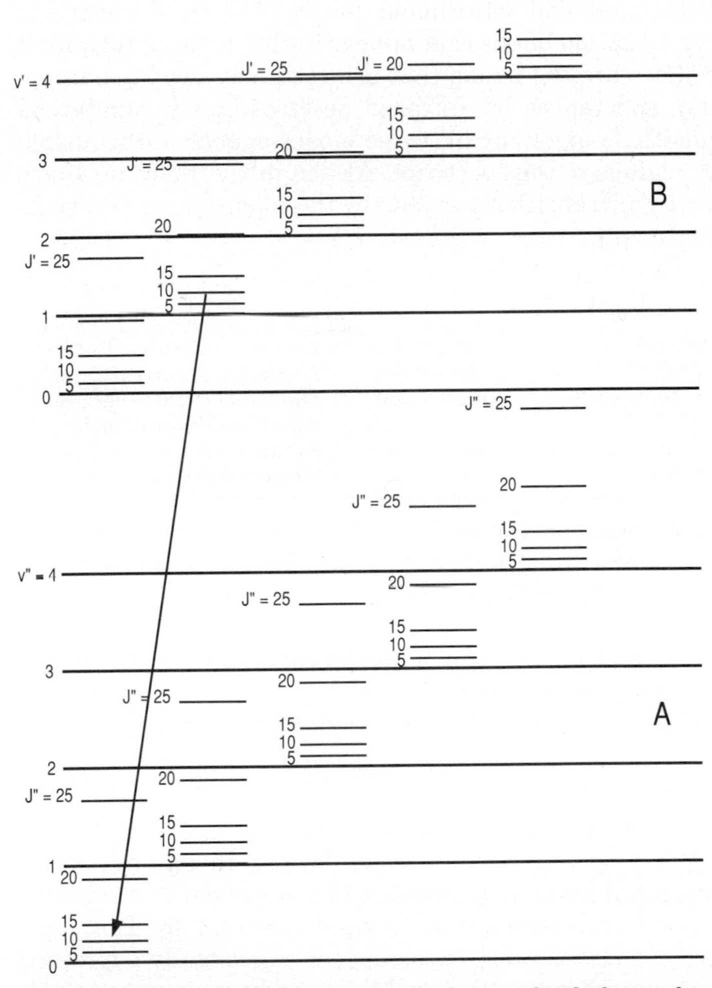

Figure 3.4 Schematic of ground and excited energy levels due to electronic configuration, vibration, and rotation of a diatomic molecule. *(From J. D. Graybeal, Molecular Spectroscopy, McGraw-Hill, 1988. Used by permission.)*

line in the emission spectrum that would be experimentally observed would appear at a frequency of $[\epsilon(B, 1, 10) - \epsilon(A, 0, 11)]/h$. And this is only one transition. When we imagine all of the possibilities involving many electronic states, many vibrational states within each electronic state, and many rotational states within each vibrational state, we begin to realize the origin of the complexity of the electronic spectra of molecules. Transitions that occur between rotational levels only of a fixed vibrational and electronic state give rise to pure rotational spectra in either absorption or emission, and these lines appear in the microwave region of the electromagnetic spectrum. Transitions that occur between rotational and vibrational levels of a fixed electronic state give rise to vibration bands that appear in the infrared regions of the spectrum. These are the bands that are routinely used to characterize particular substances by infrared spectroscopy in analytical chemistry. Transitions involving all three modes appear in the visible and ultraviolet regions of the spectrum. Although there are no sharp lines of demarcation, transitions appear in the electromagnetic spectrum about as follows:

	ν, s^{-1}	λ	Source
X-ray	10^{17}–10^{20}	1–0.001 nm	Transition of inner electron
Ultraviolet	10^{15}–10^{16}	400–10 nm	Transition of outer electron
Visible	4×10^{14}–8×10^{14}	0.8–0.4 μm	Transition of outer electron
Infrared	10^{11}–10^{14}	1 mm–1 μm	Vibrational transition
Microwave	10^{9}–10^{11}	100–1 mm	Rotational transition
FM and TV	100 MHz	1 m	Electronic device

Not all possible transitions are actually observed, and this has given rise to so-called selection rules that may be more or less rigorously obeyed. All such considerations are in the domain of molecular spectroscopy.†

The partition functions corresponding to these two sets of energy levels are also readily determined.

The rotational partition function is clearly

$$f_{\text{rot}} = \Sigma(2J + 1)e^{-hBJ(J+1)/kT} \tag{3.31}$$

The rotational constant may be expressed either as $B = h/8\pi^2 I$ or $B = h/8\pi^2 cI \times 10^2$. I appears in data compilations in $kg \cdot m^2$, while the rotational constant involving c appears in units of cm^{-1}, read *wave numbers*. This is common notation in infrared spectroscopy. However, in the literature of microwave spectroscopy, B is generally expressed in units of cycles per second $\times 10^{-6}$, or MHz, or dimensionally as s^{-1},

†See, for example, Graybeal.[7]

as is evident in the first expression for B above. The conversion factor is, of course, just the velocity of light, c in cm/s, or B in cm^{-1} is just B/c with B in s^{-1}. So data compilations appear that list I in kg · m^2, or B in s^{-1}, or B in cm^{-1}, and they are all perfectly equivalent. We merely must be alert to make sure that the exponent of the rotational partition function is dimensionless.

Some rotational constants are tabulated in Table 3.6, where one should note those data of extraordinary precision which have been obtained from microwave spectroscopy—up to 9 significant figures! No measurement can be made as precisely as can a frequency, so when utmost accuracy is important, one should strive to measure the phenomenon of interest in terms of a frequency. This may not always be possible. We also note a most significant point that spectroscopic measurements can be made on free radicals, ions, unstable, or otherwise transitory species which then allows us to calculate the thermodynamic properties of such species. This is important in a wide variety of practical situations including all high-temperature processes such as combustion, MHD applications, and plasmas. Calorimetric measurements on such species is, of course, impossible, so the theory provides the only approach to their thermodynamic behavior. Also note that the rotational constant is different for different isotopes, and therefore the entropies and then the Gibbs and Helmholtz free energies of isotopic species at the same conditions will be different.

Interestingly, the rotational constants in megahertz are somewhat above the TV broadcast range. VHF, i.e., channels 2 to 13, lies from

TABLE 3.6 Rotational Data on Several Representative Linear Molecules

Molecule	B (MHz) or I (kg · m^2)	Molecule	B (MHz) or I (kg · m^2)
H_2	0.473×10^{-47}	CS	24,495.592
F_2	31.8×10^{-47}	$Al^{35}Cl$	7288.73
Cl_2	113.5×10^{-47}	COS	6081.49255
Br_2	341.4×10^{-47}	HCN	44,315.9757
I_2	740.5×10^{-47}	FCN	10,544.20
$F^{35}Cl$	15,483.688	LiOH	35,342.44
$F^{37}Cl$	15,189.221	N_2O	$\{66.1 \times 10^{-47}$ 12,561.6338
HI	192,658.8	HC≡C—C≡N	4549.067
O_2	19.2×10^{-47}	C_2N_2	175×10^{-47}
N_2	13.8×10^{-47}	C_3O_2	395.2×10^{-47}
CO	57,635.970	OH	1.513×10^{-47}
$H^{35}Cl$	312,991.30		

SOURCE: Rotational data on hundreds of compounds are tabulated in the *Landolt-Börnstein Handbook, Numerical Data and Functional Relationships in Science and Technology*, Group II, vol. 6. Springer-Verlag, Berlin, 1974.

about 100 to 200 MHz and UHF from about 230 to 600 MHz. AlCl is at about 7000 MHz.

The rotational partition function of Eq. (3.31) may be readily summed, and the result is

$$f_{\text{rot}} = \int_0^\infty (2J + 1) \exp\left[-J(J+1)\frac{hB}{kT}\right] dJ = \frac{8\pi^2 IkT}{\sigma h^2} \tag{3.32}$$

provided only that the temperature is sufficiently high that integration over the quantum numbers, rather than summation, provides an accurate evaluation of the partition function. For everything other than H_2, D_2, HF, HCl, and HBr, this means temperatures greater than a fraction to a few kelvins.

The quantum-mechanical solution to the rigid-rotator problem also works out such that asymmetric linear molecules may have any value of J, while symmetric molecules may have either any odd value of J or any even value. Since the energy levels are so close together, adjacent levels make essentially the same contribution to f_{rot}, so we will introduce no real error by merely dividing the sum over all J's by 2 when we are concerned with a symmetric molecule. This gives rise to the so-called symmetry number, or σ, in the above partition function. Then σ is set equal to 1 or 2 for asymmetric or symmetric linear molecules respectively, so its assignment is readily done by simple inspection. This partition function is adequate for all linear species except H_2 and D_2 (but not HD), which present special cases because of their nonzero nuclear spin and the very low moment of inertia of the molecules. We shall omit consideration of this special and unique circumstance.

With the vibrational energy levels, Eq. (3.30), the partition function for a single vibrator, clearly becomes just a geometric progression,

$$f_{\text{vib}} = e^{-h\nu/2kT} + e^{-3h\nu/2kT} + e^{-5h\nu/2kT} + \cdots \tag{3.33}$$

which may be converted to

$$f'_{\text{vib}} = 1 + e^{-h\nu/kT} + e^{-2h\nu/kT} + e^{-3h\nu/kT} + \cdots,$$

if we agree to a scale change wherein we arbitrarily set the energy of the vibrator in its ground state ($n = 0$) equal to zero rather than its quantum-mechanical value of $\frac{1}{2}h\nu$. The effect on the partition function of changing the energy zero was discussed in the text leading to Eq. (2.15). All this is of no consequence, since absolute energy has no meaning anyway; we can establish an energy zero at our convenience.

The insignificance of this is again apparent from recalling that $U = nkT^2[\partial(\ln f)/\partial T]_V$, and, consequently, should we include the exp $(-h\nu/2kT)$ term in the vibrational partition function, it would produce merely an additive constant in U^0 of $\frac{1}{2} n_A h\nu$ J/mol. It then will not appear at all in C_V^0 (vib), nor does it appear at all in the expression for the entropy. It disappears by differentiation in the former, and it cancels in the latter. To remind us of this arbitrary choice of the energy zero, we will always write $(U^0 - U_0^0)$,$(H^0 - U_0^0)$,$(G^0 - U_0^0)$, and $(A^0 - U_0^0)$ for the four energy functions. Some data compilations use H_0^0 rather than U_0^0, but it is of no consequence, since $H_0^0 = U_0^0 + pV = U_0^0$ for an ideal gas at 0 K. The series extends over all positive integers from $n = 0$ to $n = \infty$. Forming $f'_{vib}e^{-h\nu/kT}$,

$$f'_{vib}e^{-h\nu/kT} = e^{-h\nu/kT} + e^{-2h\nu/kT} + e^{-3h\nu/kT} + \cdots$$

and subtracting this from f'_{vib}, we write

$$f'_{\text{vib}}(1 - e^{-h\nu/kT}) = 1$$

and

$$f'_{\text{vib}} = (1 - e^{-h\nu/kT})^{-1} \tag{3.34}$$

Although it is of no additional value, we note that the complete vibrational partition function with $\epsilon_0 = \frac{1}{2}h\nu$ rather than $\epsilon_0 = 0$ is $f(\text{vib}) = 2 \sinh (h\nu/2kT)$.

Relative Occupation of Energy Levels

One can readily see the effect of temperature on the occupation of vibrational energy levels by using the simple partition function of Eq. (3.34). Recall that, in general,

$$P = \frac{n_i}{n} = \frac{e^{-\epsilon_i/kT}}{f}$$

where the fraction of total molecules in level i, P, is also called the probability of that state. Then for the single vibrator of a diatomic molecule,

$$P = \frac{e^{-n_i h\nu/kT}}{1 - e^{-h\nu/kT}}$$

The relative occupation appears in Fig. 3.5. At low temperatures, the only significant occupation is $n = 0$. Note that *low temperature* is defined relative to $h\nu$, i.e., $T < h\nu/k$.

As the temperature is raised, the upper levels are progressively populated. Consider as an example iodine at room temperature, with a vibrational energy separation of 215 cm^{-1} and $h\nu/kT$ of

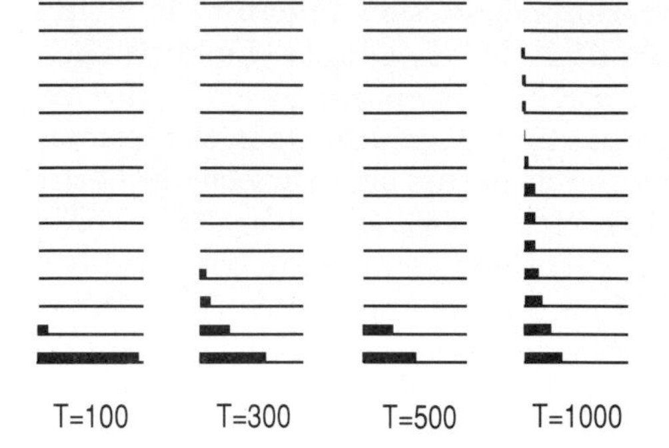

Figure 3.5 Population of vibrational states at several temperatures for I_2.

$$\frac{(6.626 \times 10^{-34}\ \text{J} \cdot \text{s})(215\ \text{cm}^{-1})(2.998 \times 10^{10}\ \text{cm} \cdot \text{s}^{-1})}{(1.381 \times 10^{-23}\ \text{J} \cdot \text{K}^{-1})(298\ \text{K})} = 1.038$$

from which we calculate $P_0 = 0.646$, $P_1 = 0.229$, $P_2 = 0.081$, $P_3 = 0.029, \ldots$ etc. at $n_i = 0,1, 2, 3$, etc. The I-I bond is weak, the frequency is low, the upper levels are significantly populated even at room temperature, and we anticipate (and properly so) that this vibrational mode will make significant contributions to the thermodynamic properties of iodine at room temperature.

Populations for all modes for all species behave similarly.

Diatomic Species

The total partition function for all diatomic species is then given by

$$f_{\text{di}} = g_0\left(\frac{2\pi mkT}{h^2}\right)^{3/2} V\left(\frac{8\pi^2 IkT}{\sigma h^2}\right)(1 - e^{-h\nu/kT})^{-1} \qquad (3.35)$$

and we will see subsequently just how easily this may be generalized to describe all linear molecules.

Rotational spectra from which we may obtain I are observed in the microwave region, vibrational spectra from which we may obtain ν are observed in the infrared, and electronic spectra are observed in the visible and ultraviolet regions of the spectrum. An exponential distribution of molecules in each mode is characterized by a particular thermodynamic temperature in kelvins, but nonequilibrium conditions can be made to exist such that these temperatures need not be the same. The rotational temperature is quickly equilibrated with the translational; that is, translation-rotation exchange is very fast. However, vibrational coupling to translation and rotation is poor, vibrational disequilibrium is very easy to produce in practice, and one can readily measure a vibrational temperature using spectroscopic techniques that is not the same as the translational temperature that would be measured with, say, a thermocouple. The electronic temperature can also be vastly different from the translational temperature. Many sorts of practical situations produce significant vibrational and electronic disequilibrium, but not necessarily with the relative occupation of these levels being exponential and thus completely describable by simply specifying a temperature. For example, this sort of nonmaxwellian vibrational or electronic excitation is necessary for the production of an infrared or a visible laser, respectively. The laser then depends for its existence on the production of disequilibrium among the modes of energy absorption of molecules.

It is interesting to note that, with little exception, chemistry as we know it is the chemistry of the electronic ground state. The chemistry of electronically excited species is sure to be distinctly different from that of the ground state particularly when there is a change in multiplicity as in, for example, ground-state $^3\Sigma$ oxygen and excited $^1\Delta$ oxygen. Electronically excited species can be made by any high-energy process, such as an electric discharge. Interestingly, electronically excited oxygen, $O_2\ ^1\Delta$, can be readily produced in large amounts from a simple chemical reaction,

$$Cl_2\ (g) + H_2O_2\ (aq.\ soln.) \rightarrow O_2\ ^1\Delta\ (g) + 2HCl\ (aq.\ soln.)$$

Here chlorine is merely sparged into the peroxide.

Some typical values for the characteristic molecular parameters of several diatomic molecules, I, ω, and D (the dissociation energy), are collected in Table 3.7.

In this table, the specific state of each species is described by a spectroscopic term symbol, e.g., H_2 is a $^1\Sigma$, O_2 is $^3\Sigma$, etc., and read *singlet sigma, triplet sigma*, etc. This shorthand notation that describes the electronic state of species has been described for atoms. It is similar for molecules, and the interested reader is referred to any of the sev-

TABLE 3.7 Spectroscopically Measured Properties of Some Typical Diatomic Molecules

Molecule	State	ω, cm^{-1}	B, MHz	D, kcal/mol
H_2	$^1\Sigma_g^+$	4395	1,774,200	103.19
F_2	$^1\Sigma_g^+$	892	26,390	37.8
Cl_2	$^1\Sigma_g^+$	565	7,895	57.06
Br_2	$^1\Sigma_g^+$	323	2,458	45.43
I_2	$^1\Sigma_g^+$	215	1,133	35.539
OH	$^2\Pi$	3735	554,660	100.2
O_2	$^3\Sigma_g^-$	1580	43,708	117.10
N_2	$^1\Sigma_g^+$	2360	60,812	225
CO	$^1\Sigma^+$	2170	58,612	233

eral textbooks noted at the end of this discussion for further explanation. Note that Table 3.7 gives data sufficient to calculate all of the properties of nonisolable OH just as it does for ordinary O_2.

All of the thermodynamic properties of all diatomics at all temperatures and pressures may now be readily evaluated. For example, consider the internal energy

$$(U^0 - U_0^0) = nkT^2 \left(\frac{\partial \ln f}{\partial T}\right)_V$$

and let us write $\ln f$ for a diatomic molecule as

$$\ln f = \frac{5}{2} \ln T - \ln (1 - e^{-h\nu/kT}) + \cdots$$

When differentiated with respect to temperature at constant volume, this becomes

$$\left(\frac{\partial \ln f}{\partial T}\right)_V = \frac{5}{2}\left(\frac{1}{T}\right) + \frac{h\nu}{kT^2}\left(\frac{e^{-h\nu/kT}}{1 - e^{-h\nu/kT}}\right)$$

and we can then write the internal energy in dimensionless form as

$$\frac{U^0 - U_0^0}{RT} = \frac{5}{2} + \frac{h\nu/kT}{e^{h\nu/kT} - 1} \tag{3.36}$$

Since the enthalpy is defined as $H = U + pV$ or $H^0 = U^0 + RT$ for an ideal gas, then,

$$\frac{H^0 - U_0^0}{RT} = \frac{H^0 - H_0^0}{RT} = \frac{7}{2} + \frac{h\nu/kT}{e^{h\nu/kT} - 1} \tag{3.37}$$

The value of the term in $h\nu/kT$ is tabulated in Table 3.8, so the evaluation of the internal energy or the enthalpy at any temperature is merely a matter of reading a value from Table 3.8 at the appropriate value of ν/T (or ω/T, see below). The heat capacity $(\partial H/\partial T)_p$ is obviously written

$$C_p^0 = R\left[\frac{7}{2} + \left(\frac{hc\omega}{kT}\right)^2 \frac{e^{-hc\omega/kT}}{(1 - e^{-hc\omega/kT})^2}\right] \tag{3.38}$$

which may also be equally well expressed as

$$C_p^0 = R\left[\frac{7}{2} + \frac{\theta_E^2 e^{\theta_E/T}}{T^2(e^{\theta_E/T} - 1)^2}\right] \tag{3.39}$$

where θ_E is called the *Einstein temperature* and is clearly just $h\nu/k$.

It has become standard practice in optical spectroscopy to give vibrations in cm^{-1}, read *wave numbers* and given the symbol ω. The wave number is just the number of cycles per centimeter length of the wave, or $\omega = \nu/c$ or frequency divided by the velocity of light, 3×10^{10} $\text{cm} \cdot \text{s}^{-1}$. For numerical calculations, it is convenient to note that

$$\frac{h\nu}{kT} = \frac{\omega hc}{kT} = 1.438831 \frac{\omega}{T}$$

where ω is in cm^{-1} and T in K. The second term in the heat capacity expression [Eq. (3.39)] is called the *Einstein heat capacity function*, the shape of which is shown schematically in Fig. 3.6. This expression was first developed by Einstein to describe the heat capacity of solids, but it was quickly supplanted by the Debye formulation. The assumption of a fixed vibrational frequency, while not reasonable for solids, is accurate for intramolecular vibrations, and we will use the Einstein formulation to calculate such vibrational contributions to all of the thermodynamic properties. For ease of calculation, Table 3.8 presents these vibrational contributions to the free energy, enthalpy, entropy, and heat capacity at values of ω/T (in $\text{cm}^{-1} \cdot \text{K}^{-1}$) at sufficiently close intervals that linear interpolation is accurate to within a maximum error of 1 percent.†

Interestingly, the spectroscopist determines the molecular parameters, B and ν, in an experiment using a real gas, and we then use these parameters to determine the properties of not the real but the ideal gas. This is not troublesome because the intramolecular modes are un-

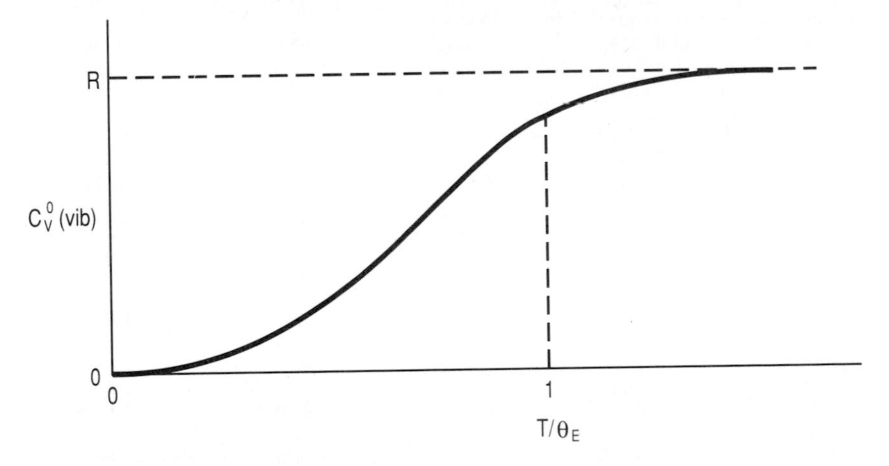

Figure 3.6 Einstein heat-capacity function.

†Closer spaced values are available in Hilsenrath and Ziegler.[8]

TABLE 3.8 Einstein Contributions to Thermodynamic Properties

ω/T, $cm^{-1} \cdot K^{-1}$	$-\left(\frac{G^0 - H_0^0}{RT}\right)$	$\frac{H^0 - H_0^0}{RT}$	$\frac{S^0}{R}$	$\frac{C_V^0}{R}$
0.10	2.00983	0.92979	2.93961	0.99828
0.15	1.63926	0.89596	2.53523	0.99613
0.20	1.38604	0.86300	2.24905	0.99313
0.25	1.19693	0.83091	2.02783	0.98929
0.30	1.04822	0.79965	1.84787	0.98461
0.40	0.82646	0.73969	1.56615	0.97285
0.50	0.66755	0.68305	1.35061	0.95796
0.60	0.54778	0.62970	1.17748	0.94014
0.70	0.45452	0.57955	1.03406	0.91958
0.80	0.38024	0.53252	0.91275	0.89654
0.90	0.32008	0.48851	0.80859	0.87124
1.00	0.27076	0.44743	0.71820	0.84397
1.10	0.22993	0.40916	0.63910	0.81501
1.20	0.19588	0.37360	0.56948	0.78464
1.30	0.16729	0.34062	0.50791	0.75314
1.40	0.14318	0.31010	0.45328	0.72081
1.50	0.12277	0.28190	0.40467	0.68789
1.60	0.10541	0.25592	0.36133	0.65466
1.70	0.09062	0.23202	0.32265	0.62136
1.80	0.07799	0.21007	0.28806	0.58820
1.90	0.06718	0.18997	0.25714	0.55541
2.00	0.05791	0.17157	0.22948	0.52315
2.10	0.04995	0.15477	0.20472	0.49159
2.20	0.04311	0.13945	0.18256	0.46088
2.30	0.03723	0.12551	0.16274	0.43112
2.40	0.03215	0.11284	0.14500	0.40241
2.50	0.02779	0.10136	0.12913	0.37483
2.60	0.02402	0.09093	0.11495	0.34846
2.70	0.02076	0.08151	0.10227	0.32331
2.80	0.01795	0.07300	0.09095	0.29943
2.90	0.01553	0.06532	0.08085	0.27680
3.00	0.01344	0.05839	0.07183	0.25545
3.10	0.01163	0.05215	0.06378	0.23536
3.20	0.01016	0.04655	0.05664	0.21650
3.30	0.00871	0.04151	0.05022	0.19885
3.40	0.00753	0.03700	0.04453	0.18237
3.50	0.00652	0.03295	0.03947	0.16701
3.60	0.00564	0.02932	0.03497	0.15275
3.70	0.00489	0.02608	0.03096	0.13952
3.80	0.00423	0.02318	0.02741	0.12727
3.90	0.00366	0.02059	0.02425	0.11596
4.00	0.00317	0.01828	0.02145	0.10553
4.10	0.00274	0.01622	0.01896	0.09593
4.20	0.00237	0.01438	0.01676	0.08712
4.30	0.00206	0.01274	0.01481	0.07902
4.40	0.00178	0.01129	0.01307	0.07162
4.60	0.00133	0.00885	0.01018	0.05865
4.80	0.00100	0.00692	0.00793	0.04786
5.00	0.00075	0.00541	0.00616	0.03892
5.10	0.00065	0.00478	0.00543	0.03506
5.20	0.00056	0.00421	0.00478	0.03156
5.30	0.00049	0.00372	0.00421	0.02839
5.40	0.00042	0.00328	0.00370	0.02552
5.60	0.00032	0.00256	0.00287	0.02058
5.80	0.00024	0.00198	0.00222	0.01655
6.00	0.00018	0.00154	0.00172	0.01328
6.50	0.00009	0.00081	0.00090	0.00759
7.00	0.00005	0.00043	0.00047	0.00431

SOURCE: Adapted from the extensive "Tables of Einstein Functions" by Hilsenrath and Ziegler.[8]

affected by intermolecular forces at pressures used by the spectroscopist. In fact, the infrared spectrum is usually only modestly changed even upon condensation. Both the constancy and the separation of the intramolecular modes makes possible empirical group contribution schemes such as UNIFAC that are popularly used to estimate activity coefficients in liquid solutions. The details of the intricate UNIFAC scheme are described in all modern textbooks on engineering thermodynamics.

All of the temperature dependency of the heat capacity then arises from the vibrational contribution, and this contribution varies from 0 to R for each vibrator in the molecule. The relative contributions of the several modes to the heat capacity of all diatomic species is obviously

$$C_V^0(\text{trans}):C_V^0(\text{rot}):C_V^0(\text{vib}) = \frac{3}{2}R:\frac{2}{2}R:0 \text{ to } R$$

We note that at high temperatures, $\omega hc/kT$ is a very small number, and $e^{-\omega hc/kT}$ can then be well replaced by $(1 - \omega hc/kT)$; the above expression for $C_V^0(\text{vib})$ becomes equal to just R, which is also just the value, but constant for all temperatures, that one would obtain from a wholly classical mechanical treatment of the harmonic oscillator (see Chap. 9). This function gives precise values of the vibrational heat capacity contribution up to temperatures at which the harmonic model becomes unreliable, and even then it is possible to make anharmonic corrections to the vibrational partition function, as we shall see subsequently.

With $x = hc\omega/kT$, we can compactly write the following expressions for the contributions to the thermodynamic properties that arise from each vibrator:

$$\frac{-(G^0 - H_0^0)}{RT} = -\ln(1 - e^{-x})$$

$$\frac{H^0 - H_0^0}{RT} = xe^{-x}(1 - e^{-x})^{-1}$$

$$\frac{S^0}{R} = xe^{-x}(1 - e^{-x})^{-1} - \ln(1 - e^{-x}) \tag{3.40}$$

$$\frac{C_V^0}{R} = x^2e^{-x}(1 - e^{-x})^{-2}$$

These are the quantities that are presented in Table 3.8. Note that the tabulated properties are dimensionless, and therefore the actual property may be obtained in any unit by multiplying by R in the desired unit. For example, at $\omega = 1.8 \times 10^3$ cm^{-1} and $T = 600$ K, S^0/R

$= 0.0718$ and $S^0 = 0.597$ J/mol · K, where we have used $R =$ 8.314333 J/mol · K. Although Table 3.8 is convenient for small or infrequent calculations, a very simple computer code is readily written to calculate the Einstein contributions for more extensive calculations or for frequent usage.

The complete entropy of all diatomic species is given by

$$\frac{S^0(T,p)}{R} = \left\{\ln\left[\left(\frac{2\pi mkT}{h^2}\right)^{3/2}\frac{kT}{p}\right] + \frac{5}{2}\right\} + \left\{\ln\left(\frac{8\pi^2 IkT}{\sigma h^2}\right) + 1\right\}$$
$$+ \left\{\frac{h\nu}{kT}(e^{h\nu/kT} - 1)^{-1} - \ln(1 - e^{-h\nu/kT})\right\} + \ln g_0 \qquad (3.41)$$

where the collections of terms in the braces represent the contribution to the entropy from translation (and you may note that it is the same as for monatomic species), rotation, vibration, and electronic excitation, respectively.

In practical units this becomes

$$\frac{S^0(T,p)}{R} = \left(\frac{3}{2}\ln M + \frac{5}{2}\ln T - \ln p + 3.4533\right)$$
$$+ (\ln T - \ln\sigma - \ln B + 10.9444) + \frac{S^0(\text{vib})}{R} + \ln g_0$$

where p is in kPa, B in MHz, T in K, and the next to last term is from Eq. (3.40) and its value is read from Table 3.8. Since we are here concerned with ideal gases, it would have been very disappointing if pressure had appeared here other than as $-R \ln p$. Further, it is well to recall that even isotopically different homonuclear diatomics are unsymmetrical and then have $\sigma = 1$. That is, $^{16}O\text{-}^{16}O$ has $\sigma = 2$, but $^{16}O\text{-}^{18}O$ has $\sigma = 1$, and the entropy of the asymmetric species will be less than that of the symmetric species by $R \ln 2$ or 1.377 cal/mol · K at all temperatures and pressures, other things being equal. The Gibbs and Helmholtz free energies will be similarly different, and the equilibrium conversion in any reaction involving O_2 will reflect this difference. One could utilize this difference in a process for isotopic enrichment, and this is the basis of the deuterium enrichment process using H_2S and isotopic exchange that is operated commercially.

As an example of the calculation of the entropy, the molecular weight of nitrogen is 28.016 g, its symmetry number is 2, its vibration frequency is 2359 cm^{-1}, and its moment of inertia is 13.81×10^{-47} kg · m^2. These data are ready to use except for the moment of inertia that must be converted to megahertz. That is,

$$\begin{aligned}
\mathrm{B(MHz)} &= B(\mathrm{cm}^{-1})c(\mathrm{cm}\cdot\mathrm{s}^{-1})10^{-6} \\
&= \left[\frac{h}{8\pi^2 cI10^2}\right]c \times 10^2 \times 10^{-6} \\
&= \frac{6.6262 \times 10^{-34} \times 10^{-6}}{8\pi^2 \times 13.81 \times 10^{-47}} \\
&= 60{,}769\ \mathrm{MHz}
\end{aligned}$$

At 298.15 K, we then calculate contributions to the entropy of 35.92, 9.80, and 0 cal/mol · K arising from the translational, rotational, and vibrational modes respectively. We note that the vibrational contribution becomes significant only at temperatures sufficiently high to excite the vibrator, that is, high enough to make ω/T sufficiently small. This does not occur for N_2 at 298 K. The total entropy of 45.72 cal/mol · K compares well with the experimental calorimetric value of 45.79 cal/mol · K wherein we note that there is a solid-phase transition in nitrogen that occurs near 35 K with a latent heat of about 55 cal/mol. This thermal effect along with that due to melting, boiling, and the heat capacities of all phases at all temperatures below the temperature of interest must, of course, be known if we are to determine a calorimetric third-law entropy.

Some comparisons of the properties of several diatomics calculated from the partition function with those from calorimetric measurements appear in Table 3.9. The agreement shown there is typical of that for all diatomic species. Molecular parameters may be more accurately measured than are calorimetric quantities; so such disagreements as occur are most likely due to inaccuracies in the calorimetric

TABLE 3.9 Calculated and Experimental Heat Capacities and Entropies of Several Diatomic Species at 101.325 kPa

		Experiment		Theory	
Species	T, K	C_p/R	S/R	C_0^0/R	S^0/R
O_2	100	3.546	21.736	3.501	20.835
	200	3.510	24.176	3.503	23.262
	300	3.537	25.602	3.534	24.687
	400	3.622	26.630	3.621	25.714
N_2	100	3.613	19.171	3.500	19.204
	200	3.515	21.625	3.501	21.631
	300	3.508	23.048	3.503	23.051
	400	3.521	24.059	3.581	24.060
CO	300	3.549	23.560	3.505	23.782
	400	3.539	24.801	3.529	24.792
	500	3.553	24.976	3.583	25.585

measurements. It is rare in science that theory is more accurate than experiment.

In work leading to the establishment of what is now known as the third law of thermodynamics, an interesting disparity was noted between the calculated and experimental entropies of mildly polar heteronuclear diatomics in that the experimental entropies from low-temperature calorimetric studies were always lower by about 4 J/mol · K than that determined using Eq. (3.41). It seems that the polarity of the molecule was not sufficient to cause all of the molecules to be aligned as the crystal grew at the freezing point. This resulted in residual disorder in the crystal at 0 K. If completely random, an ideal solid solution would have an entropy of 5.763 J/mol · K at 0 K, that is, $-R\Sigma x_i \ln x_i$ for the equimolar binary solution formed by the molecule lying in the lattice either one way or the opposite way. That is, insofar as the crystal is concerned, we can identify two species present in equal amounts if the orientation of each molecule is randomly one way or the other way. If the intermolecular forces cause the molecules in the liquid to go into the growing solid even partially preferentially aligned, then the residual entropy will be less than 5.763 J/mol · K and approach zero as the forces causing alignment in the crystal become increasingly effective. The calorimetric and statistical entropies of CO at 298.1 K and 101.3 kPa are 193.38 and 198.03 J/mol · K, respectively, whereas for HCl at the same conditions the numbers are 186.19 and 186.77 respectively. Unlike the weakly polar CO, the strong dipole of HCl causes the crystal to grow uniformly. In the case of CH_3D, there are four equivalent positions, and then the ΔS at 0 K is $-R \ln 0.25$ or 11.53 J/mol · K if the crystal is completely random. At 99.7 K, the calorimetric entropy of CH_3D is 153.64 J/mol · K, and the theoretical value is 165.23 J/mol · K. Upon adding the residual entropy of 11.53 to the experimental value, one obtains excellent agreement. It is interesting that the differences between the thermal and the statistical entropy revealed this randomness in the solid phase and thereby contributed significantly to the establishment of the third law of thermodynamics. A perfectly ordered crystal at 0 K has an entropy of zero.

The Gibbs free energy of all diatomic species as ideal gases is given by

$$-\left(\frac{G^0 - H_0^0}{T}\right) = nk \ln\left[g_0\left(\frac{2\pi mkT}{h^2}\right)^{3/2}\frac{kT}{p}\left(\frac{8\pi^2 IkT}{\sigma h^2}\right)(1 - e^{-h\nu/kT})^{-1}\right] \tag{3.42}$$

which in practical units is

$$-\left(\frac{G^0 - H_0^0}{RT}\right) = \frac{7}{2}\ln T + \frac{3}{2}\ln M + \ln g_0 - \ln p - \ln \sigma - \ln B$$
$$- \ln\left(1 - e^{-h\nu/kT}\right) + 10.8977$$

where T is in K, M is the molecular weight, g_0 is the ground-state multiplicity, p is the pressure in kPa, B is the rotational constant in MHz, the next to the last term is from Eq. (3.40) and is read from the table of Einstein contributions (Table 3.8) at the appropriate value of ω/T, and the last constant arises from the combined physical constants and numbers that appear in Eq. (3.42). Note that the symmetry number σ will be either 2 or 1, depending on whether the diatomic molecule is homo- or heteronuclear, respectively. The multiplicity g_0 is usually equal to 1 for stable species, but it will be 2 for all free radicals having one unpaired electron. There are a few free radicals that are chemically stable at room temperature, NO and O_2, for example, where g_0 is 2 and 3, respectively. In any event, the multiplicity of the electronic ground data g_0, as well as that of each excited electronic state, like the energy level itself, must be provided from experimental observation by the spectroscopist. Also note again that data compilations typically tabulate values of $(G^0 - H_0^0)/T$ obtained using the relationships shown here—that is, from the partition function in which the zero-point vibrational energy has been omitted, or, in other words, the zero of energy has been set at the zero-point vibration.

Multiatom Linear Species

The total energy of any molecule in three-dimensional space, regardless of its character, may be expressed as a sum of $3n$ kinetic energy terms, where n is the number of atoms. That is, each atom may have kinetic energy in the x, y, and z directions. Now it may be more convenient to separate and group these to talk about the kinetic energy of the center of mass of the assembly together with the rotation and vibration of the assembly itself. But mechanics insists that we still specify $3n$ kinetic energy terms to describe the molecular assembly. The molecule is said to have $3n$ degrees of freedom, and for all linear molecules, three of these describe the translational motion of the center of mass, two more describe the energy of rotation with its two (and equal) moments of inertia, and there will be then $3n - 5$ characteristic vibrational frequencies that contribute to the partition function of all linear molecules. Thus there will be four frequencies for triatomic species. The complete partition function in the rigid-rotator, harmonic-oscillator approximation for all linear triatomic molecules is then

$$f = g_0\left(\frac{2\pi(m_A + m_B + m_C)kT}{h^2}\right)^{3/2} V \frac{8\pi^2 IkT}{\sigma h^2}(1 - e^{-h\nu/kT})^{-1}$$
$$(1 - e^{-h\nu_2/kT})^{-1}(1 - e^{-h\nu_3/kT})^{-2} \quad (3.43)$$

where one vibrational mode is always doubly degenerate; that is, the molecule vibrates with the same characteristic frequency in two of its four modes. In this model, the four vibrators are taken to act independently of each other. This is an excellent approximation. As was the case for diatomic species, the entire temperature dependency of the heat capacity arises from the vibrational contributions,

$$C_p^0 = \frac{3}{2}R + \frac{2}{2}R + \sum_{i=1}^{i=4} C_V^0(\text{vib}) + R$$

where we now must read the contribution of each of the four vibrators using its particular ω/T and Table 3.8. Agreement with experiment is excellent. We infer again that spectroscopic data on transitory or unstable species allow one to calculate, with confidence, the species' thermodynamic properties at all temperatures and pressures. This is particularly important for free radicals and ions where calorimetric measurements are impossible.

Similarly to Eq. (3.41) for all diatomic molecules, the absolute entropy of all linear triatomic molecules is then

$$\frac{S^0}{R} = \ln\left[g_0\left(\frac{2\pi(m_A + m_B + m_C)kT}{h^2}\right)^{3/2}\frac{kT}{p}\right] + \frac{5}{2} + \ln\frac{8\pi^2 IkT}{\sigma h^2} + 1$$
$$+ \sum_{i=1}^{i=4}\left[\frac{h\nu_i}{kT}\left(\frac{e^{-h\nu_i/kT}}{1 - e^{-h\nu_i/kT}}\right) - \ln(1 - e^{-h\nu_i/kT})\right] \quad (3.44)$$

where the first two terms are those contributions to the entropy that arise from translation (and electronic ground-state degeneracy), the second two terms arise from rotation, and the final summation arises from vibration. If the molecule is symmetric, $\sigma = 2$ and if asymmetric, $\sigma = 1$. Comparisons of absolute entropies from this equation with that from low-temperature third-law calorimetric studies reveal excellent agreement as may be inferred from Table 3.10.

In all instances where there is a significant discrepancy, the calculated entropy is higher because of residual disorder in the crystal at 0 K that was omitted in the experimental assumption of zero entropy at 0 K. In every instance, this randomness is an artifact of the intermolecular forces that are operative as the crystal grows, as we have already discussed.

While certainly not all triatomic molecules are linear, there are a

TABLE 3.10 Entropies of Some Linear Triatomic Molecules†

Species	T, K	S^0(calc), J/mol · K	S^0(exp), J/mol · K
CO_2	194.7	198.95	199.1
CS_2	318.4	241.00	240.5
N_2O	298.2	220.00	215.2
HCN	298.2	201.8	200.5

†The estimated error for most calorimetric entropies is about 0.5 J/mol · K.

rapidly decreasing number of linear molecules that possess 4, 5, 6, etc. atoms. Dicyanoacetylene, NCC≡CCN, is an interesting linear molecule of 6 atoms. For each of these, and for all linear molecules regardless of complexity, the complete partition function is

$$f = g_0\left(\frac{2\pi mkT}{h^2}\right)^{3/2} V \frac{8\pi^2 IkT}{\sigma h^2} \prod_{i=1}^{3n-5} (1 - e^{-h\nu_i/kT})^{-1} \qquad (3.45)$$

and we can write

$$C_p^0 = \frac{3}{2}R + \frac{2}{2}R + \sum_{i=1}^{3n-5} C_V^0(\text{vib}) + R \qquad (3.46)$$

and
$$\frac{S^0}{R} = \ln\left[g_0\left(\frac{2\pi mkT}{h^2}\right)^{3/2}\frac{kT}{p}\right] + \frac{5}{2} + \ln\left(\frac{8\pi^2 IkT}{\sigma h^2}\right) + 1 + \sum_{i=1}^{3n-5}\left[\frac{h\nu_i}{kT}\left(\frac{e^{-h\nu_i/kT}}{1 - e^{-h\nu_i/kT}}\right) - \ln(1 - e^{-h\nu_i/kT})\right] \qquad (3.47)$$

and
$$-\left(\frac{G^0 - H_0{}^0}{RT}\right) = \ln\left[g_0\left(\frac{2\pi mkT}{h^2}\right)^{3/2}\frac{kT}{p}\right] + \ln\left(\frac{8\pi^2 IkT}{\sigma h^2}\right) - \sum_{i=1}^{i=3n-5} \ln(1 - e^{-h\nu_i/kT}) \qquad (3.48)$$

and similarly for all of the other thermodynamic properties. In practical units, the entropy of all linear molecules may be expressed from Eq. (3.47) as

$$\frac{S^0}{R} = \left[\frac{3}{2}\ln M + \frac{5}{2}\ln T + \ln g_0 - \ln p + 3.4533\right] + (\ln T - \ln\sigma - \ln B + 10.9444) + \sum_{i=1}^{i=3n-5} \frac{S_i{}^0(\text{vib})}{R}$$

where, as before, M is the molecular weight, T is the temperature in K, p is the pressure in kPa, B is the rotational constant in MHz, and the contribution of each vibrational frequency to the entropy is read from Table 3.8 at each value of ω/T in $cm^{-1} \cdot K^{-1}$.

For completeness, we also here write the Gibbs free energy function for any linear molecule as

$$\left(\frac{G^0 - H_0^0}{RT}\right)_{elect} = -\ln g_0$$

$$\left(\frac{G^0 - H_0^0}{RT}\right)_{trans} = \frac{5}{2} - \frac{3}{2}\ln M - \frac{5}{2}\ln T + \ln p - 3.4533$$

$$\left(\frac{G^0 - H_0^0}{RT}\right)_{rot} = 1 - \ln T + \ln \sigma + \ln B - 10.9444$$

$$\left(\frac{G^0 - H_0^0}{RT}\right)_{vib} = \sum_{i=1}^{i=3n-5} \ln\left(1 - e^{-h\nu_i/kT}\right)$$

and the enthalpy is

$$\frac{H^0 - H_0^0}{RT} = \frac{7}{2} + \sum_{i=1}^{i=3n-5} \left[\frac{H^0 - H_0^0}{RT}\right]_{vib}$$

where, as before, the vibrational contribution to the free energy and enthalpy may be conveniently read from the table of Einstein functions at each value of ω/T.

A convenient source of vibrational data on molecules is *Tables of Molecular Vibration Frequencies*.[9] The *Landolt-Börnstein Handbook*[10] also provides a useful compilation.

Example Problem 3.1 Calculate the enthalpy of acetylene at 450 K and 10 atm. $I = 15.22$ amu $\cdot$ Å^2 and the frequencies are: 3374, 1974, 3287, 612(2), and 729(2) cm^{-1}. (Recall that 1 atomic mass unit = 1×10^{-3} kg/Avogadro's number or 1.66057×10^{-27} kg.)

Equation (3.37) was developed for a diatomic species, but it is obviously expanded for any linear species as:

$$\frac{H^0 - H_0^0}{RT} = \frac{7}{2} + \sum_{i=1}^{i=3n-5} \frac{h\nu_i/kT}{e^{h\nu_i/kT} - 1}$$

The seven vibrational contributions for C_2H_2 can be read from Table 3.8:

ω/T	$(H^0 - H_0^0)_{vib}/RT$	ω/T	$(H^0 - H_0^0)_{vib}/RT$
7.50	–	1.36†	0.32231
4.39	0.03745	1.62†	0.25114
7.30	–		

†Doubly degenerate.

$$\text{then}\quad H^0 - H_0^0 = 8.314(450)[3.5 + 0.03745 + 2(0.32231) + 2(0.25114)]$$
$$= 17{,}526 \text{ J/mol}$$

An estimate of the nonideality of C_2H_2 is most conveniently obtained by using any of the several principles of corresponding states (see Chap. 5). For this acetylene, $T_r = 0.82$, $p_r = 0.21$, $z_c = 0.184$, and, from the tables of Hougan et al. (see Chap. 5),

$$H^0 - H \cong 2400 \text{ J/mol}$$

and the enthalpy of the real gas is

$$H - H_0^0 = 17{,}500 - 2400 = 15{,}100 \text{ J/mol}$$

Concept of Temperature

The several modes of energy absorption and energy content in a molecular system are loosely connected to each other. That is, energy that is placed in a vibration will leak out into rotation and into translation at rates that depend on the particular species and the particular collision partners with which the species interacts. That is, different partners will exert different sorts of force fields on the species, which will then stimulate relaxation of the energy at varying rates (see Chap. 10).

Now the absolute temperature is a number that characterizes a particular exponential distribution of the occupation numbers of one of the modes of energy content of the species. In other words, the particular set of n_J in Table 3.11 corresponds to a temperature T that we may legitimately call the *rotational temperature*. That is, this one number will give back each of the population numbers from the basic Boltzmann equation,

$$\frac{n_J}{n} = \frac{(2J + 1)\exp[-J(J+1)\,h^2/8\pi^2 IkT]}{8\pi^2 IkT/\sigma h^2}$$

TABLE 3.11 Occupation of Rotational Energy Levels for HF

This set of occupation numbers is specified by stating the single parameter T = 300 K.

Rotational level	n_J/n	$\epsilon_J(\text{J}\cdot\text{mol}^{-1})$
$J = 0$	0.090	0
$J = 1$	0.246	502
$J = 2$	0.274	1,506
$J = 3$	0.210	3,012
$J = 4$	0.120	5,020
$J = 5$	0.054	7,530
$J = 6$	0.018	10,543

$$\frac{n_J}{n} = \frac{(2J + 1)\sigma\theta_R}{T} \exp\left[\frac{-J(J + 1)\theta_R}{T}\right] \tag{3.49}$$

where θ_R is the characteristic rotational temperature defined as $h^2/8\pi^2 Ik$. For HF, $I = 1.335 \times 10^{-47}$ kg · m^2 or $\theta_R = 30.2$ K. From Table 3.11, 93 percent of the rotators have $J \leq 4$ at $T = 300$ K.

Note that T has no meaning, it is indeed undefined, for other than a Boltzmann distribution of occupation numbers.

By stretching the definition of temperature somewhat, we can consider not all but only two levels from Table 3.10 and their relative occupation numbers will specify a "temperature" of these two levels. At $J = 3$ and 4, for example,

$$\frac{n_{J=4}}{n_{J=3}} = \frac{n_{J=4}/n}{n_{J=3}/n} = \frac{9}{7} \exp\left(-\frac{\epsilon_4 - \epsilon_3}{kT}\right)$$

and

$$\frac{0.120}{0.210} = \frac{9}{7} \exp\left(-\frac{5020 - 3012}{1.381 \times 10^{-23} \times T \times 6.023 \times 10^{23}}\right)$$

or $T = 300$ K. Conversely, we see that the ratio of these two occupation numbers is fixed by stating that $T = 300$ K. Figure 3.7 is a plot of

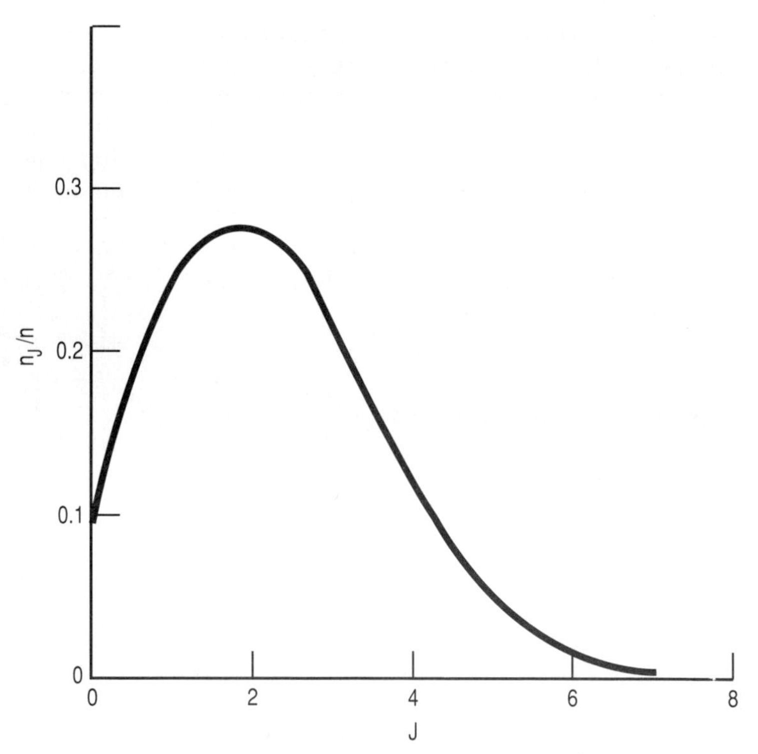

Figure 3.7 Rotational occupation numbers for HF at 300 K.

n_J/n versus J for HF. The point of maximum n_J/n may be obtained by merely differentiating Eq. (3.49) with respect to J and setting this derivative equal to 0. The result is $J = (T/2\theta_R)^{1/2} - \frac{1}{2}$, which for HF at 300 K corresponds to $J = 1.7$. The point of maximum J then depends on the product of the absolute temperature and the moment of inertia of the rotator, and as either increases, the maximum J also increases.

By any of many possible rapid energy input processes, it is possible to momentarily produce occupation numbers that are vastly different from the Boltzmann distribution. For example, in the present case of HF, if n at $J = 4$ is equal to n at $J = 3$, the implication is clearly that T = infinity. If n at $J = 4$ is greater than n at $J = 3$, then clearly the absolute temperature must be negative. It is interesting that lasers require such inverted populations (producing this inversion is called *pumping the laser*), and thus lasers typically operate at "negative absolute temperatures." The use of this rather provocative-appearing phrase requires, however, a significant extrapolation of the conventional meaning of the word temperature. The important thing from all of this is to realize again the connection between temperature and the Boltzmann distribution, and, the importance of that most probable distribution notwithstanding, it is nonetheless readily possible to produce very nonboltzmannian distributions for which temperature has no meaning in the conventional sense.

Because of the finite rates of energy transfer between modes, it is readily possible to produce molecular systems with, say, very different translational and vibrational temperatures. The former could be measured with a thermocouple while the latter could be measured spectroscopically by determining the occupation numbers. For example, the CO_2 laser may require a vibrational population of CO_2 that is so "hot" that the CO_2 is at negative kelvins, while the optical cavity containing the working laser medium of CO_2, He, and N_2 may be only warm to the touch. That is, the translational temperature is modest. Within a brief time after the source of energy is turned off, energy is transferred among the modes, a state of thermodynamic equilibrium is produced, and the occupation of all levels of all modes is fixed by one parameter—the absolute thermodynamic temperature. The required time is called the relaxation time. It is both species- and environment-specific, and for vibration-to-translation relaxation, this time is of the order of 10^{-4} s. Relaxation times between other energy modes are usually much shorter.

Nonlinear Species

One moment of inertia of both symmetric and asymmetric linear molecules will be zero and that about the other two axes will be equal. Thus there will only be one numerical value of the moment of inertia for all linear molecules, whether symmetric or not. By contrast, nonlinear molecules will, of course, have moments of inertia about all three principle axes; the energy levels are readily obtained from quantum-mechanical arguments as was the case for linear species, and the resulting rotational partition function is

$$f_{\text{rot}} = \frac{\pi^{1/2}(8\pi^2 kT)^{3/2}(I_A I_B I_C)^{1/2}}{\sigma h^3} \tag{3.50}$$

The symmetry number σ has the same meaning for nonlinear molecules as it did for linear molecules, and its value can be determined from inspection once the basic structure, but not angles and distances, of the molecule is known. Some typical values for moments, products of moments, or rotational constants appear in Table 3.12.

TABLE 3.12 Rotational Data on Several Representative Nonlinear Molecules

Molecule	B, MHz, or I, kg · m^2	Molecule	B, MHz, or I, kg · m^2
C — C ‖ ‖ C C N H	9130.610 9001.343 4532.083	NO_2	3.486×10^{-47} 63.76×10^{-47} 67.25×10^{-47}
C — C ‖ ‖ C C O	9447.153 9246.775 4670.846	CH_4 SiH_4	5.47×10^{-47} 9.78×10^{-47}
$(CH_3)_2NH$	34242.22 9334.03 8215.98	C_3H_4	98.276 98.276 5.499
$C{=}CCl^{35}$	56840.21 6029.96 5445.29	*cis*-C_4H_8	59.977 142.431 191.886
CH_2O	38835.369 34003.282	*trans*-C_4H_8	25.675 224.40 239.556
SO_2	60778.516 10317.963 8799.808	*cy*-C_6H_{12} NH_3	$I_1 \times I_2 \times I_3 = 12.583 \times 10^{-138}$ 2.782×10^{-47} 2.782×10^{-47} 4.33×10^{-47}
SO_2	13.3×10^{-47} 86.2×10^{-47} 99.4×10^{-47}		

SOURCE: *Landolt-Börnstein Handbook, Numerical Data and Functional Relationships in Science and Technology*, Group II, Vol. 6, Springer Verlag, Berlin, 1974.

The complete partition function for the most simple nonlinear molecule, that is, a bent triatomic species such as water, is

$$f = g_0\left(\frac{2\pi(m_A + m_B + m_C)kT}{h^2}\right)^{3/2} V\frac{\pi^{1/2}(8\pi^2kT)^{3/2}(I_AI_BI_C)^{1/2}}{\sigma h^3}$$
$$\times (1 - e^{-h\nu_1/kT})^{-1}(1 - e^{-h\nu_2/kT})^{-1}(1 - e^{-h\nu_3/kT})^{-1} \quad (3.51)$$

which should be compared with that for a linear triatomic species as presented in Eq. (3.43). We should note that, since there are here three degrees of rotational freedom, rather than two as for the linear triatomic species, there are here then only three vibrational contributions. For this partition function, the Gibbs free energy becomes

$$-\left(\frac{G^0 - H_0{}^0}{T}\right) = nk \ln\left[[g_0\left(\frac{2\pi mkT}{h^2}\right)^{3/2}\frac{kT}{p}\frac{\pi^{1/2}(8\pi^2kT)^{3/2}(I_AI_BI_C)^{1/2}}{\sigma h^3}\right.$$
$$\left.\times \sum_{i=1}^{i=3}(1 - e^{-h\nu_i/kT})^{-1}\right] \quad (3.52)$$

where we have again ignored the zero-point vibrational energies of each vibrator.

Molecules of whatever the complexity can be handled by the same sorts of techniques as we have now seen in detail for mono-, di-, and triatomic species. The calculations become more tedious, and an additional phenomenon, free rotation of groups about bonds, can occur and must be properly treated.

As we saw for linear molecules, the frequencies of nonlinear molecules may be degenerate, i.e., several can have the same value, depending on the symmetry of the molecule. For example, all symmetric tetraatomic pyramidal molecules have, of course, $3n - 6$, or 6, characteristic frequencies, of which two are always doubly degenerate. Such molecules are NH_3, PH_3, BF_3, $BiCl_3$, etc., and the frequencies of several of these molecules are summarized in Table 3.13. Pentatomic symmetric top molecules, such as $CHCl_3$, CH_3Cl, etc. have $3n - 6$, or 9, characteristic frequencies, of which three are doubly degenerate. The frequencies for chloroform are 3033, 667, 364, 1205 (2), 760 (2), and 262 (2). Of the 30 frequencies in benzene, 10 are doubly degener-

TABLE 3.13 Vibration Frequencies of Several Pyrimidal Molecules

Compound	ω_1	ω_2	$\omega_3(2)$	$\omega_4(2)$
NH_3	3337	950	3414	1628
PF_3	890	531	840	486
$BiCl_3$	288	130	242	96

ate, and the remaining 10 are nondegenerate. And so on. But whatever the pattern of vibrational frequencies, it must be determined by the spectroscopist, and the partition function for all nonlinear molecules is

$$f = g_0 \left(\frac{2\pi mkT}{h^2}\right)^{3/2} V \left[\frac{\pi^{1/2}(8\pi^2 kT)^{3/2}(I_A I_B I_C)^{1/2}}{\sigma h^3}\right] \prod_{i=1}^{i=3n-6} (1 - e^{-h\nu_i/kT})^{-1} \tag{3.53}$$

Electronic part Translational part Rotational part Vibrational part

which may be compared with Eq. (3.45) for all linear molecules. We recall that the general expression for enthalpy in terms of the partition function is

$$\frac{H^0 - H_0^0}{T} = RT\left(\frac{\partial \ln f}{\partial T}\right)_V + R \tag{3.54}$$

Then only that part of f which depends on temperature need be included in the differentiation of the logarithm. Thus,

$$\ln f = \frac{3}{2}\ln T + \frac{3}{2}\ln T - \sum_{i=1}^{i=3n-6} [\ln(1 - e^{-h\nu_i/kT}) + \cdots]$$

and on differentation,

$$\frac{\partial \ln f}{\partial T} = \frac{3}{2}\left(\frac{1}{T}\right) + \frac{3}{2}\left(\frac{1}{T}\right) + \sum_{i=1}^{i=3n-6} \left[\frac{e^{-h\nu_i/kT}(h\nu_i/kT^2)}{(1 - e^{-h\nu_i/kT})}\right] \tag{3.55}$$

and then

$$\frac{\mathrm{H}^0 - \mathrm{H}_0{}^0}{\mathrm{T}} = 4\mathrm{R} + \mathrm{R}\sum_{i=1}^{3n-6}\left[\frac{h\nu_i}{kT}(e^{h\nu_i/kT} - 1)^{-1}\right] \tag{3.56}$$

As before, we must evaluate each of the $(3n - 6)$ vibrational contributions at its particular ω/T using Table 3.8 or Eq. (3.40), and the simple calculation of the enthalpy function for any molecule at any temperature is then evident.

The heat capacity $C_p{}^0$ is just the derivative of $(H^0 - H_0{}^0)$ with respect to temperature, which may be written

$$C_p^0 = C_V^0(\text{trans}) + C_V^0(\text{rot}) + \left[\sum_{i=1}^{i=3n-6} C_V^0(\text{vib})\right] + (C_p^0 - C_V^0)$$

or

$$C_p^0 = \frac{3}{2}R + \frac{3}{2}R + \left[\sum_{i=1}^{i=3n-6} C_V^0(\text{vib})\right] + R \tag{3.57}$$

Table 3.8 contains C_V^0(vib) as a function of ω/T, so the heat capacity of any species may be readily determined at any temperature. We note that the entire temperature dependency arises from the vibrational contributions.

The entropy, which depends on $\ln f$ as well as the derivative of $\ln f$, may be written in practical units as

$$\left(\frac{S^0}{R}\right)_{\text{elect}} = \ln g_0$$

$$\left(\frac{S^0}{R}\right)_{\text{trans}} = \frac{3}{2}\ln M + \frac{5}{2}\ln T - \ln p + 3.4533$$

$$\left(\frac{S^0}{R}\right)_{\text{rot}} = \frac{3}{2}\ln T - \frac{1}{2}\ln B_x B_y B_z - \ln \sigma + 16.9890$$

$$\left(\frac{S^0}{R}\right)_{\text{vib}} = \sum_i \left[\frac{h\nu_i}{kT}(e^{h\nu_i/kT} - 1)^{-1} - \ln(1 - e^{-h\nu_i/kT})\right] \tag{3.58}$$

Table 3.8 lists the value of this $S^0(\text{vib})/R$ as a function of ω/T. All of the other thermodynamic properties may be expressed as simple combinations of those detailed above. It is perhaps useful to write the expression for the Gibbs free energy in practical units. First remembering

$$\frac{G^0 - H_0^0}{RT} = \frac{H^0 - H_0^0}{RT} - \frac{S^0}{R}$$

we write

$$\left(\frac{G^0 - H_0^0}{RT}\right)_{\text{elect}} = -\ln g_0$$

$$\left(\frac{G^0 - H_0^0}{RT}\right)_{\text{trans}} = \frac{5}{2} - \frac{3}{2}\ln M - \frac{5}{2}\ln T + \ln p - 3.4533$$

$$\tag{3.59}$$

$$\left(\frac{G^0 - H_0^0}{RT}\right)_{\text{rot}} = \frac{3}{2} - \frac{3}{2}\ln T + \frac{1}{2}\ln B_x B_y B_z + \ln \sigma - 16.9890$$

$$\left(\frac{G^0 - H_0^0}{RT}\right)_{\text{vib}} = \sum_{i=1}^{i=3n-6} \ln(1 - e^{-h\nu_i/kT})$$

It is useful to note that we can see in these expressions a molecular interpretation or meaning of the term *high temperature*. A temperature is high in a vibrational sense if ω_i/T is sufficiently low to yield a significant contribution to the properties. Clearly the lower-frequency vibrator will be more active at lower temperatures than will the higher-frequency vibrator in the same molecule. Similarly, because the energy levels are so closely spaced, all temperatures are high from the viewpoint of rotation. And since the levels are so far apart, electronic levels are rarely populated, and, in situations of chemical interest then, all temperatures are low from the perspective of electronic excitation. Interesting rotational phenomena are observed for very light molecules (H_2) at very low temperatures (20 K), but this is too unimportant from a practical perspective to be included here.

Example Problem 3.2 Calculate the C_p^0 of C_2H_4 at 293.55, 368.10, and 464.00 K. The calorimetric values are 10.266, 11.900, and 14.151 cal/mol · K, respectively.[11] It is important to note that the molecule is rigid; i.e., there is no internal rotation and the selected "best" molecular data on ethylene are:

Molecular weight	28.0536
Symmetry number	4
Rotational constants, cm^{-1}	4.86596 1.001329 0.828424
Vibrational frequencies, cm^{-1}	3026, 1623, 1342, 1023 3103, 1236, 949, 943, 3106, 826, 2989, 1444

Therefore,

$$C_p^0 = \frac{3}{2}R + \frac{3}{2}R + \sum_1^{12} C_V(\text{vib}) + R$$

for the translational, rotational, vibrational, and $(C_p^0 - C_V^0)$ contributions respectively. Then, using Table 3.8, we find

	$T = 293.55$		$T = 368.10$		$T = 464.00$	
ω, cm^{-1}	ω/T	$C_V^0(\text{vib})/R$	ω/T	$C_V^0(\text{vib})/R$	ω/T	$C_V^0(\text{vib})/R$
3106	10.581	0.00006	8.438	0.00079	6.694	0.00632
3103	10.571	0.00006	8.430	0.00079	6.688	0.00636
3026	10.308	0.00008	8.221	0.00102	6.522	0.00745
2989	10.182	0.00009	8.120	0.00115	6.442	0.00825
1623	5.529	0.02222	4.409	0.07097	3.498	0.16735
1444	4.919	0.04233	3.923	0.11351	3.112	0.23308
1342	4.572	0.06034	3.646	0.15658	2.892	0.27857
1236	4.211	0.08623	3.358	0.18918	2.664	0.33264
1023	3.485	0.16925	2.775	0.30431	2.205	0.45948
949	3.232	0.21059	2.578	0.35413	2.045	0.50885
943	3.212	0.21424	2.562	0.35839	2.032	0.51295
826	2.814	0.29621	2.244	0.44766	1.780	0.59477
		$\Sigma = 1.10168$		$\Sigma = 1.98848$		$\Sigma = 3.11607$

With the vibrational contributions in hand, we can immediately calculate C_p^0 = 10.138, 11.900, and 14.141 cal/mol · K, which all compare very well with the corresponding calorimetrically measured values of C_p^0 = 10.266, 11.900, and 14.151 cal/mol · K at 293.55, 368.10, and 464.00 K, respectively.

It is worthwhile to note in passing that these values of C_p^0, together with similar data on H_2 and on solid graphite, enable us to calculate ΔH_f^0 and ΔG_f^0 of ethylene at any temperature provided only that we have a measured heat of formation at some one temperature. The heat of formation cannot be deduced from statistical thermodynamics, but it can, in principle, be calculated quantum-mechanically, and we will discuss subsequently a particularly useful technique for doing this. For the formation reaction,

$$2C\ (\text{graphite}) + 2H_2\ (g) \rightarrow C_2H_4\ (g)$$

we have

$$\Delta H_f^0(T) = \Delta H_f^0(T_{\text{exp}}) + \int_{T\,\text{exp}}^{T} \Delta C_p^0\, dT$$

or, equivalently,

$$\Delta H_f^0(T) = \Delta H_f^0(T_{\text{exp}}) + \Sigma[(H^0 - H_0^0)_T - (H^0 - H_0^0)T_{\text{exp}}]$$

The "best" value of ΔH_f^0 for C_2H_4 is 12.50 kcal/mol at 298.15 K[11] and we can then make the calculation summarized below.

T, K	$(H^0 - H_0^0)_{C_2H_4}$	$(H^0 - H_0^0)_{H_2}$	$(H - H_0)_{\text{graph}}$	ΔH_f^0, kcal/mol	$S^0_{C_2H_4}$, cal/mol · K	$S^0_{H_2}$	S_{graph}	ΔG_0^f
298.15	2.525	2.024	0.2516	12.50	52.39	31.211	1.3609	16.31
400.00	3.711	2.731	0.5028	11.75	55.74	33.221	2.081	17.73
500.00	5.117	3.430	0.8210	11.10	58.81	34.773	2.788	19.30

Again, we should note that we can calculate the heat of formation and the free energy of formation at any temperature using only the statistically calculated properties and the single additional datum of the heat of formation at any one temperature.

Example Problem 3.3 Calculate the entropy of ethylene as an ideal gas at its normal boiling point of 169.40 K. The experimental value was determined as shown in the following table.†

†Data summarized by Chao and Zwolinski.[11]

0–15 K	Debye extrapolation	0.25
15–103.95 K	Graphical integration	12.226
800.8/103.95	Fusion	7.704
103.95–169.40 K	Graphical integration	7.924
3237/169.40	Vaporization	19.11
	Entropy of real gas	47.25
	Correction to ideal	0.15
	Ideal entropy	47.36 cal/mol · K

From the data presented in Example Problem 3.2, the translational entropy is given by

$$\left(\frac{S^0}{R}\right)_{\text{trans}} = \frac{3}{2}\ln M + \frac{5}{2}\ln T - \ln p + 3.4533$$

$$= \frac{3}{2}\ln 28.0536 + \frac{5}{2}\ln 169.4 - \ln 101.325 + 3.4533$$

$$= 16.6668$$

The rotational entropy is given by

$$\left(\frac{S^0}{R}\right)_{\text{rot}} = \frac{3}{2}\ln T - \frac{1}{2}\ln B_xB_yB_z - \ln \sigma + 16.9890$$

But before we can use this relationship, the rotational constants given in wave numbers, cm^{-1}, must be converted to MHz:

$$B_x(\text{MHz}) = B_x(\text{cm}^{-1}) \times 10^2 \times c$$

$$= 4.86596 \times 10^2 \times 2.9979250 \times 10^8 \times 10^{-6}$$

$$= 145{,}878$$

and similarly B_y = 30,019.09 MHz, and B_z = 24,835.5 MHz. Then the rotational entropy is

$$\left(\frac{S^0}{R}\right)_{\text{rot}} = \frac{3}{2}\ln 169.4 - \frac{1}{2}\ln\,(145{,}878)\,(30{,}019.09)\,(24{,}835.5) - \ln 4 + 16.9890$$

$$= 7.1410$$

Finally, the vibrational contribution is formed by summing the contribution from each of the twelve vibration frequencies:

T = 169.40 K			T = 169.40 K		
ω, cm^{-1}	ω/T	S^0/R	ω, cm^{-1}	ω/T	S^0/R
3106	18.335	0	1342	7.922	0
3103	18.318	0	1236	7.296	0
3026	17.863	0	1023	6.039	0.002
2989	17.645	0	949	5.602	0.003
1623	9.581	0	943	5.567	0.003
1444	8.524	0	826	4.876	0.007
					0.015

And now we can state the total entropy as

$$S = (16.6668 + 7.1410 + 0.015)8.314333 = 198.07 \text{ J/mol} \cdot \text{K}$$

which we may compare with the experimental calorimetric value of 47.36 cal/mol · K or 198.15 J/mol · K which is, of course, excellent agreement.

Note that all of the vibrational contribution to the entropy arose from the three lowest frequencies and that the total vibrational entropy of 0.015(8.314) or 0.125 J/mol · K was only 0.06 percent of the total entropy of 198.06 J/mol · K. On the other hand, the vibrational contribution to the heat capacity at this same temperature may be calculated to be 0.824 J/mol · K in a total C_p^0 of 34.081 J/mol · K, or the vibrational modes contributed 2.4 percent of the total. We see then that the heat capacity is much more sensitive to the quality of the spectroscopic data. This is always true.

The essential correctness of the statistical thermodynamic formalism as revealed in these and many other comparisons with experimental data allows a powerful inference. That is, many species that appear in chemical reactions have a transitory or short-lived existence. For example, free radicals such as OH, CH, and NH_2, are important species in calculations of the properties of a flame gas that is passing through a gas turbine. It is then important to know the heat capacity, say, of such species; they cannot be measured; and so we calculate all of the thermodynamic properties of, say, hydroxyl, OH, with great confidence that the numbers are physically accurate ones.

The model of molecules as a rigid rotator and the internal vibrations as a collection of harmonic oscillators can be improved on, as we shall see subsequently.

Symmetry Number

The symmetry number arises naturally from the symmetry of the molecule in the quantum-mechanical treatment of its rotation, for symmetrical molecules do not have as many allowed rotational energy levels as do unsymmetrical molecules. The symmetry number is, however, nothing more than the number of indistinguishable ways that the molecule can be oriented in space. For example, for all homonuclear diatomics the symmetry number is 2, while for heteronuclear diatomics it is clearly 1. For a molecule like acetone, each methyl group may rotate about its C-C bond. It is convenient to separate these rotations and first calculate the contribution to the thermodynamic properties from the rotation of the molecule as a whole and then calculate the contribution arising from these internal rotations. The symmetry number for the rotation of the molecule as a whole is deduced for the molecule taken to be rigid, or $\sigma = 2$ for acetone. The symmetry number of each rotating methyl group is 3, for there are 3 equivalent positions of the methyl relative to the remainder of the molecule. Contributions to the properties from such internal rotations require a separate partition function as we shall see. By exactly similar reasoning, the symmetry number for benzene is 12. The symmetry number for a symmetric tetrahedral molecule like CH_4 or CCl_4 is 12, for there are three equivalent positions for the base of the pyramid for each atom that we may select as the apex. And there may be four such equivalent apex atoms.

For a rigid C_2H_6, $\sigma = 2$. On the other hand, the symmetry number for internal rotation is 3, as is evident by merely rotating the molecule about its C-C axis three times by 120°. The fact that $\sigma = 3$ for rigid i-C_4H_{10} is evident by imagining the entire molecule to rotate three times about the axis formed by the C-H bond of the center carbon.

Symmetry numbers for several rigid molecules appear in Table 3.14. With some practice and geometric imagination, it is possible to write the symmetry number for any molecule.

Some molecules can have very large symmetry numbers. Consider footballene[12] C_{60}, which may have the structure

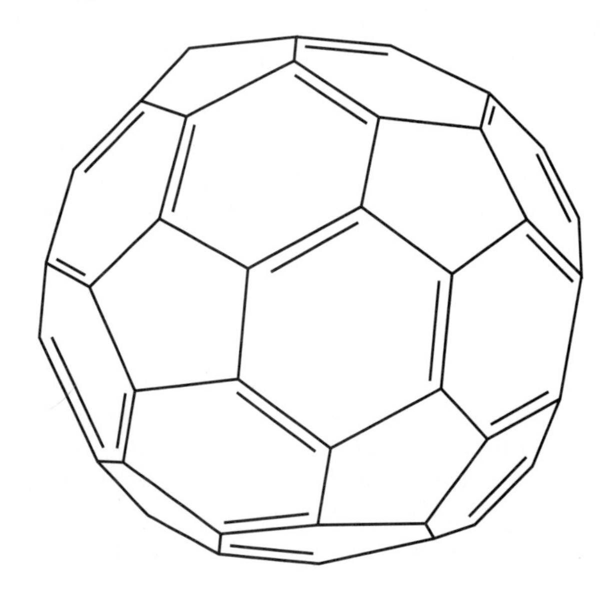

from which the origin of its name is obvious. This molecule is also called buckyball after the geodesic domes popularized by Buckminster Fuller. Assuming equal bond lengths of about 0.14 nm, footballene is a spherical shell of about 0.7 nm diameter. Its interior will be electron-rich due to the inward pointing lobes of orbitals of the delocalized Π bonds. Some species could be trapped inside in a structure reminiscent of crown-ether complexes. Footballene could be imagined as 20 hexagonal rings with each sharing a side with three other such hexagonal rings, and then $\sigma = 20 \times 3$, or 60. Or the mole-

TABLE 3.14 Symmetry Numbers of Several Molecules

Compound	σ	Compound	σ
C_2H_2	2	i-C_4H_{10}	3 (no rotation)
CO_2	2	n-C_4H_{10}	2 (no rotation)
CO	1	C_2H_6	2 (no rotation)
H_2	2	CCl_4	12
CH_4	12	C_2H_4	4
CH_3Cl	3	UF_6	24
$(CH_3)_2CO$	2 (no rotation)	NH_3	3
C_6H_6	12	SF_6	24
BF_3 (planar)	6	*trans cy*-C_6H_{12} (chair form)	6
		cis (boat form)	2
H_2O	2	C_3H_8	2 (no rotation)
		BF_4^- (tetrahedral)	4

cule could be seen as 12 pentagonal rings, and then $\sigma = 12 \times 5$ or 60. According to Eq. (3.57), the entropy will have a term of $-R \ln \sigma$ or $-$ 34.0 J/mol · K. Its entropy will be low, other things being equal. If it is assumed to have a uniform mass distribution about a sphere of 0.35-nm radius, its moment of inertia will be

$$I = \frac{2}{5}mr^2 = \frac{2}{5}\left[60 \times 12 \times \frac{1}{6.023 \times 10^{23}} \times \frac{1}{10^3}\right][0.35 \times 10^{-9}]^2$$

$$= 5.8 \times 10^{-44} \text{ kg} \cdot \text{m}^2$$

and it is, of course, the same about all three axes. The rotational constant is $B_x = 10^{-6}h/8\pi^2 I$ or 145 MHz. The rotational contribution to the entropy is, from Eq. (3.57),

$$\left(\frac{S^0}{R}\right)_{\text{rot}} = \frac{3}{2}\ln T - \frac{1}{2}\ln B_x B_y B_z - \ln \sigma + 16.9890$$

$$= \frac{3}{2}\ln T + 5.43$$

and similarly for the other properties. Although they are unknown, footballene will have 174 vibrational frequencies with high degeneracy among them. C_{60} is intensely interesting, for it may open a whole new area of organic chemistry based upon round molecules with properties quite different from any now known. Buckyballs are readily produced from a high current arc between graphite electrodes.

If internal rotation of a group is possible, we can subdivide the problem into three cases as we shall see in more detail in the next chapter. These are torsional oscillation, hindered rotation, and free rotation. If the temperature is so low that rotation is frozen, the spectrum will reveal low-level torsional vibrations (small ω) which we will include in the $3n - 6$ vibrational contributions to the thermodynamic functions in the usual way. On the other hand, if the temperature is so high that the internal rotation is free, there will be $\frac{1}{2}RT$ contributed to $(U^0 - U_0^0)$ and $(H^0 - H_0^0)$ and $\frac{1}{2}R$ to C_p^0 and C_V^0 for each such internal rotation. That these conclusions are so is qualitatively evident from consideration of a typical molecule where free rotation is possible as, for example, in *i*-butane,

$$\begin{array}{ccccc} & & H & & \\ & & | & & \\ CH_3 & — & C & — & CH_3 \\ & & | & & \\ & & CH_3 & & \end{array}$$

If each methyl is freely rotating about its C-C axis, it may rotate only about that axis rather than about all three of its axes as would be the case for an unbound methyl radical. We had earlier found molecular rotation to contribute $\frac{1}{2}RT$ per degree of rotational freedom to $U^0 - U_0^0$ and to $H^0 - H_0^0$ and then $\frac{1}{2}R$ per degree of rotational freedom to both C_V^0 and C_p^0. Therefore for each of the constrained one-dimensional rotators, that is, the bound methyl groups of i-C_4H_{10}, we expect a similar contribution of $\frac{1}{2}RT$ and $\frac{1}{2}R$ to each energy function and heat capacity, respectively. The only assumption in this argument is that the internal energy levels of the bound rotator will be as closely spaced as they were for the free rotator. This allows integration rather than summation of terms in the partition function.

The entropy and then the free energy function that depend on entropy depend directly on $\ln f$ as well as on the derivatives of $\ln f$. This somewhat more tedious evaluation appears in Chap. 4.

There will be $3n - 6 - r$ vibrational frequencies, where r is the number of freely rotating groups, and these will form the vibrational contribution to the thermodynamic properties. The contribution due to rotation of the molecule as a whole to the entropy is calculated in the usual way. The symmetry number appearing in the partition function for rotation of the molecule as a whole is that for the rigid molecule. The symmetry number characterizing the internal rotator appears explicitly in the partition function for that motion (see Chap. 4).

The intermediate region where the internal rotator appears neither as a low-frequency torsional oscillation nor as a free rotator is the region of "hindered rotation." The partition function is complex, and its reflection in the thermodynamic properties is discussed in detail in Chap. 4.

In molecular spectroscopy, molecular symmetry is expressed in terms of the symmetry groups and their character tables.[13] For example, molecules like H_2O or SO_2 possess C_{2v} symmetry (read C-2-V), NH_3 is C_{3v}, C_6H_6 is D_{6h}, etc. But none of this is significant for our applied purposes.

Group Contribution Methods

Our theoretical developments to this point allow us to understand the basis of all schemes for the correlation of thermodynamic properties that depend on so-called group contributions. The contributions due to the vibrations of, say the amine group, NH_2, will not be too different whether the group finds itself on a benzene ring (aniline) or attached to a methyl group (methyl amine). But obviously such similarities are limited.

Disequilibrium among the Modes

Several phenomena in chemistry depend on a disequilibrium among the modes of energy absorption in a molecule. The most notable of these is the laser.

Rotation-translation interaction or relaxation is fast. That is to say, molecular rotations and translation rapidly assume a Maxwell-Boltzmann distribution which is characterized, of course, by the single parameter of temperature. But because of poor coupling, it is possible to produce vibrational or electronic excitation far above the ambient translation-rotation equilibrium for sufficient lengths of time to be useful. Thus it is reasonable to talk about vibrational temperatures or electronic temperatures far in excess of the translational temperature that one would measure with a thermocouple. Many simple flames are luminous, indicating the presence of electronically excited species even though the adiabatic flame temperature is far below that which would excite any electronic level. Evidently the dynamics of the chemical reaction is such as to produce an initial electronically excited species.

To produce a laser, one must also produce vibrational or electronic excitation far above equilibrium. This excitation is referred to as *pumping the laser*. This disequilibrium is in fact a population inversion, for now more species are in the higher energy states than would be allowed by

$$n_i = \frac{ne^{-\epsilon_i/kT}}{f}$$

Such inversions are short-lived, and the rate of relaxation back to the Maxwell-Boltzmann distribution is a question in kinetics (see Chap. 10). The inversion of electronic levels may be produced by flash photolysis as in the neodymium:yttrium-aluminum-garnet (Nd:YAG) laser or by electric discharge as in the excimer laser. Inversion of the vibrational levels can be produced by chemical reaction in the so-called chemical laser and by other means.

The inverted vibrational population will relax to a Maxwell-Boltzmann distribution by rapid exchange between the vibrators. This resulting equilibrium distribution may still be characterized by a vibrational temperature far above that in the translation/rotation heat bath. This will, in turn, relax to an overall equilibrium distribution that will be characterized by a single temperature. At last, the molecular system is at a state of thermodynamic equilibrium. The HF laser is a well-known example of a chemical laser which can be achieved in the apparatus of Fig. 3.8. Here fluorine atoms are produced by electric discharge in SF_6 or F_2 in low concentration in an inert carrier. This gas mixture containing F atoms is accelerated to supersonic velocity by expanding through an array of converging-diverging nozzles. At the exit plane of the nozzles, H_2 is injected into the flow whereupon the chain reactions

$$H_2 + F \rightarrow HF^* + H$$
$$H + F_2 \rightarrow HF^* + F$$

occur. These reactions are exothermic, and one might wonder wherein lies this energy as the products recede from the collision event. Is this release of energy in translation, rotation, vibration, or electronic excitation? The chemical dynamics of these particular encounters is such that the energy appears as excess vibration, and the rate of appearance of vibrationally excited HF is higher than the rate of appearance of HF in its ground state. In subsequent collisions, this excess vibrational energy is exchanged, and the vibrators quickly assume a Maxwell-Boltzmann distribution. Both the rate of appearance of vibrationally hot HF and the rate of vibrational relaxation may be described using the techniques of molecular theory. All of this is discussed in Chap. 10.

The chemistry is evident in Fig. 3.9, where the F atom reaction produces HF's in $n = 1$, $n = 2$, and $n = 3$ in proportions of 0.3 to 1.0 to 0.5. The exothermicity of 31.6 kcal/mol preferentially appears as vibrational excitation (see Chap. 10). This excitation is clearly not present in an equilibrium or thermodynamic distribution which would demand that the lowest level be most occupied and all other levels be exponentially less occupied. Lasing from the H atom reaction is observed from $6 \rightarrow 5$ and on down the vibrational ladder. The axis of the optical cavity is at right angles to the high-speed gas flow so that HF's in an inverted population are continuously passing through the cavity. Such devices are called gas dynamic lasers.

Many examples of chemical lasers have been reported.

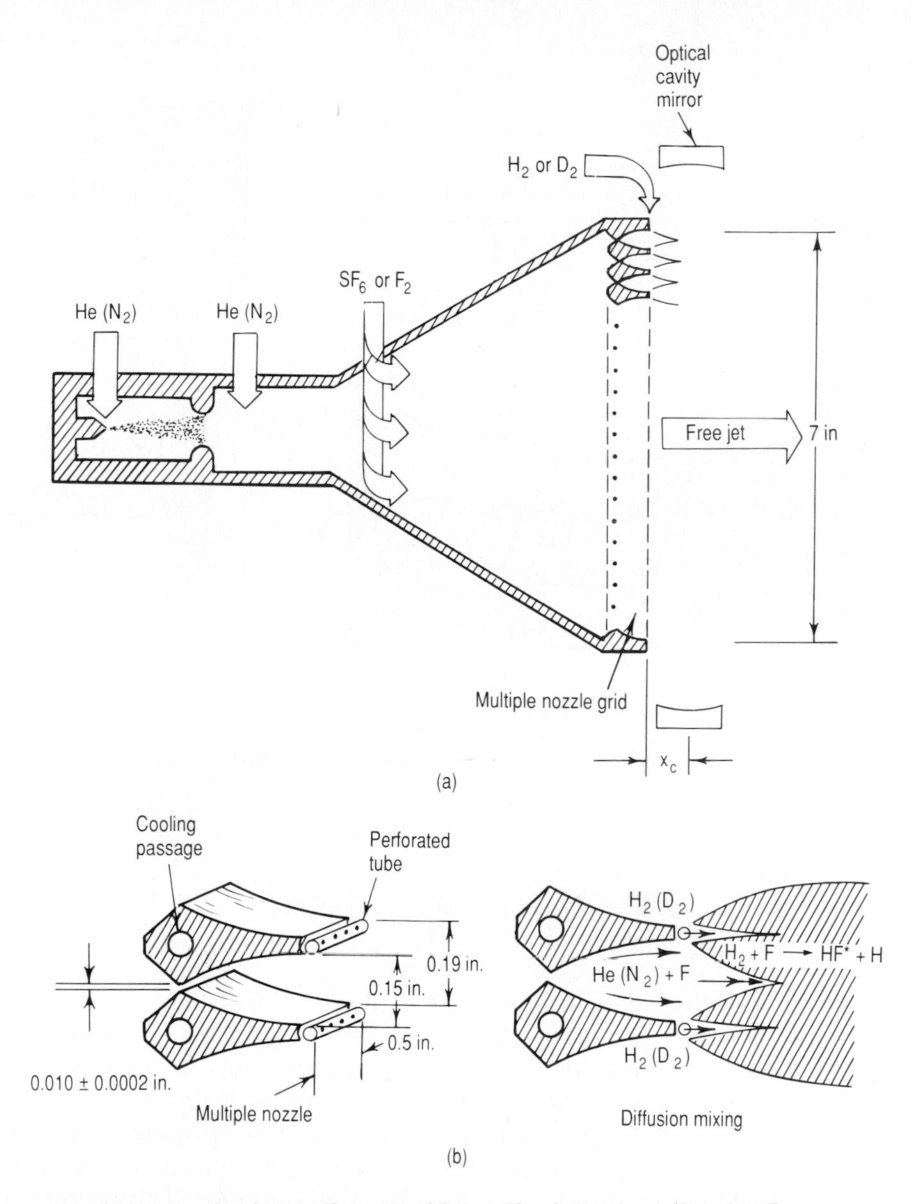

Figure 3.8 Schematic of a chemically pumped laser. The device is really a fast-flow reactor in which the product is a beam of photons rather than some compound. *(From R. W. F. Gross and J. F. Bott (eds.), Handbook of Chemical Lasers, Wiley-Interscience, New York, 1976. Used by permission.)*

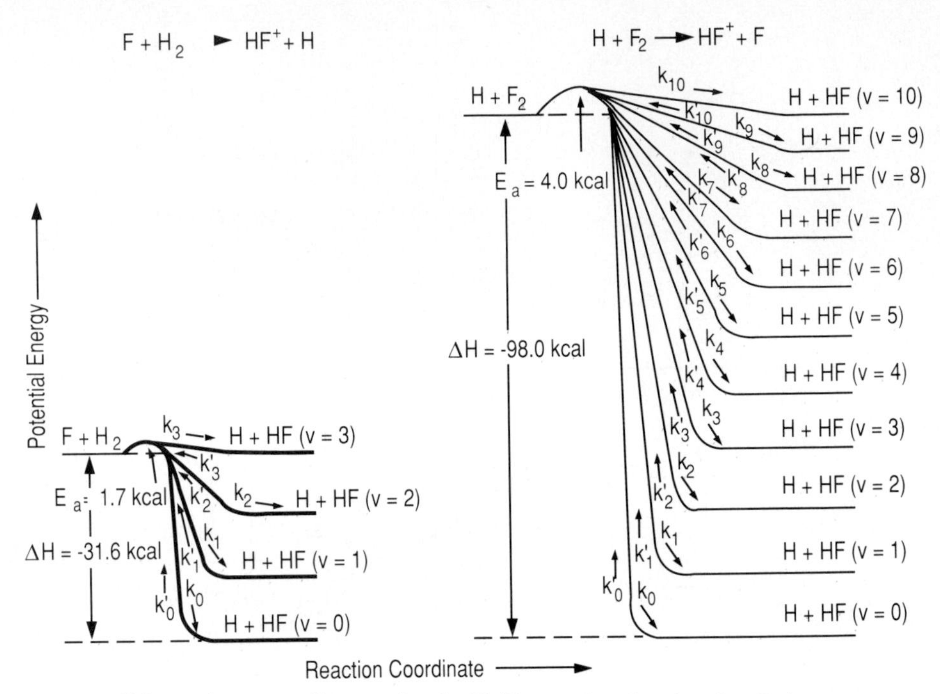

Figure 3.9 Schematic energy diagram for the H_2/F_2 reaction showing detailed rate coefficients in forming HF in specific vibrational energy levels. *(From R. W. F. Gross and J. F. Bott (eds.), Handbook of Chemical Lasers, Wiley-Interscience, New York, 1976. Used by permission.)*

A Better Model

What about the physical reasonableness of the four partition functions for the four rigidly compartmentalized energy modes, each with its particular set of energy levels? The translational energy can go to infinity as prescribed in the integrated partition function. The rotational energy must have a limit, for centrifugal force will ultimately rupture the rotator, yet our partition function has involved an integration over rotational energies that increase to infinity. Also, chemical bonds usually dissociate with two dozen or so quanta of vibrational energy, and yet our vibrational partition function is the sum of an infinite series up to infinite vibrational energies. These problems are all merely artifacts of the idealized rigid rotator, harmonic oscillator that we have selected as a model. They represent no particular problems, for at temperatures where the model is physically reasonable, these nonphysical high-energy states cannot be populated anyway; that is, ϵ_i is much larger than kT and the high-energy terms therefore cannot contribute to the partition function. This series must, however, be properly terminated in the case of the electronic partition function, and we have seen how to do this in our earlier discussion of the thermodynamic properties of plasmas. But it is rare in chemical processes that temperatures are such as to populate other than the ground electronic state anyway. Remember our earlier argument that temperatures of 4500 K were required to populate the first excited state of lithium to only 1 percent.

It is an experimental fact that molecular vibrations are increasingly anharmonic with increasing quantum number, and we find empirically that observed spectroscopic data for vibrational transitions can be well-fitted by a power series in the vibrational number $(n + \frac{1}{2})$

$$\epsilon_{\text{vib}} = (n + \tfrac{1}{2})h\nu_e - x(n + \tfrac{1}{2})^2 h\nu_e \ldots \tag{3.60}$$

where ν_e is the vibration frequency at minimum energy, i.e., the harmonic vibration at $n = 0$, and x is a constant of small magnitude. These and many other sorts of data reveal that the actual intermolecular potential that governs vibration is as shown schematically in Fig. 3.10. We also superimpose the harmonic approximation, which

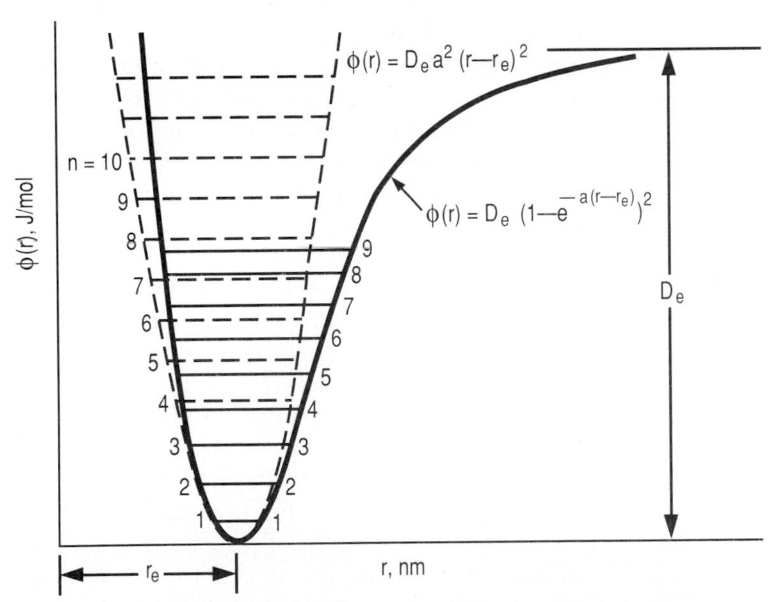

Figure 3.10 Schematic of potential energy functions depicting regions of harmonic and anharmonic vibrations. The dashed curve, a parabola, is the potential function for a harmonic oscillator for which quantum mechanics reveals energy levels of $(n + \frac{1}{2})h\nu$. The solid curve is the empirical potential function due to Morse [Eq. (3.61)], for which the quantum mechanical energy levels are given by $(n + \frac{1}{2})h\nu - (n + \frac{1}{2})^2 h\nu x$. The lines have been drawn approximately to scale for HCl, where $\omega = 2994$ cm^{-1} and $\omega x = 53.6$ cm^{-1}.

is clearly more in keeping with the experimental data at lower vibrational levels. By merely curve-fitting this shape, we find that it can be well-represented by

$$\phi(r) = D_e(1 - e^{-a(r-r_e)})^2 \tag{3.61}$$

which is the so-called Morse function in which a is an empirical constant, D_e is the depth of the potential well including the zero-point vibrational energy $\frac{1}{2}h\nu_e$, and r_e is the internuclear separation at the bottom of the potential well. Note that the experimental dissociation energy D is $D_e - \frac{1}{2}h\nu_e$.

It is interesting that the Schrödinger equation for a quantum mechanical oscillator moving under the influence of this anharmonic Morse potential can be solved to give energy levels

$$\epsilon_{\text{vib}} = \left(n + \frac{1}{2}\right)h\nu_e - \left(n + \frac{1}{2}\right)^2 \frac{(h\nu_e)^2}{4D_e} \tag{3.62}$$

which agrees in form with the experimentally observed fit of the data, Eq. (3.60), with $x = h\nu_e/4D_e$.

Similarly, a rapidly rotating diatomic molecule is subject to centrifugal forces which move the atoms further apart, thus increasing its moment of inertia and lowering its energy levels. Experimental spectroscopic data reveal that a more accurate less-than-rigid rotator will have energy levels that can be fitted by the first two terms of a simple power series in the rotational number $J(J + 1)$ and a third term to account for the greater internuclear separation with increasing n if the vibrator is anharmonic. These two effects lead respectively to two correction terms in the rotational energy levels and two additional experimental constants that must be determined by the spectroscopist from detailed analyses of spectral data. These data may be used to determine corrections to the thermodynamic properties due to anharmonicity and nonrigid rotation. The revised partition functions have been manipulated through the usual relationships to yield improved values for the thermodynamic properties. These have been expressed as corrections to the rigid-rotator, harmonic-oscillator (RRHO) results in the form of an easy-to-use table of values for the rather complex functions that result. These should be thought of as corrections to the tables of Einstein functions. The evaluation of the thermodynamic properties involves no new concepts, but rather it requires mathematical sophistication to manipulate the now more complex partition function through the formalism. However, using the results is simple. Some examples of use of the improved model appear in Table 3.15, where it is clear that the corrections are of no importance.

TABLE 3.15 Gibbs Free Energy of Several Diatomic Gases as Predicted by the RRHO Model and the Improved Model

		$-(G^0 - H_0^0)/T$, cal/mol · K	
Species	T, K	RRHO	Improved
Na_2	1000	56.99	57.12
HBr	1600	52.42	52.45
CO	1000	48.89	48.88
N_2	2000	52.45	52.46

SOURCE: From E. A. Moelwyn-Hughes, *Physical Chemistry*, Pergamon, London, 1957.

The uncertainties in the rigid-rotator, harmonic-oscillator approximation are, of course, reflected in inaccuracies in the thermodynamic properties. These inaccuracies are small, and, since we are frequently concerned with calculations of ΔG for a reaction in attempts to predict equilibrium, even these small errors tend to cancel, for ΔG(reaction) = $\Sigma\Delta G_f$(products) − $\Sigma\Delta G_f$(reactants).

The needed experimental data are obtained with a difficulty that increases with increasing molecular complexity. There are only a few molecules where the anharmonic constants for the energy levels are reasonably well-known, so our improved molecular model is of little practical significance, but we at least see the avenue of approach to a more accurate expression. Fortunately, the anharmonic effect on the thermodynamic properties is very small even at rather high temperatures. For example, consider a typical ω of 1000 cm^{-1} and a typical bond dissociation energy of 350 kJ/mol. If we take the vibration to be harmonic over the lower one-third of its energy levels, simple calculations with the Boltzmann distribution function quickly reveal that temperatures well in excess of 1000 K will be required to produce significant populations above this lower third of the levels. Usually events of much greater significance, principally dissociations, have become of concern before the rigid rotator harmonic oscillator model itself becomes questionable.

Table 3.16 is a data sheet for Si_2 taken from the large and very useful Joint Army Navy Air Force (JANAF) tables published by the National Institute of Science and Technology. The techniques and concepts from this chapter now allow understanding of this and all similar sorts of tables.

TABLE 3.16 Data Sheet on Si_2 from the JANAF Table†

Silicon, diatomic (Si_2)

(Ideal gas) GFW = 56.172

	Gibbs/mol			kcal/mol			
T, K	C_p^0	S^0	$-(G^0 - H^0_{298})/T$	$H^0 - H^0_{298}$	ΔHf^0	ΔGf^0	$\log K_p$
0	0.000	0.000	INFINITE	-2.214	140.324	140.324	INFINITE
100	7.029	46.655	61.829	-1.517	140.893	136.410	-298.124
200	7.692	51.715	55.632	-0.783	141.065	131.827	-144.064
298	8.234	54.895	54.895	0.000	141.000	127.315	-93.325
300	8.242	54.946	54.895	0.015	140.997	127.230	-92.687
400	8.658	57.375	55.224	0.861	140.829	122.665	-67.021
500	9.079	59.352	55.857	1.747	140.627	118.149	-51.643
600	9.499	61.045	56.584	2.676	140.430	113.671	-41.404
700	9.567	62.537	57.330	3.645	140.219	109.230	-34.103
800	10.151	63.875	58.066	4.647	140.021	104.614	-28.634
900	10.343	65.082	58.779	5.672	139.822	100.426	-24.387
1000	10.453	66.178	59.465	6.713	139.615	96.061	-20.994
1100	10.502	67.177	60.122	7.761	139.393	91.716	-18.222
1200	10.507	68.091	60.748	8.812	139.154	87.392	-15.916
1300	10.485	68.932	61.346	9.861	138.893	83.087	-13.968
1400	10.447	69.707	61.916	10.909	138.610	78.506	-12.302
1500	10.401	70.427	62.460	11.951	138.303	74.545	-10.861
1600	10.354	71.096	62.979	12.989	137.970	70.306	-9.603
1700	10.309	71.723	63.475	14.022	113.628	66.298	-8.523
1800	10.268	72.311	63.950	15.050	113.356	63.523	-7.713
1900	10.232	72.865	64.404	16.075	113.081	60.761	-6.989
2000	10.200	73.389	64.841	17.097	112.803	58.017	-6.340
2100	10.174	73.886	65.260	18.116	112.522	55.284	-5.753
2200	10.153	74.359	65.663	19.132	112.238	52.563	-5.222
2300	10.135	74.810	66.050	20.146	111.952	49.957	-4.737
2400	10.122	75.241	66.425	21.159	111.665	47.165	-4.295
2500	10.112	75.654	66.785	22.171	111.377	44.482	-3.889

2600	10.106	76.050	67.134	23.189	111.088	41.513	− 3.515
2700	10.101	76.432	67.472	24.192	110.798	39.151	− 3.149
2800	10.099	76.799	67.798	25.202	110.508	36.507	− 2.849
2900	10.099	77.153	68.115	26.212	110.219	33.570	− 2.552
3000	10.100	77.496	68.422	27.222	109.928	31.239	− 2.276
3100	10.103	77.827	68.720	28.232	109.638	29.619	− 2.018
3200	10.107	78.148	69.009	29.242	109.348	26.014	− 1.777
3300	10.111	78.459	69.291	30.251	109.059	23.414	− 1.551
3400	10.117	78.761	69.565	31.265	108.771	20.823	− 1.338
3500	10.124	79.054	69.832	32.277	108.483	19.239	− 1.139
3600	10.131	79.339	70.092	33.289	− 75.727	20.161	− 1.225
3700	10.139	79.617	70.346	34.303	− 75.815	22.945	− 1.349
3800	10.147	79.887	70.594	35.317	− 75.903	25.515	− 1.467
3900	10.156	80.151	70.635	36.332	− 75.992	25.161	− 1.579
4000	10.165	80.408	71.071	37.349	− 76.080	30.556	− 1.686
4100	10.175	80.660	71.302	38.365	− 76.169	33.523	− 1.787
4200	10.185	80.905	71.528	39.383	− 76.255	36.200	− 1.884
4300	10.195	81.145	71.749	40.402	− 76.340	38.579	− 1.976
4400	10.206	81.379	71.965	41.422	− 76.426	41.560	− 2.064
4500	10.218	81.609	72.177	42.444	− 76.508	44.249	− 2.149
4600	10.229	81.833	72.384	43.466	− 76.590	46.934	− 2.230
4700	10.241	82.053	72.588	44.489	− 76.673	49.621	− 2.307
4800	10.254	82.269	72.787	45.514	− 76.752	52.306	− 2.382
4900	10.266	82.481	72.983	46.540	− 76.830	54.995	− 2.453
5000	10.279	82.688	73.175	47.567	− 76.907	57.683	− 2.521
5100	10.293	82.892	73.363	48.596	− 76.982	60.381	− 2.588
5200	10.306	83.092	73.548	49.626	− 77.056	63.074	− 2.651
5300	10.320	83.288	73.730	50.657	− 77.127	65.772	− 2.712
5400	10.335	83.481	73.909	51.690	− 77.198	68.467	− 2.771
5500	10.349	83.671	74.085	52.724	− 77.263	71.161	− 2.828
5600	10.364	83.858	74.258	53.760	− 77.336	73.964	− 2.883
5700	10.379	84.041	74.428	54.797	− 77.403	76.571	− 2.936
5800	10.395	84.222	74.595	55.836	− 77.470	79.269	− 2.987
5900	10.410	84.400	74.760	56.876	− 77.536	81.965	− 3.036
6000	10.424	84.575	74.922	57.914	− 77.600	84.570	− 3.084

SOURCE: JANAF Table, Dec. 31, 1960; Dec. 31, 1962; Mar. 31, 1967.

TABLE 3.16 Data Sheet on Si_2 from the JANAF Table† (*Continued*)

Silicon, diatomic (Si_2)	ideal gas
Ground state configuration $^3\Sigma_g^-$	$\Delta H_f^0(0\text{ K}) = 140.3 \pm 3$ kcal/mol
$S^0(298.15\text{ K}) = 54.895$ gibbs/mol	$\Delta H_f^0(298.15\text{ K}) = 141 \pm 3$ kcal/mol
Gram formula weight = 56.172	$\omega_e = 510.98\text{ cm}^{-1}$
$\sigma = 2$, $r_e = 0.2246$ nm	$B_e = 0.2390\text{ cm}^{-1}$

Heat of formation

The selected value of ΔH_f^0 (298.15) = 141.000 kcal/mol, based upon spectroscopic and equilibrium data, corresponds to a dissociation energy at 0 K of 73.0 kcal/mol. Below the melting point of 1685 K, the tabulated values correspond to the formation of Si_2 as an ideal gas from solid silicon, then to the boiling point of 3513.8 K ideal Si_2 is formed from liquid silicon, and above this temperature, Si_2 is formed as an ideal gas from the real vapor of whatever composition. At all temperatures, $\Delta G_f^0 = \Delta H_f^0 - T\,\Delta S_f^0$.

Heat Capacity and Entropy

Not all of the electronic states of Si_2 have been experimentally observed. Comparisons with molecules that are isoelectronic with Si_2 such as C_2, BN, BeO, etc. allows one to postulate these states and to estimate their energies. These estimates appear in the electronic partitition function and produce uncertainties then in the calculated values of C_p^0, S^0, etc.

$\log K_p$

Tabulated values of $\log K_p$ are (base 10) logarithms of the equilibrium constant of formation of Si_2 as an ideal gas from the elements at the indicated temperature. $\Delta G_f^0 = -2.303RT \log K_p$.

Equation of State

With an expression for the Helmholtz free energy in terms of the partition function, we can evaluate the equation of state. The Helmholtz free energy is $A = U - TS$, and then

$$dA = dU - T\,dS - S\,dT = -p\,dV - S\,dT$$

which implies that

$$\left(\frac{\partial A}{\partial V}\right)_T = -p$$

and since $A = nkT[\ln(f/n) + 1]$, the pressure may be written in terms of the partition function as

$$p = nkT\left(\frac{\partial \ln f}{\partial V}\right)_T$$

We have considered only ideal gases, and there only the translational partition function depends on V. When differentiated, the result is

$$p = nkT\left(\frac{1}{V}\right)$$

and we would have been terribly upset if the result had been anything else. The partition function for neither rotational nor vibrational motions depends on V, and therefore this will be the proper expression for the pressure of all ideal gases. Again, we would be upset if the results were anything else.

The intermolecular forces between real molecules will affect the translational energy of a species, and hence its translational partition function, and hence yield a correspondingly revised equation of state. This is an enormously complex problem that has not been solved for other than very simple assumptions, but we will look at the essential components of this problem subsequently.

References

1. H. A. McGee, Jr. and G. Heller, *ARS Journal, 32*, 203 (1962).
2. D. R. Inglis and E. Teller, *Astrophys. J., 90*, 439 (1939).
3. H. Margenau and M. Lewis, *Rev. Mod. Phys., 31*, 569 (1959).
4. P. W. Gilles, *J. Chem. Phys., 19*, 129 (1951).
5. H. M. Spencer, "Chemical Thermodynamics," unpublished notes, Department of Chemistry, University of Virginia, 1983.

6. C. E. Moore, "Atomic Energy Levels," NSRDS-NBS 35, Washington, D.C., December 1971.
7. J. D. Graybeal, *Molecular Spectroscopy*, McGraw-Hill, New York, 1988.
8. J. Hilsenrath and G. G. Ziegler, monograph no. 49, National Bureau of Standards, Washington, D.C., July 1962.
9. T. Shimanouchi, "Tables of Molecular Vibration Frequencies," NSRDS-NBS, Washington, D.C., 1972.
10. *Landolt-Börnstein Handbook, Numerical Data and Functional Relationships in Science and Technology*, Springer Verlag, New York, 1974.
11. J. Chao and B. J. Zwolinski, *J. Phys. Chem. Ref. Data, 4*, 251 (1975).
12. A. D. J. Haymet, *J. Am. Chem. Soc., 108*, 319 (1986).
13. H. Eyring, J. Walter, and G. E. Kimball, *Quantum Chemistry*, John Wiley and Sons, New York, 1944.

Further Reading

1. G. N. Lewis and M. Randall, revised by K. S. Pitzer and L. Brewer, *Thermodynamics*, McGraw-Hill, New York, 1961.
2. D. R. Stull, E. F. Westrum, Jr., and G. C. Sinke, "Chemical Thermodynamics of Organic Compounds," John Wiley and Sons, New York, 1969.
3. E. A. Moelwyn-Hughes, *Physical Chemistry*, Pergamon, London, 1957.
4. D. A. McQuarrie, *Statistical Thermodynamics*, Harper and Row, New York, 1973.
5. E. U. Condon and G. H. Shortley, *The Theory of Atomic Spectra*, Cambridge University Press, England, 1953.

Chapter

4

Thermodynamic Properties of Nonrigid Molecules

Since we are taking atoms to be point masses, there will be no energy of internal rotation of a linear group of atoms in a molecule, for there is no moment of inertia about this axis of the molecule. However, if that group has any off-axis mass, if the group is joined to the remainder of the molecule by a single bond, and if steric hindrance is unimportant, then internal rotation may occur. Consider, for example, CH_3OH wherein the hydroxyl hydrogen is not in line with the C–O bond. Rotation of the OH group about this C–O bond can occur, and the rotation would be hindered as the hydroxyl proton passed over each of the three hydrogens on the methyl group. We expect that an activation energy for this rotation should exist such that the hydroxyl would pass from a rocking motion, that is, a weak harmonic oscillation, between two of the methyl hydrogens to a "three-bump" rotation that will become increasingly a free rotation as the temperature is increased. Whether such barriers to free rotation occur as a result of steric hindrance or van der Walls interactions, or whatever, is here unimportant. Quantum-mechanical calculations indicate that there is no barrier to rotation about single bonds, and therefore the experimentally observed barrier must result from some force of interaction between the parts of the molecule. This barrier to free rotation is usually a few kilojoules in height, but the barrier is not normally large enough to stabilize structural isomers of a species. Much higher barriers can produce rotational isomers, and many are evident. For example, 1,3-butadiene exists in both cis and trans forms where the energy is 10.15 kJ/mol lower for the trans than for the cis, as is revealed from spectroscopic data. The barrier varies from negligible as in $Cd(CH_3)_2$, to which we will return in a moment, to extremely large as in C_2H_4, where only torsional motion about the C═C double bond may occur without the entire character of the molecule changing. Between these two extremes, where the barrier is neither much smaller nor much larger than kT, molecules experience hindered or restricted rotation. Although the barrier is small, the thermodynamic properties may be nonetheless significantly influenced. As an aside, we note that the rate of reaction will also be influenced by the possibility of hindered rotation, since the activated couplex may rotate in this way, and its properties, and thus the rate, will be influenced. Hindered rotation is also important in considerations of the flexibility of both natural and synthetic polymers.

It has long been known that molecules experience rotation about single bonds, but that higher-order bonds allow twisting but not complete rotation of the molecule. About 1930, it was noted that the experimental entropy of ethane was between the value calculated by assuming free rotation and that by assuming no rotation. This disagreement between theory and experiment was explained by assuming that the rotation was not completely free, and a potential energy barrier to rotation of 13.180 kJ/mol led to thermodynamic conclusions that were in agreement with the calorimetric data. In ethane the hydrogens are 120° apart, and thus the molecule possesses two possible configurations: a low-energy staggered form and a high-energy eclipsed form. Figure 4.1 illustrates the low-energy orientation of the methyl groups and also the potential energy of interaction as a function of methyl group orientation. If one of the hydrogen atoms on each carbon is replaced by a halogen such as in 1,2-dichloroethane, a potential curve such as that shown in Fig. 4.2 results. In this case steric hindrance of the chlorine atoms becomes significant, in addition

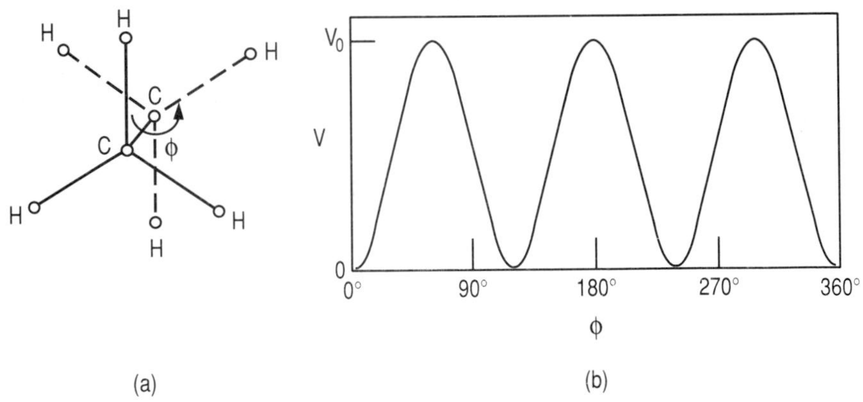

Figure 4.1 (*a*) Orientation of CH_3 groups in ethane. (*b*) Potential energy function. This energy is zero when the H atoms are staggered as shown in (*a*) and V_0 when they are opposite.

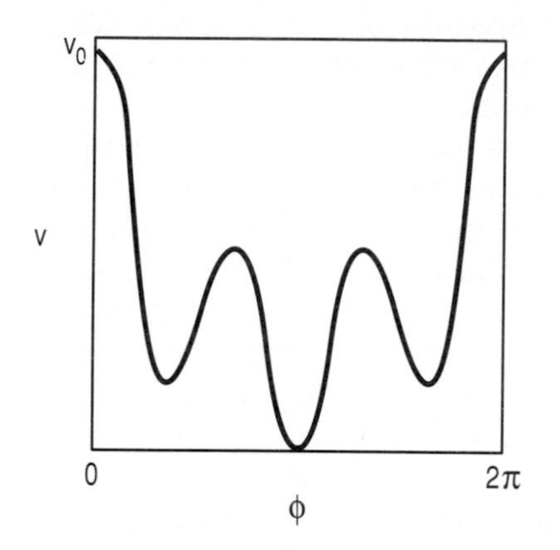

Figure 4.2 Potential curve for internal rotation of 1,2-dichloroethane.

to electrostatic repulsion. Simple qualitative reasoning tells us that the maximum occurs in the cis configuration and the other two maxima occur in the remaining eclipsed configurations. The lowest minimum occurs in the trans and the two remaining minima occur in the gauche configurations. In dimethyl cadmium, the hydrogen atoms on the two methyl groups are much farther apart than in ethane, and thus free rotation results because of the negligible interaction of these hydrogen atoms.

A consideration of the quantum-mechanical problem of the free, nonhindered internal rotator leads to energy levels of

$$\epsilon = \frac{h^2K^2}{8\pi^2 I_r} \qquad K = \ldots, -3, -2, -1, 0, 1, 2, 3, \ldots \tag{4.1}$$

where K is a quantum number having all integral values positive, negative, and zero, and where I_r is called the reduced moment of inertia of the internal rotation. This quantity, like the other molecular data, must be deduced from the molecular architecture by the spectroscopist. Replacing the sum with an integral and carrying through the mathematics leads to a partition function for unhindered free rotation of

$$f_{\text{int rot}} = \frac{2.7928}{n}(10^{45} I_r T)^{1/2} \tag{4.2}$$

where I_r is in units of kg · m^2, T in K, and n is a sort of symmetry number that is equal to the number of equivalent orientations of the rotating group relative to the molecule as a whole. The factor $1/n$ arises because in molecules in which either one or both of the rotating groups are symmetrical, only $1/n$th of the quantum numbers are allowed for n equivalent orientations. For example, since each rotating group in ethane has three equivalent orientations, then $n = 3$. Thus the contributions of free unhindered internal rotation to the thermodynamic properties at any temperature may be calculated as soon as we know I_r. The value of n will be obvious by inspection, and the number of vibration frequencies will always be $3n - 6 - r$, where r is the number of rotating groups. Since $f_{\text{int rot}}$ varies as $T^{1/2}$, free internal rotation of a single group will always contribute ½R to the heat capacity.

It was mentioned earlier that dimethyl cadmium is a molecule which experiences free rotation since the methyl groups are so far apart. The calculation of the entropy of this molecule reveals that the statistical-mechanical value agrees well with the experimental value determined from low-temperature heat-capacity measurements. The entropy arising from free internal rotation was obtained simply by substituting the expression for the partition function for free rotation, Eq. (4.2), into the general expression for the entropy of nonlocalized molecules:

$$S = R\left[\ln\left(\frac{f}{n}\right) + 1\right] - T\left(\frac{\partial \ln f}{\partial T}\right)_V$$

which may be expanded to

$$S = R\left[\ln\left(\frac{f_{\text{trans}} f_{\text{rot}} f_{\text{vib}} f_{\text{int rot}}}{n}\right) + 1\right] - T\frac{\partial}{\partial T}(\ln f_{\text{trans}} f_{\text{rot}} f_{\text{vib}} f_{\text{int rot}})_V$$

or $\quad S = S_{\text{trans}} + S_{\text{rot}} + S_{\text{vib}} + S_{\text{int rot}}$

where clearly

$$S_{\text{int rot}} = R \ln f_{\text{int rot}} + RT\left(\frac{\partial \ln f_{\text{int rot}}}{\partial T}\right)_V$$

With the $f_{\text{int rot}}$ of Eq. (4.2), one obtains

$$S_{\text{int rot}} = R\left[\ln \frac{2.7928}{n} + \frac{45}{2} \ln 10 + \frac{1}{2} \ln I_r + \frac{1}{2} \ln T + \frac{1}{2}\right] \tag{4.3}$$

and similarly for several other important properties for free rotators,

$$\left(\frac{H^0 - H_0^0}{RT}\right)_{\text{int rot}} = \frac{1}{2}$$

and

$$\left(\frac{C_V^0}{R}\right)_{\text{int rot}} = \frac{1}{2}$$

and

$$\left(\frac{G^0 - H_0^0}{RT}\right)_{\text{int rot}} = \frac{1}{2} - \frac{S_{\text{int rot}}^0}{R}$$

TABLE 4.1 Entropy of Dimethyl Cadmium at 298.16 K

Translation and rotation	253.80	J/mol K
Vibration ($3n - 7 = 20$ frequencies)	36.65	
Free internal rotation	12.26	
Total, calculated	302.71	

SOURCE: J. C. M. Li, *J. Am. Chem. Soc.*, *78*, 1081 (1956).

For $Cd(CH_3)_2$, $n = 3$, $I_r = 2.70 \times 10^{-47}$ kg · m^2, and at $T = 298.16$ K, we deduce $S_{\text{int rot}}$ as 12.26 J/mol · K. Thus, as is shown in Table 4.1, we deduce the total entropy to be 302.71 J/mol · K, which agrees with the calorimetric value, the calculation of which is summarized in Table 4.2. Conversely, one could reasonably compare the calorimetric entropy of 302.92 J/mol · K with that calculated due to translation and rotation of 253.80 J/mol · K and that due to vibration of 36.65 J/mol · K. This calculation takes the C-Cd-C structure to be linear and the product of the moments for the rotations of the molecule as a whole to be 6.929×10^{-136} kg^3 · m^6. The contribution due to internal rotation by difference is then 12.47 J/mol · K, which may be contrasted with the entropy of 12.26 J/mol · K calculated by using the partition function for free rotation and Eq. (4.3.) With this agreement, we conclude that the rotation is free.

One might wonder why there are not two internal rotators in $Cd(CH_3)_2$. However, insofar as rotational motion is concerned, two independently rotating methyl groups are exactly equivalent to one stationary group with the other rotating about it at the difference of the angular frequencies of the two rotations about a fixed Cd atom. In this sense the molecule is no different from ethane.

We note also that molecules with free internal rotations will have significantly higher entropies than would be the case should that ro-

TABLE 4.2 Calculation of Entropy of $Cd(CH_3)_2$, cal/mol · K, from Thermal Data

0 to 14 K ($\theta_D = 90$)	0.705	
14 to 254.35 K (solid II)	34.592	
Transition 363.5/254.34	1.429	
254.35 to 270.48 K (solid I)	1.539	
Fusion 1873/270.48	6.925	
270.48 to 298.16 K (liquid)	3.056	
Liquid at 298.16 K	48.246	± 0.10
298.16 to 291.5 K	−0.712	
Vaporization 9153/291.5 ($p^s = 2.27$ cm Hg)	31.400	
Compression (2.27 cm to 1 atm)	−6.973	
291.5 to 298.16 K ($C_p^0 = 19.6$ cal/mol K)	0.442	
Ideal gas at 298.16 K and 1 atm	72.40	± 0.20 (302.92 J/mol K)

tation rather appear as a low-frequency torsional oscillation. For a typical torsion of 300 cm^{-1}, ω/T is 1 at room temperature and S^0 is about 1.4 cal/mol · K. S^0 due to free rotation in $Cd(CH_3)_2$ is about 3 cal/mol · K at room temperature. Also the contribution to heat capacity can be no more than ½R for the free rotator, but it may have a maximum value of R for the vibrator.

On the other hand, if free rotation is impossible, as in C_2H_4, a low-frequency torsion will be evident in the spectrum, and this will contribute to the thermodynamic properties through the Einstein function, as would any other vibration. For example, for ethane this torsion occurs at 826 cm^{-1}.

Between completely free rotation and a torsional vibration, there is the possibility of hindered rotation, and the quantum-mechanical treatment of this class of problem is now much more complicated, since it must account for the potential barrier which may be of variable height, width, and shape. Each molecule is then a special case to be handled individually. We use, however, the identical approach to the problem as in the freely rotating case. We will here represent the potential by a Fourier series where experience suggests that the first two terms alone usually give an excellent approximation, or,

$$V = \tfrac{1}{2}V_0(1 - \cos n\phi) \tag{4.4}$$

where V_0 is the barrier height, n is again the number of equivalent orientations (3 in ethane), and ϕ is the angle of rotation away from the potential minimum (where V is zero). This function describes the potential energy of Fig. 4.1*b*, and it would be accurate for all molecules where this symmetry about the internal axis of rotation is apparent. Manipulating this barrier function through the quantum-mechanical formalism to obtain the energy levels, and then summing to evaluate the partition function, and then finally developing the contributions to the thermodynamic properties from this partition function are all difficult tasks. All of this has, however, been done for the model potential of Eq. (4.4), and the results have been presented in convenient and easy-to-use tables that have been prepared for molecules with rigid frames with symmetrical rotating groups. These tables allow quick, if sometimes approximate, determination of the thermodynamic contributions due to internal rotation. The contributions to the thermodynamic properties appear in terms of the reduced barrier height V/RT and the reciprocal of the partition function, $1/f_{\text{int rot}}$, if the rotation were completely free rather than hindered. These data appear in Tables 4.4 to 4.7, and, to use these tables, we need only I_r and V. In these tables, if V/RT is zero, the rotation is free, $(H^0 - H_0^0)/T$ is ½R and C_p^0 is ½R, and both are in-

TABLE 4.3 Some Typical Barriers to Free Internal Rotation

Compound	Barrier, kJ/mol	Compound	Barrier, kJ/mol
CH_3CH_3	13.18	CH_3CH_2F	13.81
CH_3CF_3	14.56	Cis-$CH_3CH{=}CHF$	4.42
C_6H_5OH	13.76	Trans-$CH_3CH{=}CHF$	9.23
CH_3OH	4.49	BH_3PF_3	13.56
CH_3-cy-propane	11.97	$(CH_3)_2CO$	4.18
CH_3CHO	4.88	Styrene	9.20
CH_3NH_2	8.27	CH_3CCl_3	12.41
CH_3CF_3	14.56	$CH_3{-}C{\equiv}C{-}CH_3$	0

SOURCE: *Landolt-Börnstein Handbook, Numerical Data and Functional Relationships in Science and Technology*, Springer-Verlag, New York, 1974 and G. Herzberg, *Molecular Spectra and Molecular Structure*, Vol. II, Van Nostrand, New York, 1945.

dependent of temperature. Some examples of barriers to free internal rotation appear in Table 4.3.

A generalization from many such data is that when the two groups on the axial bond suggest a barrier of sixfold maxima, then rotation will be free, and contributions to the thermodynamic properties are evaluated using Eq. (4.2). The height of the potential barrier that one deduces to explain experimental data is also a function of the shape of that assumed barrier, that is, forms other than Eq. (4.4) will yield different thermodynamic properties, so it is best to think of these values as qualitative "equivalent" barriers on which we will not attach too much physical significance. Only this case of the cosine potential, Eq. (4.4), has proved to be sufficiently useful to justify a tabulation of the thermodynamic properties when using it in the partition function. The tables assume one group to be symmetrical about the axis of rotation, but the other group need not be. The tables are more or less valid for more complex molecules, where, for example, one would expect good results with *p*-xylene, but not so good with *o*-xylene where interaction or a sort of gearing of the rotations may occur. Methods for treating molecules with more than two rotating parts have been presented,[1–4] but their utility does not justify discussion here. We have implicitly assumed that internal rotation and rotation of the molecule as a whole are independent activities. This is strictly true only when the moments of inertia of the molecule are unchanged by the internal rotation, and this is true only for the rotation of what is called a *symmetric top* like, for example, the CH_3 groups in $(CH_3)_2Cd$. It is certainly not true for the rotation of the C_2H_5 group in butanone, but we will assume this separation of rotational motions anyway. More accurate treatment of such cases is sometimes possible, but there are no generally useful techniques. Fortunately the internal moment is frequently small compared to that of the molecule as a whole, and little error is

TABLE 4.4 Heat Capacity, C_V^0(int rot), cal/mol · K

V/RT	$1/f_{\text{int rot}}$																			
	0.0	0.05	0.10	0.15	0.20	0.25	0.30	0.35	0.40	0.45	0.50	0.55	0.60	0.65	0.70	0.75	0.80	0.85	0.90	0.95
0.0	0.994	0.994	0.994	0.994	0.994	0.994	0.994	0.994	0.994	0.994	0.994	0.994	0.994	0.994	0.994	0.994	0.994	0.994	0.994	0.994
0.2	1.0035	1.003	1.003	1.002	1.001	1.000	0.999	0.998	0.998	0.998	1.000	1.000	1.000	1.000	1.000	1.000	1.000	0.999	0.999	0.999
0.4	1.0328	1.033	1.032	1.030	1.028	1.025	1.024	1.021	1.019	1.017	1.018	1.017	1.015	1.013	1.012	1.010	1.008	1.007	1.005	1.004
0.6	1.0801	1.080	1.079	1.076	1.073	1.068	1.065	1.060	1.056	1.051	1.049	1.046	1.041	1.036	1.031	1.026	1.021	1.017	1.014	1.011
0.8	1.1435	1.143	1.141	1.138	1.133	1.128	1.121	1.114	1.106	1.099	1.092	1.084	1.075	1.067	1.058	1.049	1.040	1.031	1.025	1.020
1.0	1.2203	1.219	1.217	1.212	1.206	1.199	1.190	1.180	1.169	1.157	1.144	1.131	1.118	1.105	1.091	1.078	1.065	1.052	1.040	1.031
1.5	1.4508	1.449	1.444	1.435	1.423	1.408	1.391	1.370	1.348	1.324	1.299	1.273	1.247	1.218	1.192	1.165	1.141	1.115	1.090	1.070
2.0	1.6778	1.695	1.687	1.673	1.655	1.632	1.606	1.574	1.541	1.505	1.465	1.424	1.382	1.341	1.300	1.258	1.218	1.180	1.146	1.113
2.5	1.9213	1.917	1.908	1.888	1.866	1.840	1.801	1.756	1.717	1.670	1.619	1.562	1.504	1.448	1.393	1.341	1.289	1.238	1.190	1.146
3.0	2.0989	2.095	2.082	2.062	2.033	1.996	1.952	1.900	1.846	1.794	1.732	1.663	1.597	1.532	1.466	1.401	1.337	1.276	1.217	1.164
3.5	2.2226	2.218	2.204	2.180	2.146	2.106	2.054	1.995	1.934	1.869	1.803	1.727	1.654	1.580	1.506	1.432	1.361	1.293	1.226	1.165
4.0	2.2989	2.294	2.276	2.249	2.213	2.168	2.110	2.048	1.980	1.907	1.834	1.754	1.674	1.593	1.513	1.435	1.359	1.286	1.215	1.148
4.5	2.3358	2.330	2.312	2.280	2.238	2.190	2.129	2.062	1.990	1.911	1.832	1.749	1.664	1.578	1.496	1.413	1.333	1.259	1.185	1.115
5.0	2.3447	2.338	2.318	2.285	2.241	2.186	2.120	2.056	1.972	1.890	1.808	1.718	1.631	1.543	1.457	1.373	1.292	1.214	1.140	1.068
6.0	2.3158	2.307	2.283	2.245	2.192	2.130	2.059	1.979	1.893	1.803	1.711	1.614	1.520	1.429	1.342	1.255	1.173	1.096	1.022	0.954
7.0	2.2650	2.256	2.228	2.185	2.126	2.055	1.973	1.883	1.787	1.688	1.588	1.487	1.390	1.296	1.207	1.120	1.040	0.962	0.890	0.826
8.0	2.2160	2.205	2.174	2.125	2.058	1.979	1.888	1.788	1.684	1.576	1.468	1.366	1.262	1.164	1.074	0.988	0.908	0.834	0.765	0.704
9.0	2.1762	2.164	2.130	2.074	1.999	1.909	1.808	1.699	1.587	1.474	1.362	1.250	1.144	1.048	0.956	0.869	0.789	0.717	0.652	0.593
10.0	2.1457	2.133	2.094	2.033	1.951	1.854	1.745	1.630	1.507	1.382	1.262	1.151	1.045	0.943	0.850	0.765	0.688	0.618	0.556	0.499
12.0	2.1053	2.089	2.043	1.972	1.877	1.763	1.636	1.502	1.365	1.233	1.107	0.989	0.877	0.774	0.682	0.600	0.528	0.463	0.407	0.358
14.0	2.0813	2.063	2.009	1.923	1.814	1.686	1.546	1.400	1.254	1.112	0.978	0.855	0.744	0.644	0.554	0.479	0.411	0.352	0.303	0.262
16.0	2.0657	2.044	1.983	1.887	1.764	1.622	1.468	1.311	1.156	1.009	0.873	0.749	0.639	0.542	0.457	0.387	0.324	0.272	0.229	0.194
18.0	2.0547	2.031	1.961	1.853	1.717	1.562	1.397	1.232	1.070	0.919	0.780	0.657	0.549	0.456	0.378	0.312	0.259	0.215	0.175	0.144
20.0	2.0465	2.020	1.944	1.827	1.678	1.510	1.333	1.158	0.991	0.837	0.701	0.580	0.477	0.389	0.316	0.256	0.208	0.168	0.135	0.109

SOURCE: From *Thermodynamics*, revised by K. S. Pitzer and L. Brewer, McGraw-Hill, New York, 1961.

TABLE 4.5 Heat Content $(H^o - H_0^0)/T$ due to Internal Rotation, cal/mol · K

V/RT	$1/f_{\text{int rot}}$																			
	0.0	0.05	0.10	0.15	0.20	0.25	0.30	0.35	0.40	0.45	0.50	0.55	0.60	0.65	0.70	0.75	0.80	0.85	0.90	0.95
0.0	0.994	0.994	0.994	0.994	0.994	0.994	0.994	0.994	0.994	0.994	0.994	0.994	0.994	0.994	0.994	0.994	0.994	0.994	0.994	0.994
0.2	1.1824	1.142	1.106	1.074	1.050	1.032	1.022	1.015	1.008	1.004	1.000	0.996	0.994	0.994	0.994	0.992	0.992	0.991	0.990	0.989
0.4	1.3515	1.300	1.249	1.200	1.151	1.106	1.073	1.051	1.036	1.025	1.015	1.006	0.999	0.994	0.992	0.990	0.988	0.988	0.986	0.985
0.6	1.5013	1.437	1.374	1.311	1.251	1.190	1.138	1.099	1.072	1.049	1.030	1.014	1.004	0.995	0.990	0.987	0.984	0.982	0.980	0.979
0.8	1.6326	1.556	1.482	1.411	1.340	1.272	1.211	1.157	1.114	1.077	1.048	1.026	1.009	0.996	0.984	0.980	0.976	0.974	0.972	0.971
1.0	1.7463	1.660	1.576	1.495	1.418	1.344	1.275	1.211	1.155	1.106	1.065	1.038	1.014	0.996	0.982	0.972	0.965	0.962	0.960	0.959
1.5	1.9610	1.856	1.753	1.654	1.561	1.472	1.385	1.306	1.230	1.164	1.103	1.059	1.019	0.987	0.962	0.945	0.932	0.922	0.916	0.915
2.0	2.0937	1.971	1.854	1.742	1.636	1.536	1.440	1.350	1.265	1.190	1.120	1.057	1.005	0.962	0.928	0.904	0.886	0.873	0.864	0.860
2.5	2.1660	2.031	1.900	1.779	1.662	1.550	1.448	1.351	1.260	1.179	1.104	1.032	0.972	0.922	0.882	0.850	0.827	0.811	0.801	0.796
3.0	2.1974	2.049	1.909	1.777	1.651	1.535	1.426	1.321	1.224	1.140	1.060	0.988	0.924	0.870	0.828	0.791	0.763	0.744	0.732	0.728
3.5	2.2033	2.043	1.893	1.753	1.621	1.497	1.382	1.275	1.176	1.088	1.006	0.933	0.868	0.811	0.765	0.727	0.697	0.676	0.663	0.659
4.0	2.1947	2.024	1.864	1.715	1.577	1.448	1.329	1.221	1.121	1.030	0.947	0.872	0.806	0.749	0.701	0.661	0.630	0.609	0.595	0.590
4.5	2.1791	1.998	1.829	1.673	1.529	1.394	1.273	1.162	1.061	0.968	0.884	0.810	0.744	0.687	0.638	0.599	0.567	0.545	0.531	0.526
5.0	2.1610	1.971	1.794	1.631	1.481	1.344	1.218	1.104	1.002	0.909	0.824	0.750	0.685	0.628	0.580	0.540	0.508	0.485	0.470	0.465
6.0	2.1264	1.918	1.727	1.552	1.392	1.247	1.115	0.999	0.893	0.799	0.714	0.644	0.580	0.523	0.476	0.437	0.406	0.383	0.368	0.361
7.0	2.0987	1.875	1.670	1.484	1.315	1.164	1.029	0.908	0.802	0.708	0.624	0.554	0.491	0.437	0.392	0.354	0.324	0.302	0.286	0.279
8.0	2.0784	1.840	1.623	1.427	1.251	1.095	0.955	0.833	0.725	0.631	0.549	0.480	0.420	0.368	0.326	0.290	0.261	0.239	0.223	0.215
9.0	2.0637	1.811	1.583	1.379	1.196	1.035	0.892	0.768	0.661	0.569	0.488	0.421	0.363	0.312	0.273	0.240	0.211	0.191	0.176	0.168
10.0	2.0529	1.787	1.548	1.335	1.147	0.982	0.838	0.715	0.608	0.515	0.437	0.370	0.314	0.269	0.231	0.200	0.174	0.154	0.140	0.132
12.0	2.0385	1.749	1.492	1.264	1.067	0.896	0.745	0.624	0.519	0.431	0.356	2.296	0.244	0.202	0.170	0.143	0.121	0.101	0.091	0.084
14.0	2.0295	1.717	1.441	1.202	0.997	0.823	0.672	0.551	0.450	0.365	0.297	0.240	0.195	0.158	0.127	0.103	0.084	0.072	0.062	0.056
16.0	2.0232	1.690	1.401	1.150	0.937	0.760	0.613	0.493	0.394	0.314	0.249	0.198	0.157	0.127	0.098	0.076	0.061	0.051	0.044	0.038
18.0	2.0185	1.666	1.363	1.102	0.886	0.707	0.561	0.443	0.347	0.271	0.211	0.164	0.128	0.099	0.077	0.060	0.047	0.036	0.029	0.026
20.0	2.0150	1.646	1.329	1.061	0.841	0.660	0.515	0.399	0.307	0.236	0.181	0.138	0.105	0.080	0.061	0.047	0.036	0.028	0.022	0.018

SOURCE: From *Thermodynamics*, revised by K. S. Pitzer and L. Brewer, McGraw-Hill, New York, 1961.

TABLE 4.6 Free Energy $-(G^0 - H_0^0)/T$ due to Internal Rotation, cal/mol · K

	$1/f_{int\ rot}$														
V/RT	0.25	0.30	0.35	0.40	0.45	0.50	0.55	0.60	0.65	0.70	0.75	0.80	0.85	0.90	0.95
0.0	2.754	2.392	2.086	1.821	1.587	1.377	1.190	1.014	0.856	0.710	0.575	0.443	0.323	0.208	0.102
0.2	2.710	2.359	2.061	1.803	1.574	1.368	1.182	1.009	0.852	0.707	0.570	0.441	0.321	0.207	0.101
0.4	2.623	2.296	2.014	1.765	1.543	1.342	1.164	0.997	0.842	0.699	0.565	0.438	0.318	0.206	0.099
0.6	2.518	2.208	1.944	1.708	1.498	1.309	1.136	0.974	0.826	0.687	0.555	0.431	0.315	0.204	0.097
0.8	2.406	2.106	1.856	1.636	1.442	1.266	1.099	0.947	0.804	0.670	0.543	0.424	0.310	0.200	0.096
1.0	2.296	2.004	1.764	1.559	1.379	1.214	1.056	0.912	0.777	0.647	0.526	0.411	0.302	0.195	0.094
1.5	2.040	1.770	1.548	1.370	1.210	1.069	0.937	0.815	0.700	0.588	0.481	0.379	0.277	0.178	0.084
2.0	1.819	1.563	1.360	1.193	1.052	0.927	0.817	0.713	0.615	0.521	0.428	0.338	0.249	0.160	0.074
2.5	1.630	1.389	1.197	1.043	0.912	0.802	0.705	0.616	0.534	0.454	0.375	0.298	0.219	0.141	0.063
3.0	1.473	1.240	1.059	0.914	0.793	0.695	0.608	0.530	0.458	0.390	0.324	0.258	0.191	0.122	0.053
3.5	1.340	1.117	0.943	0.802	0.694	0.603	0.525	0.457	0.395	0.336	0.278	0.222	0.165	0.105	0.042
4.0	1.225	1.013	0.847	0.713	0.613	0.527	0.455	0.393	0.339	0.288	0.239	0.190	0.140	0.088	0.034
4.5	1.133	0.925	0.764	0.637	0.543	0.463	0.398	0.340	0.290	0.247	0.205	0.162	0.117	0.074	0.027
5.0	1.053	0.849	0.696	0.577	0.483	0.408	0.347	0.297	0.253	0.214	0.177	0.139	0.102	0.063	0.020
6.0	0.919	0.728	0.586	0.477	0.393	0.325	0.273	0.230	0.193	0.161	0.131	0.103	0.074	0.045	0.012
7.0	0.819	0.636	0.503	0.402	0.325	0.267	0.218	0.181	0.149	0.123	0.100	0.078	0.056	0.032	0.008
8.0	0.735	0.564	0.440	0.346	0.275	0.221	0.179	0.145	0.118	0.096	0.078	0.060	0.042	0.024	0.005
9.0	0.667	0.504	0.388	0.300	0.235	0.186	0.149	0.120	0.095	0.078	0.062	0.047	0.032	0.019	0.004
10.0	0.610	0.456	0.345	0.264	0.203	0.159	0.124	0.100	0.079	0.063	0.049	0.037	0.026	0.015	0.002
12.0	0.521	0.380	0.280	0.209	0.157	0.120	0.092	0.071	0.054	0.042	0.033	0.025	0.018	0.010	0.001
14.0	0.452	0.321	0.232	0.169	0.124	0.092	0.069	0.052	0.038	0.030	0.023	0.016	0.012	0.007	0.000
16.0	0.396	0.276	0.195	0.139	0.100	0.072	0.053	0.039	0.028	0.021	0.016	0.012	0.008	0.004	0.000
18.0	0.351	0.240	0.166	0.117	0.082	0.058	0.042	0.030	0.022	0.016	0.012	0.008	0.006	0.003	0.000
20.0	0.315	0.211	0.144	0.098	0.068	0.047	0.033	0.023	0.017	0.012	0.009	0.006	0.004	0.002	0.000

SOURCE: From *Thermodynamics*, revised by K. S. Pitzer and L. Brewer, McGraw-Hill, New York, 1961.

introduced in using Tables 4.4 to 4.7 for molecules such as toluene. If multiple rotations are possible as in *p*-xylene, we take the partition function as representative of each, and we will merely add the contribution of each to determine the total contribution to the thermodynamic properties.

A number of experimental techniques have been employed to determine the barrier height in restricted rotations. Nuclear magnetic resonance (NMR) may be used with good results only when the barrier height is high (greater than about 30 kJ/mol) and this usually occurs when the rotation is about a bond with some double-bond character. Microwave spectroscopy is the most accurate method for determining barrier heights, for the absorption spectrum that results from rotation is dependent on the moments of inertia of the molecule. Bond lengths and bond angles may be accurately determined by studying different isotopically substituted structures. The coupling of internal rotation with overall rotation results in the splitting of the absorption lines into different components, and the magnitude of this splitting is in-

TABLE 4.7 Entropy S due to Internal Rotation, cal/mol · K

V/RT	$1/f_{\text{int rot}}$														
	0.25	0.30	0.35	0.40	0.45	0.50	0.55	0.60	0.65	0.70	0.75	0.80	0.85	0.90	0.95
0.0	3.748	3.386	3.079	2.814	2.580	2.371	2.182	2.009	1.850	1.703	1.567	1.438	1.316	1.203	1.097
0.2	3.743	3.382	3.076	2.811	2.578	2.369	2.180	2.003	1.848	1.701	1.563	1.433	1.312	1.196	1.091
0.4	3.730	3.370	3.065	2.801	2.568	2.359	2.170	1.996	1.837	1.691	1.555	1.428	1.307	1.193	1.085
0.6	3.709	3.347	3.043	2.780	2.547	2.340	2.151	1.980	1.823	1.677	1.541	1.415	1.295	1.184	1.076
0.8	3.679	3.318	3.013	2.750	2.519	2.315	2.125	1.957	1.800	1.654	1.523	1.399	1.284	1.171	1.068
1.0	3.638	3.279	2.974	2.714	2.485	2.279	2.094	1.928	1.774	1.629	1.499	1.377	1.262	1.153	1.052
1.5	3.512	3.156	2.854	2.600	2.376	2.173	1.997	1.833	1.685	1.552	1.428	1.310	1.201	1.094	1.000
2.0	3.355	3.004	2.709	2.458	2.241	2.048	1.874	1.718	1.578	1.450	1.332	1.224	1.122	1.024	0.936
2.5	3.180	2.836	2.548	2.303	2.091	1.907	1.739	1.589	1.456	1.335	1.224	1.126	1.031	0.942	0.860
3.0	3.008	2.667	2.380	2.138	1.933	1.756	1.576	1.456	1.330	1.217	1.114	1.021	0.936	0.855	0.779
3.5	2.838	2.500	2.218	1.978	1.782	1.610	1.458	1.323	1.206	1.100	1.004	0.919	0.841	0.769	0.703
4.0	2.678	2.343	2.069	1.834	1.643	1.475	1.328	1.199	1.087	0.988	0.901	0.821	0.748	0.683	0.623
4.5	2.528	2.199	1.926	1.698	1.511	1.348	1.209	1.086	0.978	0.884	0.804	0.730	0.662	0.607	0.551
5.0	2.396	2.068	1.798	1.579	1.392	1.233	1.097	0.982	0.881	0.794	0.716	0.648	0.588	0.535	0.486
6.0	2.166	1.844	1.585	1.370	1.192	1.040	0.915	0.808	0.715	0.637	0.568	0.509	0.457	0.412	0.372
7.0	1.983	1.665	1.411	1.204	1.033	0.891	0.774	0.672	0.588	0.516	0.453	0.401	0.357	0.319	0.285
8.0	1.830	1.519	1.272	1.071	0.906	0.770	0.660	0.566	0.486	0.422	0.366	0.320	0.281	0.248	0.220
9.0	1.703	1.397	1.156	0.962	0.804	0.674	0.570	0.483	0.407	0.350	0.300	0.258	0.223	0.195	0.171
10.0	1.593	1.295	1.060	0.872	0.719	0.596	0.496	0.414	0.348	0.293	0.248	0.211	0.180	0.154	0.134
12.0	1.417	1.125	0.904	0.728	0.588	0.476	0.388	0.315	0.255	0.213	0.176	0.146	0.122	0.101	0.084
14.0	1.275	0.994	0.783	0.620	0.492	0.388	0.309	0.247	0.196	0.157	0.126	0.100	0.084	0.069	0.056
16.0	1.157	0.890	0.688	0.533	0.414	0.322	0.251	0.196	0.155	0.119	0.092	0.075	0.059	0.048	0.038
18.0	1.058	0.801	0.609	0.464	0.353	0.270	0.205	0.158	0.121	0.093	0.072	0.056	0.042	0.034	0.026
20.0	0.975	0.727	0.542	0.405	0.303	0.228	0.170	0.129	0.097	0.073	0.056	0.042	0.032	0.024	0.018

SOURCE: From *Thermodynamics*, revised by K. S. Pitzer and L. Brewer, McGraw-Hill, New York, 1961.

versely proportional to the barrier height. The frequency of torsional vibration may also be available from microwave spectroscopy which may also be used to determine the barrier height. A notable characteristic of this torsional frequency, and one that complicates calculations, is its very large anharmonicity. Remember that this vibration becomes a hindered rotation, and this conversion then could be viewed as an increasingly anharmonic vibration, for clearly the restoring force will finally not bring the group back toward its equilibrium position at all at the onset of hindered rotation. And much less is the restoring force proportional to the square of the displacement. This state of affairs is somewhat reminiscent of the increasingly anharmonic vibrations of the molecules of a solid until a point is reached at which the restoring force is completely overrun, and we say the solid has melted. The frequency of torsional vibration, determined from the microwave spectrum of the species, may be used to calculate the barrier height.[5] It is possible to calculate these heights for molecules such as trimethylamine that have multiple rotating tops and where we as-

TABLE 4.8 Some Measured Torsional Frequencies and the Corresponding Barrier Heights to Free Rotation†

Molecule	Torsional frequency, cm^{-1}	Barrier, kJ/mol
CH_3CHO	150	4.94
CH_3CH_2Cl	243	14.98
CH_3CHF_2	222	13.43
$HCOOCH_3$	130	4.87
$(CH_3)_2CO$	109	3.47
$(CH_3)_2O$	242	10.98
$(CH_3)_3N$	269	18.45

SOURCE: From W. A. Fateley and F. A. Miller, *Spectrochim. Acta, 18*, 977 (1962).
†Multiple groups are taken to be oscillating independently.

sume each group to be rotating independently of the others. From a more detailed perspective, one must worry about assymetric tops and about interactions between rotating groups on a single frame, but there is no general solution to such issues. Some torsional frequencies and the corresponding barriers appear in Table 4.8.

The oldest method of determining the barrier height is to compare the calculated ideal gas entropy with that from experimental third-law studies. Any discrepancy is due to hindered internal rotation, and we merely select a barrier height that makes the calculated entropy agree with the third-law entropy. But this method is obviously very sensitive to experimental error in measuring the low-temperature heat capacities and in the assignment of vibration frequencies from spectral data, for it depends on small differences in two large numbers. Also, in practical calculations, our problem is inevitably one of estimating thermodynamic properties from other more readily available molecular properties, and this method of determining the barrier height is then self-defeating.

These interrelationships are evident in dimethyl ether, $(CH_3)_2O$,[6] where after subtracting the contributions to the entropy due to all modes from the experimental calorimetric entropy at 200 K, there remains 11.09 J/mol · K which may be assigned to the internal rotation of the two methyl groups. Assuming barriers of the form of Eq. (4.4), one can then deduce a barrier height of 13 kJ/mol for both rotations from the measured ΔS(int rot) of 11.09 J/mol · K. Frequencies characteristic of very weak bonds of 160 cm^{-1} and 300 cm^{-1} (among the $3n - 6$ total frequencies) have been reported for $(CH_3)_2O$, and these will contribute 11.00 J/mol · K to the entropy at 200 K, or essentially perfect agreement with the calorimetric value. These frequencies also correctly predict C_p^0 and S^0 at higher temperatures as well. These data were the first example of agreement between calorimetric and statistical thermodynamic properties when a torsional oscillator was involved.

These methods have been extended for use with unsymmetrical tops and in the general case for molecules in which there are multiple internal rotations. These cases will not be considered here for they require that many approximations be made in order to use the tables, the accuracy diminishes, and the results then are of less value from an application perspective.

Now that we have developed an understanding of the nature of internal rotation and recognize the sources of the experimental data necessary to calculate the contributions of internal rotation to the thermodynamic properties, it will be useful to consider a few examples and thus see how well the theory agrees with the actual experimental data. The barrier height for ethane is 13.18 kJ/mol. The statistical calculation of the absolute entropy of ethane at 355 K, assuming the incorrect molecular model of free rotation, gives an entropy of 245.77 J/mol · K, while the calculation assuming hindered rotation gives an entropy of 238.74 J/mol · K. The experimental third-law entropy is 239.32 ±0.8 J/mol · K. The hindered rotation model gives very good results.

Figure 4.3 shows a plot of the contribution of internal rotation to the heat capacity as a function of barrier height and absolute temperature. As mentioned before, when there is no potential barrier to rotation, the molecule rotates freely, and there is a contribution to the heat capacity of ½R that is independent of temperature, as illustrated in Figure 4.3. When there is a potential barrier, as the molecule is heated from absolute zero it begins to rock about the rotational axis and this contributes more and more to the total heat capacity as described by the Einstein function. For a torsional oscillation of 800 cm^{-1}, the contribution to C_p^0 at 200 K would be 2.5 J/mol · K as compared to 4.14 J/mol · K if the rotation were free. When such temperature is reached that the molecule begins to rotate, the contribution to the heat capacity reaches a maximum of about R, with molecules possessing higher potential barriers reaching this maximum at higher temperatures. If the temperature is increased still further, the molecule begins to rotate more and more freely and the heat-capacity contribution decreases, approaching that of the free rotator of ½R. If the rotation is entirely hindered, the contribution to the heat capacity will vary from 0 to R just as for any other Einstein vibrator. All of this behavior as it relates to experimental data on ethane is evident in Fig. 4.4. Data from several different calorimetric investigations are much better represented as a restricted rotation than as a harmonic oscillation. To obtain the data appearing in Fig. 4.4, one subtracts $(C_p - C_V)$, the calculated heat capacity due to translation, rotation of the mole-

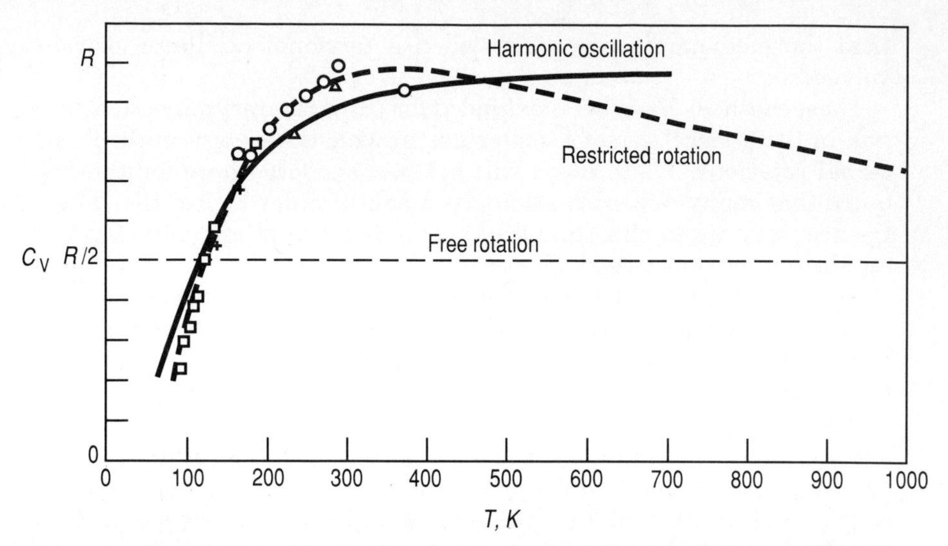

Figure 4.3 Contribution of torsional mode to molar heat capacity of ethane. *(From S. M. Blinder, Advanced Physical Chemistry, Macmillan, 1969. Used by permission.)*

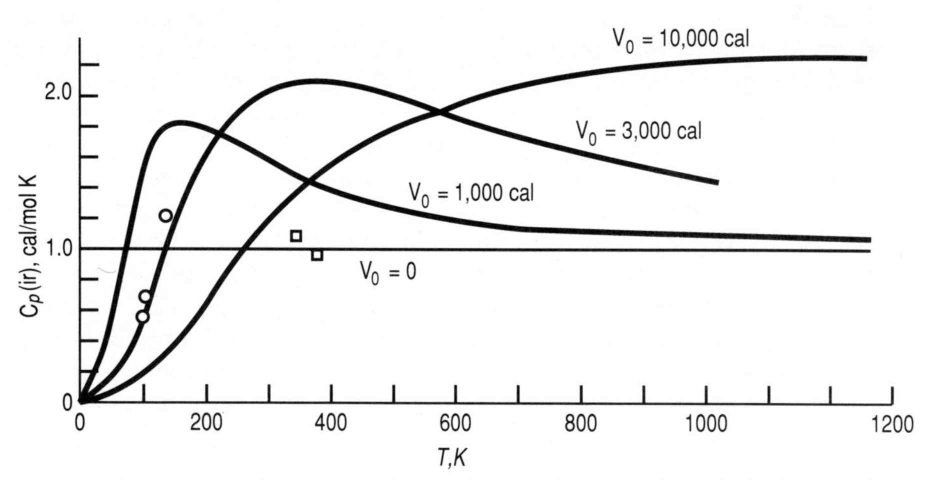

Figure 4.4 The contribution of internal rotation to the heat capacity of C_2H_6 for several assumed barrier heights. The actual barrier is 3150 cal (13.2 kJ) as given by Table 4.3. *(From G. Herzberg, Molecular Spectra and Molecular Structure II: Infrared and Raman Spectra of Polyatomic Molecules, Van Nostrand Reinhold, 1945. Used by permission.)*

cule as a whole, and the $3n - 7$ ordinary Einstein vibrators from the measured calorimetric value. Figure 4.4 is a plot of this difference. The barrier to rotation in ethane is 13.18 kJ/mol. Certainly the rotation is not free, but the hindered rotor model that well fits these data does predict a gradual decline toward free rotation as the temperature increases.

For CH_3CCl_3, the barrier height is 12.41 kJ/mol, $I_r = 5.25 \times 10^{-47}$ kg · m^2, $\sigma = 3$, the product of the moments of inertia for the whole molecule is 6.14×10^{-134} kg^{3}m^6, the $3n - 7$ vibration frequencies are 2954, 1383, 1075, 526, 344, 3017 (2), 1456 (2), 1089 (2), 725 (2), 351 (2), and 239 (2) cm^{-1}, and the torsional oscillation about the C—C bond is 214 cm^{-1}. Suppose we wish to calculate the entropy of this compound at 286.53 K. The translational, rotational, and vibrational contributions to the entropy are all calculated in the usual way wherein we consider all $3n - 7$ vibrational modes. To determine the contribution from internal rotation, we calculate $f_{\text{int rot}}$ and V/RT as

$$f_{\text{int rot}} = \frac{2.7928}{3}[(10^{45})(5.25 \times 10^{-47})(286.53)]^{1/2} = 3.6106$$

or

$$\frac{1}{f_{\text{int rot}}} = 0.2769$$

and

$$\frac{V}{RT} = \frac{12,410}{(8.3143)(286.53)} = 5.211$$

Direct double linear interpolation in Table 4.7 gives a value of 9.00 J/mol · K for the internal rotational contribution to the entropy of methylchloroform, and we note from Table 4.9 that we have calculated an entropy that agrees well with the calorimetric value. This provides a very good check of the theory since thermal data were not used in evaluating any of the molecular constants.

As another example, we might determine the internal rotational contribution to the entropy of dimethyl cadmium, which was calculated before directly from the partition function for the free rotator, but here by simply using Table 4.7. We have

TABLE 4.9 The Entropy of Methylchloroform at 286.53 K

Translation and rotation	276.06 J/mol · K
Vibration	33.76
Internal rotation	9.00
Total, calculated	318.82
Experimental, third-law	318.90 ± 0.67

SOURCE: K. S. Pitzer and J. Hollensberg, *J. Am. Chem. Soc.*, *75*, 2219 (1953).

$$f_{\text{int rot}} = \frac{2.7928}{3}[(10^{45})(2.70 \times 10^{-47})(298.16)]^{1/2} = 2.6413$$

and $\dfrac{1}{f_{\text{int rot}}} = 0.3786$ and $\dfrac{V}{RT} = 0$

Direct linear interpolation in Table 4.7 gives $S_{\text{int rot}}$ = 12.26 J/mol · K, as was obtained before. Interestingly, rotation of the CH_3 groups in dimethyl acetylene also appears to be free.

Ethyl cyanide is an interesting example wherein a barrier height of 21.757 kJ/mol was found from the frequency of the torsional vibration using the microwave spectroscopic method discussed earlier. The contributions due to internal rotation were obtained from the tables exactly as has been described, and the results appear in Table 4.10. The large barrier height is postulated to be caused by the highly polar nature of the —C≡N group. This is probably correct since the electron density around the carbon atom is much greater than that in ethane and thus more energy would be required to rotate the molecule. In other words, the bond has partial double bond character and pi bonds resist rotation.

Thermodynamic properties for molecules with more than two groups which rotate with respect to each other, such as para-xylene where both methyl groups rotate independently, as well as for molecules in which the rotating groups are unsymmetrical, as in chloro-unsymmetrical-dimethyl-hydrazine, have been discussed. If the rotating groups are far enough apart that no coupling can occur, as in *p*-xylene or trans-2-butene, we can use the theory developed above with impunity. This is not the case, however, for molecules like *o*-xylene or cis-2-butene. These situations are too complex for our discussion and would take us too far into molecular physics. However, such situations can be treated, and the interested reader should refer to the suggested "Further Reading" for more insight.

Most of the more recent work dealing with hindered internal rotation has been within the realm of theoretical physics, and it does not yet improve the computational skills of interest to the applied scien-

TABLE 4.10 Calculated Thermodynamic Functions for Ethyl Cyanide, cal/mol · K

Temperature, K	$(H^0 - H_0^0)/T$	$-(G^0 - H_0^0)/T$	C_p^0	S^0
298.16	11.814	55.998	17.225	67.812
300	11.816	56.077	17.298	67.922
400	13.705	59.734	21.324	73.439
500	15.585	62.989	24.861	78.574
600	17.416	65.992	27.962	83.408

SOURCE: N. E. Duncan and G. J. Janz, *J. Chem. Phys.*, *23*, 434 (1955).

tist. Tables 4.4 to 4.7, developed 40 years ago, remain the method of choice for calculating the thermodynamic contributions of hindered internal rotation in most engineering applications. More finesse in treating more complex molecules exists, but it requires the specialist, and its accuracy becomes more questionable.

Example Problem 4.1 Calculate the C_p^0 for propanone at 405.2 K. The selected best data are a molecular weight of 58.0798, $\sigma = 2$, $I_A I_B I_C = 1.39188 \times 10^{-135}$ $kg^3 \cdot m^6$, potential barrier = 3.47 kJ/mol for each methyl group, and internal rotational constants of 5.931 and 5.517 cm^{-1}.

$$C_p^0 = \frac{3}{2}R_{\text{trans}} + \frac{3}{2}R_{\text{rot}} + \sum_0^{22} C_V(\text{vib}) + \sum_0^{2} C_V(\text{int rot}) + (C_p^0 - C_V^0)$$

where the several terms arise from contributions due to translation, rotation of the entire molecule, 22 intramolecular vibrations, two internal rotations, and $(C_p^0 - C_V^0)$ or R. The several vibrations contribute:

ω, cm^{-1}	ω/T, $cm^{-1} \cdot K^{-1}$	$C_V^0(\text{vib})/R$
3019(2)	7.451	0
2972	7.335	0
2963	7.312	0
2937(2)	7.248	0
1731	4.272	0.0813
1454	3.588	0.1545
1435	3.541	0.1612
1426	3.519	0.1643
1410	3.480	0.1701
1364(2)	3.366	0.1880
1216	3.001	0.2552
1091	2.692	0.3253
1066	2.631	0.3407
891	2.199	0.4612
877	2.164	0.4719
777	1.918	0.5496
530	1.308	0.7506
484	1.194	0.7865
385	0.950	0.8576

and using Table 4.6, we deduce a total $C_V(\text{vib})$ of 5.906(8.31433) or 49.104 J/mol · K. The potential barrier for internal rotation of each methyl group is 3.47 kJ/mol, and the internal rotational constants are 5.931 cm^{-1} and 5.517 cm^{-1}. We need V/RT and $1/f_{\text{int rot}}$.

$$\frac{V}{RT} = \frac{3470}{8.314(405.2)} = 1.030 \qquad \text{for both rotators}$$

The internal rotational constant is commonly reported in units of cm^{-1} whereas our expression for the partition function, Eq. (4.2), requires the reduced moment of inertia in units of $kg \cdot m^2$. The relationship between I_r and B_r is exactly as defined earlier in our discussion of the rotation of the molecule as a whole, namely,

$$B_r = \frac{10^{-2}h}{8\pi^2 c I_r}$$

where h and c are in their usual units of J · s and m · s^{-1}, respectively. In the present example calculation,

$$B_r = 5.931 \text{ cm}^{-1} = \frac{10^{-2}h}{8\pi^2 c I_r}$$

or

$$I_r = \frac{10^{-2}h}{(5.931)(8)(\pi^2)(c)}$$

Then

$$f_{\text{int rot}} = \frac{2.7928}{3}\left[\frac{10^{45} \times 6.626 \times 10^{-34} \times 10^{-2} \times 405.2}{5.931 \times 8 \times \pi^2 \times 2.998 \times 10^8}\right]^{1/2}$$

or

$$\frac{1}{f_{\text{int rot}}} = 0.246$$

Double linear interpolation in Table 4.4 leads to C_V^0(int rot) = 1.213 cal/mol · K or 5.075 J/mol · K. Similarly for the other rotational constant of 5.517 cm^{-1}, we obtain $1/f_{\text{int rot}} = 0.237$, and, thus, C_V^0(int rot) = 5.079 J/mol · K.

The total heat capacity is then

$$C_p^0 = \tfrac{3}{2}R + \tfrac{3}{2}R + 49.104 + 5.075 + 5.079 + R = 92.515 \text{ J/mol} \cdot \text{K}$$

This is equivalent to C_p^0 = 22.11 cal/mol · K. The heat capacity of the vapor at ⅓ atm has been calorimetrically found to be 22.39 cal/mol · K at 405.2 K, and by using second virial coefficient data, this experimental value was corrected to that at zero pressure from

$$\left(\frac{\partial C_p}{\partial p}\right)_T = -T\left(\frac{\partial^2 V}{\partial T^2}\right)_p$$

which when integrated from p = ⅓ atm to p = 0 atm yields $C_p - C_p^0$ = 0.21 cal/mol · K, from which we conclude C_p^0(experimental) = 22.18 cal/mol · K, which compares well with our C_p^0(theoretical) = 22.11 cal/mol · K. As an alternative to the tables, the partition function due to internal rotation has been evaluated more accurately from a term-by-term summation of computed energy levels up to 17,000 cm^{-1}, and the result was C_p^0(theoretical) = 22.14 cal/mol · K—an insignificant difference.

At sufficiently low temperatures, the two methyl groups undergo low-frequency, sluggish torsional oscillations of 105 and 108 cm^{-1}. These frequencies contribute to the thermodynamic properties as would any other Einstein vibrator. As the temperature increases, this Einstein vibrator smoothly becomes a hindered rotator, which in turn smoothly becomes a free internal rotator.

Of course, the molecule may rearrange or dissociate or undergo other pervasive change which would alter completely the thermodynamic properties. Even here, our statistical thermodynamic insights will provide guidance to the effort to predict the macroscopic properties of the species (see later).

Example Problem 4.2 Calculate the entropy of propanone as an ideal gas at 298.15 K and 101.325 kPa (1 atm), and compare with the calorimetric third-law entropy.

The contribution to the entropy from the electronic modes is

$$\left(\frac{S^0}{R}\right)_{\text{elect}} = \ln g_0$$

which is equal to zero since $g_0 = 1$ for propanone.

The contribution to the entropy from translation is

$$\left(\frac{S^0}{R}\right)_{\text{trans}} = \frac{3}{2}\ln M + \frac{5}{2}\ln T - \ln p + 3.4533$$

or $$S^0_{\text{trans}} = \frac{3}{2}R\ln 58.0798 + \frac{5}{2}R\ln 298.15 + R\ln 1 - R\ln 101.325 + 3.4533R$$

$$= 50.657 + 118.429 - 38.398 + 28.712 = 159.400 \text{ J/mol}\cdot\text{K}$$

The contribution to the entropy from rotation of the molecule as a whole is

$$\left(\frac{S^0}{R}\right)_{\text{rot}} = \frac{3}{2}\ln T - \frac{1}{2}\ln B_xB_yB_z - \ln\sigma + 16.9890$$

We have been given the product of the three principal moments of inertia in units of $\text{gm}^3\cdot\text{cm}^6$, so this must be converted to the product of three rotational constants each in units of megahertz. So,

$$B_xB_yB_z = \frac{h^3(10^{-6})^3}{(8)^3(\pi^2)^3(I_xI_yI_z)} = \frac{(6.626\times10^{-34})^3(10^{-18})}{(8)^3(\pi)^6(1.39188\times10^{-135})}$$

$$= 0.424597\times10^{12}\text{ MHz}^3$$

As a cross-reference on this conversion, another source[7] reports rotational constants for propanone of 10,165.01, 8,514.88, and 4,910.25 MHz for a product of $0.425001\times10^{12}\text{ MHz}^7$.

Returning to the calculation of S^0_{rot},

$$S^0_{rot} = \tfrac{3}{2}R\ln 298.15 - \tfrac{1}{2}R\ln 0.424597\times10^{12} - R\ln 2 + 16.9890R$$

$$= 71.058 - 111.306 - 5.763 + 141.252 = 95.241 \text{ J/mol K}$$

The contributions to the entropy from the two internal rotations are determined from Table 4.7 by using the partition function due to internal rotation and the potential barrier to free rotation:

$$f_{\text{int rot}} = \frac{2.7928}{3}\left[\frac{10^{45}\times6.626\times10^{-34}\times10^{-2}\times298.15}{5.931\times8\times\pi^2\times2.998\times10^8}\right]^{1/2}$$

or $$\frac{1}{f_{\text{int rot}}} = 0.287 \qquad \text{for } B_r = 5.931\text{ cm}^{-1}$$

and $$\frac{1}{f_{\text{int rot}}} = 0.276 \qquad \text{for } B_r = 5.517\text{ cm}^{-1}$$

The value of V/RT of $3470/8.314\times298.15$ or 1.400 is the same for both rotors. We obtain 13.697 and 14.024 J/mol · K for B_r of 5.931 and 5.517 cm^{-1}, respectively.

The contribution to the entropy that arises from intramolecular vibrations is calculated in the usual way for each of the separate 22 Einstein vibrators:

ω, cm^{-1}	ω/T, $cm^{-1} \cdot K^{-1}$	S^0/R
3019(2)	10.126	0
2972	9.968	0
2963	9.938	0
2937(2)	9.851	0
1731	5.806	0.0022
1454	4.877	0.0072
1435	4.813	0.0078
1426	4.783	0.0081
1410	4.729	0.0087
1216	4.078	0.0195
1364(2)	4.575	0.0105
1091	3.659	0.0326
1066	3.575	0.0361
891	2.988	0.0729
877	2.941	0.0772
777	2.606	0.1142
530	1.778	0.2957
484	1.623	0.3524
385	1.291	0.5135

The total contribution from vibration is then

$$S^0_{vib} = 1.5691\ (8.314) = 13.045\ \text{J/mol} \cdot \text{K}$$

The total entropy is just the sum of all of these contributions, or

$$\begin{aligned} S^0 &= S^0_{elect} + S^0_{trans} + S^0_{rot} + S^0_{int\ rot} + S^0_{vib} \\ &= 0 + 159.400 + 95.241 + 27.721 + 13.046 \\ &= 295.408\ \text{J/mol} \cdot \text{K}\ (70.60\ \text{cal/mol} \cdot \text{K}) \end{aligned}$$

The calculated entropy[8] may be compared with the calorimetric value which is calculated as shown in the following table.

0–17.77	Debye extrapolation	0.667	
17.77–126.00	Integration, crystal II	17.778	
126.00	Transition 10.0/126.00	0.079	
126.00–178.50	Integration, crystal I	6.871	
178.50	Fusion 1380/178.50	7.731	
178.50–298.15	Integration, liquid	14.676	
	Vaporization	24.83	
	Gas imperfection at vapor pressure	0.22	
	Compression from vapor pressure (0.3026 atm) to 1 atm	- 2.38	
		70.47	± 0.25 cal/mol · K

There is a solid-phase transition at 126 K. And clearly the agreement is excellent.

It is interesting that the rotational constants for fully deuterated propanone are 8469.40, 6419.60, and 4011.28 MHz as compared to 10,165.20, 8515.27, and 4910.15 MHz for the normal species. The spectroscopic data also reveal a barrier for internal rotation of 274 cm^{-1} for the normal species vs. 258 cm^{-1} for the deuterated species, and detailed analysis suggests that this difference is real. The paper from which these data were taken[9] determined $V = 274\ cm^{-1}$, which is equivalent to 3.28 kJ/mol and somewhat different from the "selected" value of $V = 3.47$ kJ/mol used in the above calculation.

There are many compilations of computed thermodynamic properties of ideal gases. The techniques and models of Chaps. 3 and 4 are all that is required to calculate every number in every such table.

References

1. K. S. Pitzer and W. D. Gwinn, *J. Chem. Phys.*, *10*, 428 (1942).
2. K. S. Pitzer and W. D. Gwinn, *J. Chem. Phys.*, *14*, 239 (1946).
3. K. S. Pitzer and W. D. Gwinn, *J. Chem. Phys.*, *17*, 1064 (1949).
4. K. S. Pitzer and W. D. Gwinn, *J. Chem. Phys.*, *5*, 469 (1937).
5. D. R. Lide and D. E. Mann, *J. Chem. Phys.*, *28*, 572 (1958).
6. K. S. Pitzer, *J. Chem. Phys.*, *10*, 605 (1942).
7. *Landolt-Börnstein Handbook, Numerical Data and Functional Relationships in Science*, Vol. 4, *Molecular Constants from Microwave Spectroscopy*, Springer Verlag, New York, 1967.
8. R. A. Pennington and K. A. Kobe, *J. Am. Chem. Soc.*, *79*, 300 (1957).
9. J. D. Swalen and C. C. Costain, *J. Chem. Phys.*, *31*, 1562 (1959).

Further Reading

1. Frankiss and Green, Specialist Periodical Report, *Chemical Thermodynamics*, vol. I, The Chemical Society, 1973.
2. S. Mizushima, *Structure of Molecules and Internal Rotation*, Academic Press, New York, 1954.
3. W. J. Orville-Thomas, *Internal Rotation in Molecules*, John Wiley and Sons, New York, 1974.
4. G. J. Janz, *Estimation of Thermodynamic Properties of Organic Compounds*, Academic Press, New York, 1958.
5. D. G. Lister et al., *Internal Rotation in Molecules*, Academic Press, New York, 1978.

Chapter

5

Real Gases

The development of all partition functions to this point has implicitly assumed that the molecules have no energy because of the nearness of other molecules. But molecules do have size and shape, they are deformable and compressible, and forces of attraction exist between them. The existence of size and attraction has its influence on the total energy of a molecule, and, of course, anything that affects this energy will affect the partition function, and through that, all of the thermodynamic properties. Intuitively we sense that the effects of size and attraction will depend on how close together the molecules are, which is to say, the influence of size and attraction is expected to be density-dependent. The problem of calculating macroscopic properties in terms of molecular parameters will be greatly complicated if the partition function is to become density-dependent, for heretofore the partition function has been only temperature-dependent, with the exception of the simple linear dependence of the translational partition function on volume.

We should be forewarned that the treatment of real gases is much more complex than that of ideal gases, and, most importantly, the formalism does not yet exist to allow accurate calculations of macroscopic quantities of technical interest in general. So we here do not expect the feeling of completeness that has characterized much of our previous discussion.

A proper discussion of the properties of real gases and liquids demands the concept of ensembles. Arguments are tedious and long before reaching practical or useful results. Ensemble ideas are omitted here, but the essence of the problem is nonetheless apparent from our perspective on isolated or single-molecule systems. We gain essence but at the expense of elegance and accuracy. For the latter, the reader is referred to any one of the numerous books on statistical mechanics cited throughout this book.

The most nonideal gas of all is a liquid. A real gas is mostly space with some molecules here and there, but a liquid is just the opposite. Liquids are mostly molecules with some empty space, or holes, here and there. Because of their complexity, we will not discuss the highly nonideal gases called *liquids*. There are, however, many so-called hole theories. Suffice to say here, we are well able to handle gases, i.e., systems of complete disorder, and we are well able to handle solids (see Chap. 6), i.e., systems of complete order. We are unable to well handle the intermediate state called *liquid* that exhibits both order and disorder.

Excluded Volume

Let us think first only about the finite size of molecules and delay any consideration of intermolecular attractions. Further, let us simplify the problem to its very basic elements by considering the molecules to be billiard balls or rigid spheres. This state of affairs is depicted in Figure 5.1*a*, where we see that each molecule, say the shaded one in the figure, has available to it not the volume of the system, V, but rather a smaller volume because all of the other $(n - 1)$ molecules occupy a part of V and this part may not be entered by the shaded molecule. If the molecules have a radius of σ, no part of any other molecule may enter a surrounding sphere of volume $\frac{4}{3}\pi\sigma^3$. And if n of these molecules occupy volume V, the thereby excluded volume is $n(\frac{4}{3}\pi\sigma^3)$. This excluded volume is the sum of the "interaction" of one molecule with n other molecules, that is, the result of n pairwise interactions. Then only one-half this effect should be ascribed to the one molecule, and we assert that the volume available to any one molecule, its free volume V_f, is

$$V_f = V - n\left(\frac{2}{3}\pi\sigma^3\right) = V - nb$$

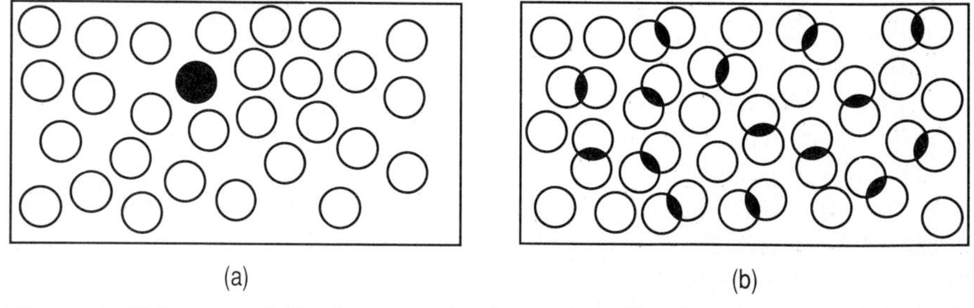

Figure 5.1 Volume available for one molecule is reduced by the volume excluded by all the other molecules.

The idea of "free volume" is prominent in liquid solution theory. With V_f, the partition function is

$$f = \left(\frac{2\pi mkT}{h^2}\right)^{3/2}(V - nb)$$

where, for this rigid-sphere model, we have omitted the usual parts of the partition function that arise from electronic, rotational, and vibrational motions of a real molecule. Now the pressure is rigorously given by

$$p = nkT\left[\frac{\partial \ln f}{\partial V}\right]_T$$

or, with the excluded volume model,

$$p(V - nb) = nkT$$

and if n is Avogadro's number, the equation of state is

$$p(V - b') = RT \tag{5.1}$$

where V is clearly the molar volume and $b' = n_A b$, that is, $b' = \frac{2}{3}\pi n_A \sigma^3$. The problem of excluded volume was recognized first by van der Waals, and Eq. (5.1) is half of the famous van der Waals equation of state.

It is possible to improve on this very simple idea. Improvements can occur from recognizing that molecules are not really hard spheres, but rather they have complex shapes and they are soft and deformable. Also this simple model overestimates the excluded volume at high densities, for then the molecules interpenetrate each other's "sphere of influence" as shown in Fig. 5.1*b*, and it is necessary to then determine the degree of this overlap if we are to expect a reasonable calculation of the pressure. This overlapping will necessarily impact the molecular vibrations and rotations as well, so even this part of the partition function will be density-dependent, and a thorough approach must consider this phenomenon as well. So we recognize many avenues of improvement in the idea of a free volume.

Intermolecular Attractions

In addition to occupying space, molecules also attract each other, for all gases, even helium, condense to form liquids. Intermolecular attractive forces are called long-range, or van der Waals, forces, as may be contrasted with a second class of forces that are referred to as short-range, or chemical or valence, forces. These latter forces give rise to chemical bonds. The long-range forces may be described in three categories: (1) electrostatic forces, (2) polarization forces, and (3) dispersion or London forces. The electrostatic and polarization forces can be understood classically, and they result from the nearness of two charges or two permanent dipoles or higher multipoles and from the induction of polarity in an otherwise nonpolar molecule due to the nearness of a permanent charge or a permanent dipole. Dispersion forces are understandable only as quantum effects.

Molecules obviously attract each other, for otherwise gases would never condense. And these attractive forces vary greatly in magnitude. Both Ne and H_2O have about the same mass (20 vs. 18), and each has a collision radius of about 0.3 nm (average or effective in the case of water). Yet Ne boils at 27 K while H_2O boils at 373 K. The forces of attraction are then much greater between molecules of water than between molecules of neon.

What is the origin of these forces, can they be characterized, can they be either theoretically calculated or experimentally measured, and, most importantly, once these forces are known, can they be used to calculate the macroscopic properties of real gases? Let us first look at the origin and characterization of forces of intermolecular attraction.

Permanent Charges

Forces between molecules are fundamentally electrical in nature, and units of quantities that associate mechanical and electrical concepts are troublesome. Units are troublesome throughout engineering because there are so many different systems and conventional usages. The English system, centimeter-gram-second (cgs) system, and International System (SI) are in varying usage in different countries and by different technical journals. This mixed state of affairs is in secular use as well, for in the United States, the outside temperature is in °F (not °C); when on a diet, we watch our calories (certainly not our Btus or joules); and we buy milk by the quart or gallon (not the cubic meter). This book utilizes predominantly SI, with fundamental units:

Mass	Kilogram	kg
Length	Meter	m
Time	Second	s
Current	Ampere	A
Temperature	Kelvin	K

but some usages that are improper in SI are still conventional. An example is the use of wave numbers in cm^{-1} rather than m^{-1}. It is pleasant in any system of units if common quantities come out to be about unity. Thus atmospheric pressure is about 1 atm, the density of water is about 1 g/cm^3, and the heat capacity of water is about 1 cal/g · °C.

Much of this pleasantness is lost in SI, for atmospheric pressure is 10^5 N/m^2, the density of water is 10^3 kg/m^3, and the heat capacity of water is 4200 J/kg · K. This unpleasantness exists with electrical units as is evident with Coulomb's law† to connect mechanical force with electrical charge. This is, of course, analogous to our use of Newton's law to connect mechanical force to mass. In Coulomb's law,

$$\text{Force} = \frac{\text{charge} \cdot \text{charge}}{\text{distance}^2}$$

and we want charge to be in the fundamental units of ampere-seconds or the secondary units of coulombs, distance is to be in the fundamental unit of meters, and force is to be in the secondary units of newtons. Coulomb's law is then written

$$F_{ij} = \frac{(z_i C)(z_j C)}{4\pi\varepsilon_0 r_{ij}^2} \tag{5.2}$$

where z is a positive or negative integer representing the number of electronic charges of 1.6024×10^{-19} C on the species i and j. The separation of the two species is r_{ij}, in m. The constant of proportionality, for historical reasons, is written as $1/4\pi\varepsilon_0$, which has a defined value of exactly 10^{-7} times the velocity of light squared. That is,

$$\frac{1}{4\pi\varepsilon_0} = 10^{-7}c^2 = 10^{-7}\,(2.9979 \times 10^8)^2 = 8.987 \times 10^9$$

The quantity ε_0 is called the vacuum permittivity. The units of $1/4\pi\varepsilon_0$ are, from Coulomb's law, N · m^2/C^2 or equivalently V · m/C. The value and units of ε_0 are then 8.854×10^{-12} C^2 · s^2 · kg^{-1} · m^{-3}. The force acting along the line of centers between the two charged species will be in newtons. Unlike charges attract; that is, when z_i and z_j are integers of opposite sign, the attractive force will be negative in sign. Conversely repulsive forces will be positive in sign.

The force on a charged species z_iC due to another charged species z_jC may also be expressed as

$$F_{ij} = (z_i C)(E) \tag{5.3}$$

By Coulomb's law, the electric field E in Eq. (5.3) must be $z_jC/4\pi\varepsilon_0 r_{ij}^2$, which, since power is potential × current, will have units of V/m. For example, the electric field at 0.1 nm from an electron is

†By focusing a beam from a krypton fluoride excimer laser at 249 nm to a tiny spot, very large intensities (W/cm^2) have been attained. At such electric fields, the behavior of an electron about an atom is controlled by the laser field rather than by the coulombic attraction of the atomic nucleus. The implications of such processes remain unclear.

$$E = \frac{(1.602 \times 10^{-19}\ \mathrm{C})(8.987 \times 10^{9}\ \mathrm{N \cdot m^2/C^2})}{0.1 \times 10^{9}\ \mathrm{m}^2}$$

or $$E = 1.4 \times 10^{11} \quad \text{in N/C or N} \cdot \text{m/C} \cdot \text{m or J/C} \cdot \text{m}$$

or $$E = 1.4 \times 10^{11}\ \mathrm{V/m}$$

Equation (5.2) is defined for the forces between charges in a vacuum. If some medium intervenes, the force is reduced by the factor D, and we write

$$F_{ij} = \frac{(z_i C)(z_j C)}{4\pi\varepsilon_0 D r_{ij}^2} \tag{5.4}$$

where D is the relative permittivity or the dielectric constant of the medium in which the charges are immersed. It is a measure of the reduced strength of electrostatic interactions due to the presence of the intervening medium. The value of the dielectric constant is 1.000586 for air at standard conditions, and values for some other substances are listed in Table 5.1.

In physics every force is the negative of the derivative of some potential energy with respect to distance, and it will be convenient in our discussions of the thermodynamics of real gases to speak in terms of intermolecular potentials. This greater convenience occurs because energy is the more natural concept in quantum-mechanical arguments. Thus the Coulomb potential between charged species is

$$\phi(r_{ij}) = \frac{(z_i C)(z_j C)}{4\pi\varepsilon_0 D r_{ij}} \tag{5.5}$$

where we have used the boundary condition of $\phi(r) = 0$ at $r = \infty$ to de-

TABLE 5.1 Some Typical Dielectric Constants

Gases			Liquids			Solids (20°C)	
	T, °C	D		T, °C	D		D
Air	0	1.000590	C_6H_6	20	2.284	Acetamide	4.0
NH_3	0	1.0072	CH_3OH	25	32.63	Napthalene	2.52
CO_2	100	1.000985	N_2	−203	1.454	CaF_2	7.36
CF_2Cl_2	100	1.00049	HCl	−15	6.35	Urea	3.50
C_2H_4	110	1.00144	CO_2	20	1.60	NaCl	6.12
HCl	0	1.0046	$(Et)_2O$	20	4.335	Se	6.6
H_2O	0	1.0126	Stearic acid	70	2.29	Cocaine	3.10
			H_2O	25	78		

termine the constant of integration to be zero. This potential appears as shown in Fig. 5.2 with attractive unlike charges having negative potential energies and like charges having positive repulsive potential energies.

For somewhat more elaboration we might recall that gravitational force is $F = -mM/r^2$, where the negative sign accounts, of course, for the always attractive character of gravitation, and the gravitational potential is then $\phi(r) = -mM/r$. Also recall that the restoring force of an elastic material, for example a wire, is proportional to its displacement, $F = -c(r - r_e)$, and this relationship is known as Hooke's law. Of course, materials may be stretched beyond their elastic limit and Hooke's law will no longer apply. But in its region of applicability, the corresponding potential is $\phi(r) = \frac{1}{2}c(r - r_e)^2$ which is immediately recognized as parabolic in shape and characteristic of the familiar harmonic oscillator.

One can get some idea of the strength of the coulombic interaction by considering a 1 molar solution of NaCl in H_2O at 25°C. If uniformly

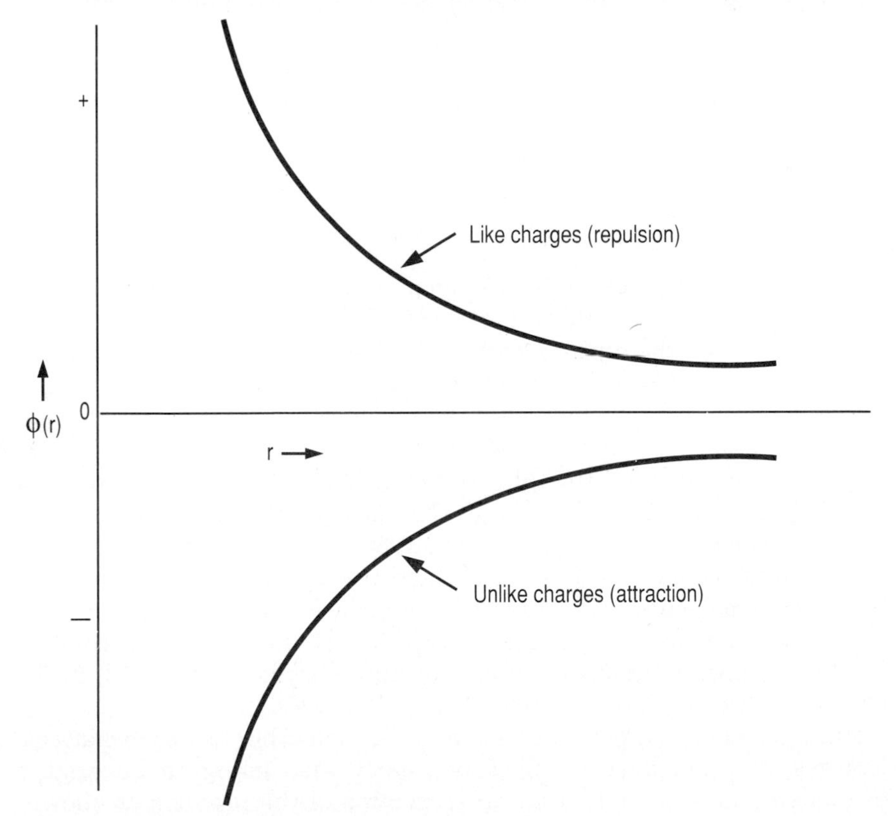

Figure 5.2 Schematic of Coulomb potential.

dispersed, the Na^+ and Cl^- ions would be approximately $[(1000\ \text{cm}^3)/2 \times 6 \times 10^{23}]^{1/3}$ or 1 nm apart. The attraction will then be a potential energy of

$$\phi(r) = \frac{(1.602 \times 10^{-19})^2(1)(-1)}{4\pi(8.854 \times 10^{-12})(1 \times 10^{-9})(78)} \cong -3 \times 10^{-21}\ \text{J}$$

At room temperature, kT is $(1.38 \times 10^{-23})(300)$ or 4×10^{-21} J, or the coulombic attraction is about the same as kT at room temperature. We conclude that the coulombic interaction is significant.

Permanent Dipoles

Some molecules are so configured electrically that there is an overall charge separation. Where there is a single space of negative charge and an equal amount of positive charge in a single other space, one has a dipole. HCl is a good example of this. Another example is the aminoacid, glycine, which has a basic and an acidic end in water,

$$H-\overset{\overset{\large NH_3^+}{|}}{\underset{\underset{\large H}{|}}{C}}-COO^-$$

When the migration of a nucleus, such as H^+ in this example, has occurred, the species is called a zwitterion. The magnitude of a dipole is characterized by the dipole moment μ, which is defined as the product of the charge and its separation, that is,

$$\mu = el \tag{5.6}$$

The common unit of the dipole moment is the debye, which is 3.338×10^{-30} C · m. If the charge that is separated is one electronic charge (1.60×10^{-19} C) and the separation is 0.1 nm, the magnitude of the dipole is about 4.8 debyes, which is very large (see Table 5.2). Symmetric molecules have no dipole moment and increasing asymmetry results in increasing moments, which is also enhanced, of course, by the presence of atoms or groups of high electronegative (Cl, F, O, etc.) or high electropositive (H^+, Na^+, NH_4^+, etc.) character.

Such permanent dipoles attract or repel each other as do permanent charges, but the overall effect is clearly also going to be angle-dependent, as is obvious from an inspection of the problem as shown schematically in Fig. 5.3. When the geometry of Fig. 5.3 is combined

TABLE 5.2 Dipole Moments for Some Typical Molecules Measured at Room Temperature as Gases or in Solution (in Debyes)

HF	1.91	C_2H_4O	1.90	CO	0.10
H_2O	1.85	$(Et)_2O$	1.15	CsCl	10.42
Hg	0	nC_6	0	NH_3	1.47
N_2O	0.17	toluene	0.36	CH_3COCH_3	2.85
UF_6	0	HgO_2	0.2	CH_3COOH	1.7
EtOH	1.70	H_2O_2	2.13		

with Coulomb's law, it may be shown that the potential energy between two dipoles is given by

$$\phi(r_{ij}, \text{angles}) = -\frac{\mu_i\mu_j}{4\pi\varepsilon_0 D r_{ij}^3}[2\cos\theta_i\cos\theta_j - \sin\theta_i\sin\theta_j\cos(\phi_j - \phi_i)] \tag{5.7}$$

For a pure polar species, $\mu_i\mu_j$ is, of course, μ^2. In keeping with one's intuition, the function of *angles* in the square brackets reveals that the repulsion is greatest when the dipoles are in a line with like charges nearest while the attraction is greatest under the same geometric circumstances but with unlike charges nearest to each other. With $\phi_i = \phi_j = \theta_i = 0$ and $\theta_j = \Pi$, the positive ends of the dipole are opposite each other and $\phi(r_{ij})$ is positive. With all four angles equal to zero, opposite charges are geometrically opposite each other and $\phi(r_{ij})$ is negative.

$$\phi(r_{ij}) = \pm\frac{2\mu_i\mu_j}{4\pi\varepsilon_0 D r_{ij}^3} \tag{5.8}$$

We can calculate the magnitude of the maximum attractive potential between molecules of, for example, water where the dipole is 1.85 debyes, where the dielectric constant is 78 with a collision radius of 0.10 nm, or about the molecular separation at room temperature. The energy becomes

$$\phi(r) = \frac{-2(1.85)^2(3.338 \times 10^{-30})^2}{4\pi(8.854 \times 10^{-12})(78)(0.1 \times 10^{-9})^3}$$

with units of $(\text{C}\cdot\text{m})^2/(\text{C}^2\cdot\text{s}^2\cdot\text{kg}^{-1}\cdot\text{m}^{-3})(\text{m}^3)$, or J. We obtain $\phi(r) = -9 \times 10^{-21}$ J or about twice kT at room temperature. The attraction from this source at this separation is strong. As with all angle-dependent potentials, a torque will also be exerted that will tend to rotate the molecules.

The overall macroscopic effect of dipole-dipole interactions of mole-

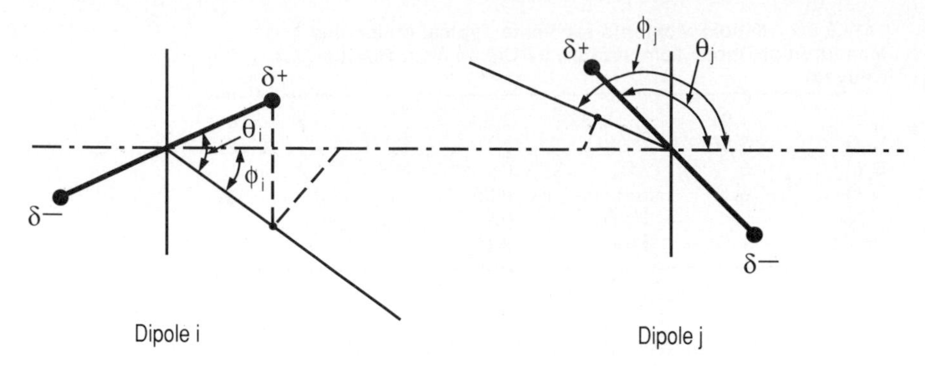

Figure 5.3 Geometric schematic of interaction between two dipoles.

cules is obtained by averaging all of the random orientations from which it may be shown that the average potential energy of interaction is attractive and varies as $\mu_i^2\mu_j^2/r^6$. Thus, for pure polar substances, the attractive force increases as the fourth power of the dipole moment, and we see that a small change in the magnitude of the dipole moment can have a very large effect on the intermolecular forces. And this effect will be even more strongly obvious if the polar molecule is small as well, in view of the r^{-6} dependence.

Polar molecules will be ordered or aligned by an applied electric field, but thermal forces will also tend to disorder this alignment. Such effects are evident in calorimetric third-law studies wherein the solid crystal of a growing phase of a highly polar species will exert sufficient forces that each molecule that condenses on the face of the crystal will be aligned, and the solid will be uniform. On the other hand, the crystal face of a weakly polar species like CO, with a dipole moment of only 0.10 debyes, does not exert forces sufficient for alignment, and there is then a certain resulting disorder in the crystal. This disorder is evident in the calorimetric entropy being less than the statistical entropy. If perfectly aligned, the residual entropy at 0 K would be zero, whereas, if the dipoles in the solid are equally misaligned in each direction, the residual entropy would be $-R[\Sigma x_i \ln x_i]$ or $-R \ln (0.5)$ or 5.76 J/mol · K. The residual entropy for CO is in between at 4.65 J/mol · K.

The experimental dipole moments for some typical molecules are summarized in Table 5.2.

The interaction between an ion and a dipole is schematically evident in Fig. 5.4. With Coulomb's law acting in the plane between the

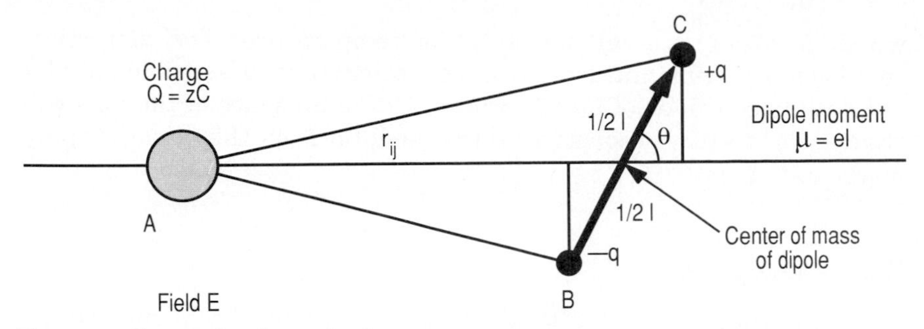

Figure 5.4 Geometric schematic of interaction between an ion and a dipole.

ion of charge zC and the two partial charges of the dipole, $\pm q$, Eq. (5.5) becomes

$$\phi(r,\theta) = -\frac{zCq}{4\pi\varepsilon_0 D}\left[\frac{1}{\overline{AB}} - \frac{1}{\overline{AC}}\right] \tag{5.9}$$

with attraction along the line $\overline{AB}$ and repulsion along the line $\overline{AC}$. The dipole is torqued by the charge.

If the separation between the two species is large compared to the intramolecular charge separation l that forms the dipole, then $\overline{AB} = r_{ij} - \frac{1}{2}l\cos\theta$ and $\overline{AC} = r_{ij} + \frac{1}{2}l\cos\theta$, and the geometry permits the simplification

$$\phi(r,\theta) = -\frac{zC\mu\cos\theta}{4\pi\varepsilon_0 Dr^2} \tag{5.10}$$

The maximum attraction or repulsion occurs when the dipole approaches the ion head-on or tail-on; that is, with $\theta = 0$ and $\cos\theta = 1$, $\phi(r, \theta)$ is negative, while with $\theta = \pi$ and $\cos\theta = -1$, $\phi(r, \theta)$ is positive. Consider Na^+ within a collision radius of about 0.25 nm with a molecule of water. At this distance of minimum separation, corresponding of course to the r of maximum interaction, the potential energy is

$$\phi(r) = -\frac{(1)(1.602\times10^{-19})(1.85)(3.338\times10^{-30})}{4\pi(8.8544\times10^{-12})(78)(0.25\times10^{-9})^2} = -2\times10^{-21}\ \text{J}$$

which is about one-half kT at room temperature. The attraction/repulsion is significant. Note that this calculation is sensible only for very dilute solutions of ions in water, for the presence of the ions will change the dielectric constant of the medium from the value for pure water used here.

Higher Multipoles

The spatial and electrostatic architecture of molecules is such that a four-part separation of charge can occur, and such species are then said to have a quadrupole moment. Similarly, an eightfold charge separation produces an octapole. Quadrupole moments are difficult to measure; they cause much weaker forces between molecules, and consequently their effect on the thermodynamic properties is usually small. For dipole-dipole interaction, the average potential as we have seen varies with separation as r^{-6}; for dipole-quadrupole interactions, the potential varies as r^{-8}; and for quadrupole-quadrupole interactions, the potential varies as r^{-10}. Thus increased multipole interactions are increasingly short-range in their effects.

Polarization

An otherwise symmetric molecule having no permanent dipole moment may have a dipole induced within it by an applied electric field. The magnitude of this induced dipole depends on the strength of the applied field and the responsiveness of the molecule, that is,

$$\mu = \alpha E \tag{5.11}$$

where μ is the moment in $C \cdot m$, E is the field strength in $V \cdot m^{-1}$, and α is a property of the molecule itself that is called the polarizability. Units are again awkward, for α must have units of $\varepsilon_0 m^3$ or $C^2 \cdot s^2 \cdot kg^{-1}$. Polarizabilities are, however, always tabulated as $\alpha/4\pi\varepsilon_0$, which has units of m^3. Typically these numbers are roughly comparable to the volume of the species itself. For example, for the H atom, the polarizability has been measured to be 0.66×10^{-30} m^3 while its volume is $\frac{4}{3}\pi a_0^3$ or 0.62×10^{-30} m^3.

The magnitude of the induced dipole will depend on the orientation of the molecule relative to the applied field, which is, of course, random in a gas or liquid due to thermal motion. So one measures an average polarizability, and some typical values are collected in Table 5.3.

Just as an applied external field may induce a dipole, so may an ion

TABLE 5.3 Polarizabilities of Some Typical Molecules

Units are 10^{-30} m^3.

He	0.20	NH_3	2.3	$CH_2{=}CH_2$	4.3
H_2	0.81	CH_4	2.6	C_2H_6	4.5
H_2O	1.48	HCl	2.6	Cl_2	4.6
O_2	1.60	CO_2	2.6	$CHCl_3$	8.2
Ar	1.63	CH_3OH	3.2	C_6H_6	10.3
CO	1.95	Xe	4.0	CCl_4	10.5

SOURCE: Adapted from J. N. Israelachvili, *Intermolecular and Surface Forces*, Academic Press, New York, 1985.

induce a dipole in an otherwise nonpolar species. A positive ion nearby a neutral species will cause a drift of the electron cloud such as to show more charge density toward the ion. The value of the electric field at the neutral species i, E_i, depends on its separation from the ion j as specified by Coulomb's law according to Eqs. (5.3) and (5.4),

$$E_i = \frac{z_j C}{4\pi\varepsilon_0 D r_{ij}^2}$$

If the polarizability of the molecule is α, the dipole moment that will be induced in the molecule is $\mu = \alpha E$ or $\alpha(zC)/4\pi\varepsilon_0 D r^2$ and the resulting potential of interaction between the ion and the induced dipole is given by Eq. (5.10) as

$$\phi(r) = -\frac{zC\mu}{4\pi\varepsilon_0 D r^2} \tag{5.12}$$

Here θ is obviously zero for the ion attracts the opposite charge along the line of centers to produce an in-line dipole. The induced dipole is αE from Eq. (5.11), and the potential of interaction becomes

$$\phi(r) = -\frac{\alpha(zC)^2}{(4\pi\varepsilon_0)^2 D^2 r^4} \tag{5.13}$$

Just as an applied external electric field or the electric field from a nearby ion can induce a dipole, so can a permanent dipole or higher multipole induce a dipole in an otherwise symmetric molecule. This induction will produce then an intermolecular potential between the permanent and the induced dipole that is not too unlike that between two permanent dipoles. We can see intuitively, however, that the inductive effect will always be attractive, since clearly the positive end of a permanent dipole will induce a negative charge in the nearest part of a neighboring molecule and vice versa. It will also be instantaneous when compared to the speed of molecular motions. The potential energy due to induction is therefore independent of temperature. An analysis of this problem in electrostatics reveals that the potential energy due to induction in a pure polar substance (like water) varies with the separation distance as r^{-6},

$$\phi(r) = -\frac{2\mu^2\alpha}{(4\pi\varepsilon_0)^2 D r^6} \tag{5.14}$$

This is sometimes called the Debye energy or the induction interaction. A similar relationship is obtained for $\phi(r)$ between different species. The derivation of Eq. (5.14) is more complicated than that shown for the interaction between an ion of charge zC and a neutral species of polarizability α that led to Eq. (5.13). There is, however, nothing new in principle, and we then merely state the end result, Eq. (5.14). The value of this energy in pure water at a collision radius of 0.1 nm is

$$\phi(r) = \frac{-2(1.85 \times 3.338 \times 10^{-30})^2(1.48 \times 10^{-30})}{(78)(0.1 \times 10^{-9})^6(4\pi\varepsilon_0)}$$

or $\phi(r) = -13 \times 10^{-21}$ J, which is 3 times kT at room temperature. The effect is large.

Dispersion

The electrostatic origin of forces between molecules that are charged, that is, between ions or between polar species, seems reasonable. But all species, even argon, will condense. Thus forces of attraction are again operative, but what is their origin?

The electron cloud that is around and within all atoms and all molecules may well have a momentary asymmetry. That is, the electrons of argon in their normal motion around the nucleus are not always absolutely spherically symmetric. An instantaneous photograph of the electron cloud would show more charge density here and less over there, and of course a series of such snapshots would reveal this asymmetry of charge density to be continually moving around on the species. Now imagine two such argon atoms near each other, and we can imagine that an instantaneous polarity in the one could induce polarity in the other and vice versa. This problem was first successfully examined by London who showed that the potential energy of interaction from this source varied with distance as r^{-6}, and, like the induction forces arising from a permanent polarity, that this instantaneous polarity was also always attractive. These forces are sometimes referred to as *London forces*, but the more common term is *dispersion forces* because a parameter in the London theory was attainable from the index of refraction of the species. The dispersion energy between two different molecules is

$$\phi(r) = -\frac{3\alpha_i\alpha_j(I_iI_j)}{2(I_i + I_j)(4\pi\varepsilon_0)^2r^6} \tag{5.15a}$$

while the dispersion energy between two identical molecules of symmetric charge distribution is

$$\phi(r) = -\frac{3\alpha^2 I}{4(4\pi\varepsilon_0)^2r^6} \tag{5.15b}$$

This is the famous result of London derived in 1930 where α is the polarizability in $C^2 \cdot s^2 \cdot kg^{-1}$ and I is the ionization potential in J. Note that $\alpha/4\pi\varepsilon_0$ has units of m^3, which may be read from compilations like Table 5.3. For asymmetric molecules, one deduces a series where the lead term varies as r^{-6}. To get some sense of the relative size of this dispersion energy, take a typical α of 1.5×10^{-30} m^3 and I of about 2×10^{-18} J, which are roughly characteristic of O_2. One can then calculate $\phi(r) = 1.4 \times 10^{-20}$ J at a separation of about 0.25 nm, which might characterize a solid noble gas, for example. This is approximately 3 times kT at room temperature; that is, it is a significant energy. This $\phi(r)$ of 1.4×10^{-20} J is also equivalent to about 8.4 kJ/mol, which is the right order of magnitude of the heat of vaporization of spherically symmetric nonpolar molecules such as methane (ΔH_{vap} of CH_4 at its normal boiling point is about 8.9 kJ/mol).

The essential results of all of this discussion of the origin of intermolecular forces are summarized in Table 5.4, where in each instance the interaction energy is in joules.

Summary

It is instructive to compare the relative size of intermolecular attractions from each of the three sources that have been discussed and to do so for several different molecules. Table 5.5 displays such a summary. The averaged energy of the dipole-dipole interaction, the so-called

TABLE 5.4 Intermolecular Interactions

Type	Interaction energy
Charge-charge	$z_iCz_jC/4\pi\varepsilon_0 r$, Coulomb energy
Dipole-dipole	$-2\mu^4/3(4\pi\varepsilon_0)^2kTr^6$, Keesom energy
Dipole-nonpolar	$-2\mu^2\alpha/(4\pi\varepsilon_0)^2r^6$, Debye energy
Nonpolar-nonpolar	$-3I\alpha^2/(4\pi\varepsilon_0)^2r^6$, London energy

TABLE 5.5 Contributions to the Intermolecular Attraction between Several Molecules

				$10^{79}\phi r^6$, J · m^6		
Molecule	μ, D	α, m^3 × 10^{30}	I, eV	$\mu - \mu$	$\mu - \alpha$	$d - d$
He	0	0.2	24.7	0	0	1.2
Ar	0	1.6	15.8	0	0	48
CO	0.12	2.0	14.3	0.0034	0.057	67.5
HCl	1.03	2.63	13.7	18.6	5.4	105
NH_3	1.5	2.21	16	84	10	93
H_2O	1.84	1.48	18	190	10	47

SOURCE: Adapted from A. W. Adamson, *Physical Chemistry*, Academic Press, New York, 1979, p. 290.

Keesom energy (see Table 5.4), was evaluated at 20°C. This energy can be dominant for very polar molecules. The dipole-induced dipole interaction, the so-called Debye energy, is calculated according to the relationship in Table 5.4, and it is generally small. The dispersion, or so-called London energy, is again given in Table 5.4, and it is significant for all molecules.

All of this understanding of the basis of intermolecular potentials is worthwhile, and we have no reason to doubt the general authenticity of the models and the arguments. However, the arguments by London and by others who have examined this problem have many approximations, and the resulting potential is then itself only an approximation. As we shall see momentarily, even if we could write down an exactly correct expression for the intermolecular potential, we still would be unable, in general, to deduce the macroscopic result of this complete microscopic understanding simply because of the mathematical complexity of the formalism that links the intermolecular potential to, for example, the empirical equation of state, $f(p, V, T) = 0$.

If we could calculate, from first principles, the magnitude of the intermolecular forces as a function of separation, this might be empirically modeled; and this model might be manipulated through the well-developed formalisms that are presented in this book to yield both the volumetric and transport properties of matter. Thus we could calculate all of the macroscopic properties of matter from first principles, and we need not experimentally measure anything. It is a fond dream and a worthy goal, but useful numerical results are not yet possible. Nonetheless, useful insights from these ideas form the locus of modern research in molecular engineering.

The theory then gives us direction and a sense of understanding, but it does not yet allow quantitative predictions of macroscopic properties in a general sense. However, much progress has been made.

A First Approximation†

We have now seen the origin of nonidealities in gases as arising both from the size and shape of the molecules and from the forces of attraction between the molecules. Let us make a very simple model of these characteristics, write the partition function for such a simple approximation, and then derive the equation of state from that partition function, as a first correction to the ideal gas relationship. We will then worry about improving the model. We will not expect superaccuracy, but we will expect an understanding of a qualitative sort and of trends.

Let us take the molecules to be repulsive as rigid spheres and to be attractive according to r^{-6} as we have seen. This is a specification of the so-called Sutherland potential which in general can have any value for the exponent r. Mathematically all of this can be written as

$$\phi(r) = \begin{cases} \infty & \text{for } r < \sigma \\ -\varepsilon\left(\dfrac{\sigma}{r}\right)^6 & \text{for } r \geq \sigma \end{cases} \tag{5.16}$$

where σ is the rigid sphere radius and ϕ is the maximum attraction when the molecules are at their closest approach. This intermolecular potential is also graphically depicted in Fig. 5.5. With this model, we can now formally write the partition function as

$$f = \sum \exp -\left[\frac{\varepsilon_{\text{trans}} + \varepsilon_{\text{rot}} + \varepsilon_{\text{vib}} + \varepsilon_{\text{elect}} + \varepsilon_{\text{intermol}}}{kT}\right]$$

where each fixed energy state composed of a translational, rotational, vibrational, and electronic component has now an additive constant, $\varepsilon_{\text{intermol}}$. This $\varepsilon_{\text{intermol}}$ is just the sum of all of the interactions of the molecule of interest with all of the other molecules and where each such pairwise interaction has energy $\phi(r)$. This partition function is more descriptively written

$$f = g_0\left[\frac{2\pi mkT}{h^2}\right]^{3/2} [(V - nb)][f_{\text{vib,rot}}]e^{\phi/2kT} \tag{5.17}$$

wherein the actual volume available to a molecule is the macroscopic total volume less the excluded volume occupied by the molecules themselves, nb. Here b is half the volume of the molecule, $\frac{2}{3}\pi\sigma^3$; that is, the exclusion of volume in a binary encounter arises equally from

†A concise overview of these simple as well as more complex models is given by Prausnitz.[1]

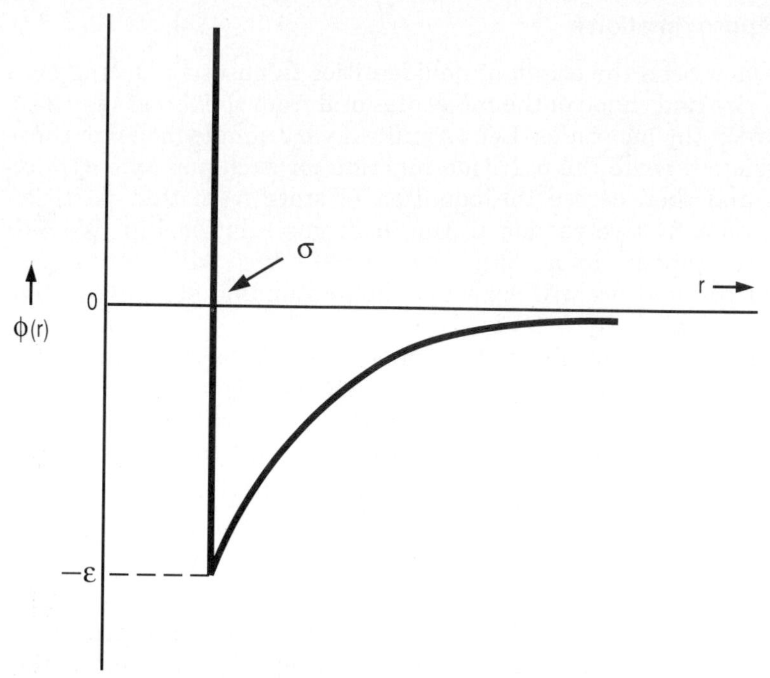

Figure 5.5 Schematic representation of a potential with rigid sphere repulsion and an r^{-6} attraction.

each molecule. The potential energy between one molecule and all the others, ϕ, is rigorously given by integrating the bimolecular potential of Eq. (5.16) over all bimolecular interactions at the density of interest:

$$\phi = \frac{n}{V}\int_0^\infty \phi(r)g(r)4\pi r^2\,dr \qquad (5.18)$$

This potential is again divided by 2 to ascribe the energy equally to each molecule. In this expression $\phi(r)$ may be any intermolecular potential function, but for now we will use that of Fig. 5.5 or Eq. (5.16). The function $g(r)$ tells us the actual molecular density at a distance r from the molecule of interest as a fraction of the average density n/V. This function $g(r)$ is referred to as the *radial distribution function*. Its implication appears, for example, in the local concentration idea of the Wilson equation, which is one of the most successful models for correlating and predicting activity coefficients in liquids. Suffice now to note that it is not unity at all r, for the intermolecular forces do produce a certain semblance of order which is density- and temperature-dependent and which can be experimentally determined. If the number density of molecules were uniform, the molecules contained in a spherical shell of volume $4\pi r^2\,dr$ would be just $(n/V)(4\pi r^2\,dr)$. The function $g(r)$ is just the ratio of the actual local number density to the average, and so the actual number of molecules in the shell is $(n/V)(g(r))(4\pi r^2\,dr)$. To determine the total energy that will be experienced by one molecule, we merely integrate the potential energy due to those molecules at r over all r. And that is, of course, exactly the expression for ϕ of Eq. (5.18). For our present attempt to make a first approximation, let us assign the radial distribution function as zero for all $r < \sigma$ (the spherical molecules are taken to be inpenetrable) and as everywhere unity for all $r > \sigma$, that is, the fluid is perfectly uniform. Then

$$\phi = \frac{n}{V}\int_0^{\sigma}(\infty)(0)(4\pi r^2\,dr) + \frac{n}{V}\int_{\sigma}^{\infty} -\varepsilon\left(\frac{\sigma}{r}\right)^6 (1)(4\pi r^2\,dr)$$

or
$$\phi = -\frac{4\pi n\varepsilon\sigma^3}{3V}$$

In the partition function of Eq. (5.17), we divided the total potential by 2, since we imagine half the energy to be associated with the one molecule and the remaining half to be associated with the remainder of the molecules. That is, we take each pair interaction as shared equally between the two species in determining the total energy of any one molecule.

These arguments about the radial distribution function are cursory, much more may be said, but it would be outside the scope of the present discussion. Continuing for now with the present argument, we recall that the equation of state that we seek is rigorously related to the partition function by

$$p = nkT\left[\frac{\partial \ln f}{\partial V}\right]_T$$

and we can write

$$\ln f = \ln(V - nb) + \frac{4\pi n\varepsilon\sigma^3}{3V(2kT)} + \cdots$$

where we have omitted all terms that do not depend on the volume. Differentiating, we obtain

$$\left[\frac{\partial \ln f}{\partial V}\right]_T = \frac{1}{V - nb} - \frac{2\pi n\varepsilon\sigma^3}{3V^2kT}$$

and the equation of state becomes

$$p = \frac{nkT}{V - nb} - \frac{2\pi n^2\varepsilon\sigma^3}{3V^2} \qquad (5.19)$$

which may be rewritten for a mole of the species (i.e., for $n = n_A$) as

$$\left[p + \frac{a}{V^2}\right][V - b'] = RT \tag{5.20}$$

where $a = 2\pi n_A^2 \varepsilon\sigma^3/3$, $b' = 2n_A\pi\sigma^3/3$, and V in Eq. (5.20) is understood to be the molar volume of the species. This is, of course, the famous equation of van der Waals that was historically the first attempt to describe the nonidealities of gases, and we have evaluated the van der Waals constants in terms of the parameters ε and σ of the intermolecular potential. Although it has no real utility from a practical perspective, such accuracy as may be possible with the van der Waals equation is better attained by fitting a and b directly to experimental pVT data. Nevertheless, the above relationship between the molecular model and the empirical pVT model is instructive.

A More General Approach

The above specification of the Sutherland potential to an inverse-sixth attraction has led us to a molecular interpretation of the van der Waals equation. But its development has required a number of assumptions. We must, of course, assume some intermolecular interaction potential; we must assume something about the radial distribution function; and finally we must assume something about the excluded volume. The assumptions that led to the van der Waals formulation were reasonable first approximations, but we made no pretense of real accuracy. We need a more general approach. One such approach is to take a different tack altogether and develop the so-called virial equation of state.

If we imagine the total potential energy between all n_A molecules to be the sum of all of the interactions between pairs of molecules, it can be rigorously shown that the compressibility is given by

$$\frac{pV}{RT} = 1 + \sum_{n=1}^{\infty} \frac{A_n(T)}{V^n} \tag{5.21}$$

This is the *virial equation of state* and the functions of temperature, $A_n(T)$, are called the *second, third, etc., virial coefficients*. This is the only equation of state that has a rigorous basis in statistical thermodynamics. These virial coefficients can be expressed in terms of the intermolecular potential function where the second virial coefficient relates to the interaction between one pair of molecules, the third to interactions of all pairs in a three-body encounter—that is, 1,2 interactions, 1,3 interactions, and 2,3 interactions—and so on for four-body encounters, etc. The rigorous relationships between the virial coefficients and the intermolecular potential are as follows:

$$A_1(T) \equiv B(T) = 2\pi n_A \int_0^\infty (1 - e^{-\phi(r)/kT}) r^2 \, dr \tag{5.22}$$

and

$$\mathrm{A}_2(\mathrm{T}) \equiv C(T) = -8\pi^2 n_A^2 \int\int\int_0^\infty (e^{-\phi(r_{12})/kT} - 1)(e^{-\phi(r_{13})/kT} - 1)$$

$$(e^{-\phi(r_{23})/kT} - 1) r_{12} r_{13} r_{23} \, dr_{12} \, dr_{13} \, dr_{23} \tag{5.23}$$

and so on with ever increasing complexity for each of the virial coefficients. This formalism assumes that all interactions are pairwise; i.e., the 1,2 interaction is unaffected by 3 being nearby. This is erroneous for molecules that tend to associate. The mathematical details of this development are unimportant; suffice it to note that we can discuss real gases if we know the intermolecular potential and if its functional form is such that these expressions for $B(T)$, $C(T)$, etc. can be evaluated.

For these expressions, $\phi(r)$ is the functional description of the potential energy between two molecules. The force between the molecules is simply related to the potential, $F = -(\partial\phi/\partial r)$, which is, of course, true for all forces and all potentials. We expect our assumption of pairwise additivity to be reasonable at moderate sorts of pressures where kinetic theory would suggest that the molecules are not in each other's force fields long enough for any sort of coordinated action to occur.

What about the effect, if any, of the rotational and vibrational modes on this intermolecular energy? Vibrations are not much affected by the intermolecular forces, as is evidenced by the infrared absorption being approximately equivalent whether one is observing a gaseous or a liquid sample, or whether the sample is pure or in some solvent. This is the basis of infrared spectroscopic analysis. Physically this means that the intermolecular forces are very small compared to the forces that form chemical bonds and give rise, through their vibration, to the characteristic infrared spectra of molecules. Similarly, with one or two notable exceptions, at moderate pressures, we will always regard molecules as rigid rotators which are behaving classically and are noninteracting, to a good approximation, with the intermolecular forces. To moderate pressures, we can with reasonable safety neglect the effect of the intramolecular forces on the equation of state.

At fixed temperature, as p increases, V decreases, and the pV product should then be more nearly constant. Fitting this slowly varying product with a power series in either V or p is an old and purely empirical idea that predates the theoretical arguments surrounding Eqs. (5.22), (5.23), etc. In volume, we have

$$\frac{pV}{RT} = 1 + \frac{B(T)}{V} + \frac{C(T)}{V^2} + \cdots \tag{5.24}$$

and in pressure

$$\frac{pV}{RT} = 1 + B^1(T)p + C^1(T)p^2 + \cdots \tag{5.25}$$

Relationships between the coefficients are made evident by solving Eq. (5.24) for p, substituting this into Eq. (5.25), and equating the coefficients of the resulting terms in V^{-1}, V^{-2}, etc. with the corresponding coefficients in Eq. (5.24). The result is

$$B^1 = \frac{B}{RT}$$

$$C^1 = \frac{C - B^2}{R^2T^2}$$

etc.

The theoretical virial coefficients from Eqs. (5.22) and (5.23) are then immediately useful in the expansion in pressure, Eq. (5.25). For example, at modest pressures,

$$\frac{pV}{RT} = 1 + \frac{B(T)p}{RT}$$

or

$$V = \frac{RT}{p} + B(T)$$

which is analogous to

$$p = \frac{RT}{V} + \frac{B(T)RT}{V^2}$$

from Eq. (5.24).

With its unique formulation in molecular theory, the virial form points the way to useful analytic forms that may be useful in correlating and predicting the pVT properties of gases. For example, the very accurate Benedict-Webb-Rubin equation of state,

$$\frac{pV}{RT} = 1 + \frac{B_0 - A_0/RT - C_0/RT^3}{V} + \frac{b - a/RT}{V^2} + \frac{a\alpha/RT}{V^5} + \frac{c/RT}{V^2T^2}\left(1 + \frac{\gamma}{V^2}\right)\exp\left(-\frac{\gamma}{V^2}\right) \tag{5.26}$$

involves reasonable second, third, and sixth virial coefficients followed by a manufactured term to fit the residuals. We will see later in this chapter that even the form of the temperature dependence of $B(T)$ seen here is suggested by our molecular arguments. The popular cubic equations of state can be cast in virial form. For example, the van der Waals equation becomes

$$z = 1 + \left(b - \frac{a}{RT}\right)\frac{1}{V} + b^2\left(\frac{1}{V}\right)^2 + b^3\left(\frac{1}{V}\right)^3 + \cdots$$

Experimental pVT data have been fitted with a polynomial in V^{-n} since Kamerlingh Onnes (1901).† So there is the possibility of evaluating virial coefficients from intermolecular potential functions which are at least qualitatively consistent with theoretical ideas and comparing them with experiment. Let us examine this idea.

At low pressures, the volume is large, and, to a good approximation,

$$B(T) = \left(\frac{pV}{RT} - 1\right)V$$

Experimental volumetric data then allows a determination of $B(T)$. Data at somewhat higher pressures similarly permits the determination of $C(T)$, and so on.

All experiments reveal that the intermolecular potential function of spherically symmetric molecules is of the schematic form shown as "Reality" in Fig. 5.6. We note that this potential has the same general shape as that between the two atoms of a chemical bond, but here the depth of the well is shallower by an order of magnitude.

Idealizing this all away to say $\phi(r)$ is zero for all r leads to the ideal gas relation, as we have seen. Let us take the molecules to be rigid spheres, i.e., model A in Fig. 5.6, for which $\phi(r) = \infty$ for all $r \le \sigma$ and $\phi(r) = 0$ for all $r > \sigma$. Then $B(T)$ may be easily evaluated as

$$\begin{aligned} B(T) &= 2\pi n_A\left[\int_0^\sigma (1)r^2\,dr + \int_\sigma^\infty (0)r^2\,dr\right] \\ &= \tfrac{2}{3}\,\pi n_A \sigma^3 \\ &= b_0 \end{aligned}$$

and similarly for the other virial coefficients,

†H. Kamerlingh Onnes is a famous name in physics and thermodynamics. He first liquefied helium, he discovered the superfluidity of helium, and he discovered the phenomenon of superconductivity.

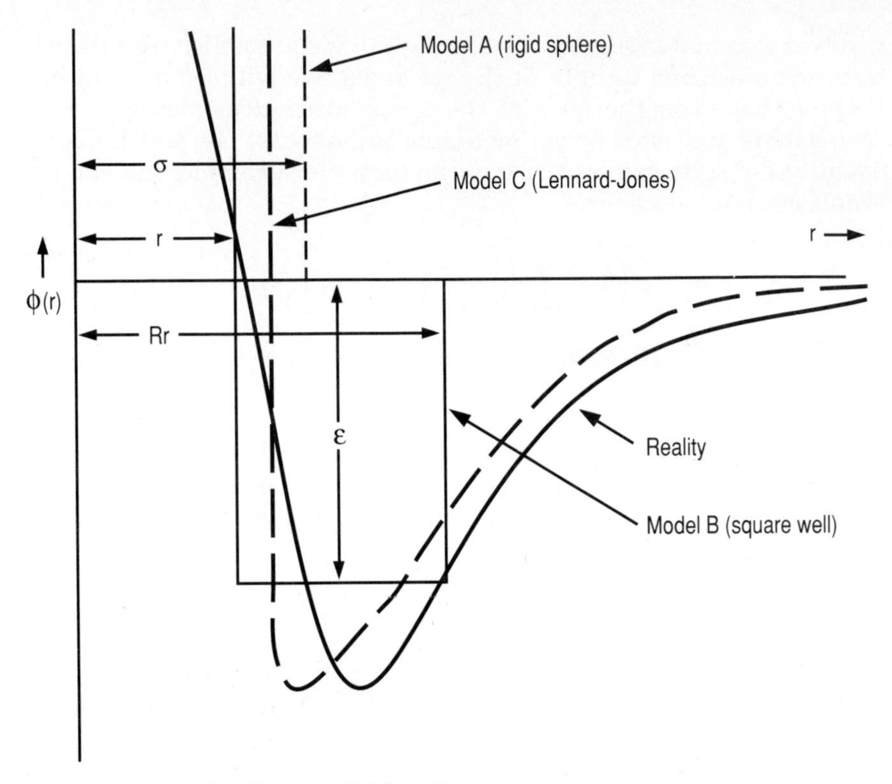

Figure 5.6 Intermolecular potential functions.

$$C(T) = 5/8\, b_0^2$$

$$D(T) = 0.2869 b_0^3$$

$$E(T) = (0.115 \pm 0.005)\, b_0^4$$

where we see that (1) the virial coefficients are not temperature dependent and (2) the difficulty of the calculation increases greatly for coefficients of increasing order. The fifth virial coefficient, even for this simplist of all approximations, can be evaluated only approximately by a Monte Carlo technique.[2] The compressibility factor is here always greater than one. This approximation has found some use at very high temperatures where intermolecular repulsion is more important than attraction. The value of σ is, of course, determined by equating experimental values of $B(T)$ to those calculated from the potential using Eq. (5.22). The evaluation of parameters in any potential function is similar.

The second and third virial coefficients may be evaluated for the square-well potential, model B of Fig. 5.6, which is clearly nearer to reality than was the rigid-sphere model. One may obtain

$$B(T) = b_0(1 - 4.832\Delta)$$

$$C(T) = b_0^2(0.625 - 2.085 + 10.118\Delta^2 - 8.430\Delta^3)$$

where b_0 is the same as previously defined, and $\Delta = \exp(\varepsilon/kT) - 1$, and where the width of the well has been taken to be 0.8, i.e., $R = 1.8$. Three adjustable parameters enable one to well-fit the second virial coefficient data of even rather complex molecules. A more general expression for $B(T)$ for the square-well potential may be written in reduced coordinates as

$$B^*(T^*) = 1 - (R^3 - 1)[\exp(T^*)^{-1} - 1]$$

Here B^* is B/b_0 and T^* is $T/\varepsilon/k$. This result is clear if one merely inserts the piecewise continuous expression for $\phi(r)$ into the general expression for the second virial coefficient and performs the mathematical operations.

Let us now return to the potential function that led us to the van der Waals equation. With the exponent of 6 in Eq. (5.16) allowed to take any value, that is, if we consider the Sutherland potential,

$$\phi(r) = \begin{cases} \infty & \text{for } r \le \sigma \\ -cr^{-\gamma} & \text{for } r > \sigma \end{cases} \tag{5.27}$$

we can evaluate the second virial coefficient as

$$B(T) = 2\pi n_A\left[\int_0^\sigma (1 - e^{-\infty})r^2\,dr + \int_\sigma^\infty \left(1 - e^{-cr^{-\gamma}/kT}\right)r^2\,dr\right]$$

or $$B(T) = \tfrac{2}{3}\pi n_A\sigma^3 - 2\pi n_A\int_\sigma^\infty (e^{-cr^{-\gamma}/kT} - 1)r^2\,dr$$

The integration may be conveniently carried out by first expanding the exponential in a Taylor series. The result is

$$B(T) = \frac{2\pi n_A\sigma^3}{3}\left[1 - \sum_{j=1}^{\infty}\frac{(-1)^j}{j!}\left(\frac{3}{j\gamma - 3}\right)\left(\frac{c}{\sigma^\gamma kT}\right)\right] \tag{5.28}$$

where we see that—depending on the specific values of c and γ and, of course, T—more or less terms in the summation will be required to obtain an accurate value of $B(T)$. The other virial coefficients may be similarly evaluated but with increasing mathematical difficulty.

There are two major problems with this approach. Because of the complex character of the intermolecular forces, it is not possible to write a rigorous expression for the intermolecular potential. And even if such a potential were in hand, mathematical difficulties would limit its utility in evaluating $B(T)$, $C(T)$, etc.

We have seen that the attractive potentials arising from dipole-dipole or dipole-induced dipole or dispersion all vary with separation as r^{-6}. A variety of experimental data reveal that molecules strongly repel each other when they are very close together. So it is perhaps reasonable that we take the total potential energy of interaction as composed of repulsive and attractive contributions as

$$\phi(r) = \frac{A}{r^n} - \frac{B}{r^m}$$

where A, B, n, and m are all positive constants and where $n > m$. This form of an intermolecular potential function was first proposed by Mie in 1903.[3] Clearly the form of this potential will be as schematically shown in Fig. 5.7. At the point of greatest attraction, $r = r_e$, the force between the molecules is zero, for

$$\text{Force} \equiv -\frac{\partial \phi}{\partial r} = 0$$

and then clearly

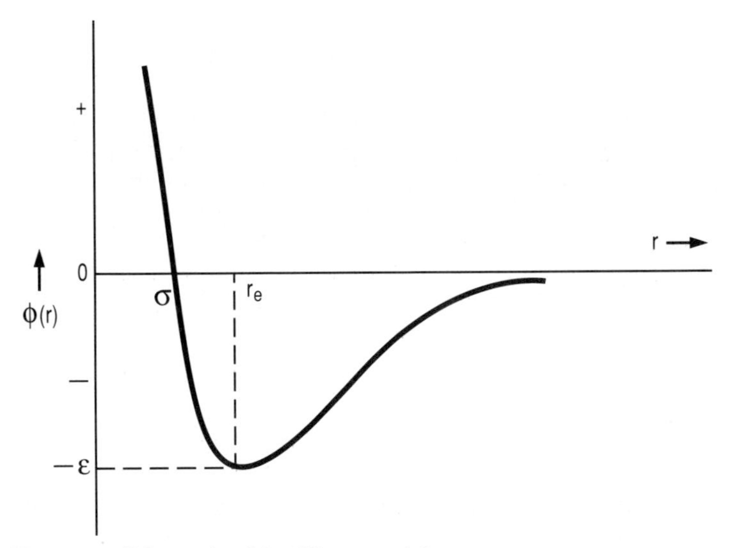

Figure 5.7 Schematic of the Mie potential.

$$r_e^{\,n-m} = \frac{An}{Bm}$$

Eliminating either A or B in the original potential gives

$$B = -\,\varepsilon r_e^m\left(\frac{n}{m-n}\right)$$

and
$$\phi(r) = -\,\varepsilon\left(\frac{n}{m-n}\right)\left(\frac{mr_e^m}{nr^n} - \frac{r_e^m}{r^m}\right)$$

But we also note that $\phi(r) = 0$ when $r = \sigma$, and then

$$\frac{r_e}{\sigma} = \left(\frac{n}{m}\right)^{1/(n-m)}$$

which allows a substitution of σ for r_e in the potential,

$$\phi(r) = \varepsilon\left(\frac{n}{n-m}\right)\left(\frac{n}{m}\right)^{m/(n-m)}\left[\left(\frac{\sigma}{r}\right)^n - \left(\frac{\sigma}{r}\right)^m\right] \tag{5.29}$$

If we set $m = 6$ for attraction and $n = 12$ for repulsion, we obtain the famous Lennard-Jones potential,

$$\phi(r) = 4\varepsilon\left[\left(\frac{\sigma}{r}\right)^{12} - \left(\frac{\sigma}{r}\right)^6\right] \tag{5.30}$$

which has been by far the most frequently used function in arguments concerning intermolecular forces because of its combination of simplicity and realism (model C of Fig. 5.6). It has been used to describe properties of species in all four states of matter.

Theory suggests the functional form of the interaction potential, and experimental data are used to determine empirically the adjustable parameters in the potential function. In the Lennard-Jones case, one merely seeks the values of σ and ε/k that will yield values of $B(T)$ that most agree with experiment.

The Lennard-Jones potential function, while not precisely accurate, nonetheless does have the character of a "real" intermolecular potential function, as has been well-revealed by a wide variety of different experimental techniques. These have included pVT measurements, transport property measurements, and molecular beam scattering measurements, to mention only three. This character-filled potential can also be manipulated through the statistical mechanical formalism to yield values for second and third virial coefficients. It will be convenient to discuss these results in terms of reduced variables which we will define as follows:

$$r^* = \frac{r}{\sigma} \qquad T^* = \frac{kT}{\varepsilon}$$

$$B^* = \frac{B}{\frac{2}{3}\pi n_A \sigma^3} = \frac{B}{b_0} \qquad C^* = \frac{C}{b_0^2}$$

$$B_k^* = T^{*k}\left(\frac{d^k B^*}{dT^{*k}}\right) \qquad C_k^* = T^{*k}\left(\frac{d^k C^*}{dT^{*k}}\right)$$

$$V^* = \frac{V}{b_0}$$

where the virial coefficients are reduced by values characteristic of the rigid sphere model and where the first, second, third, etc. derivatives B_k^* and C_k^* will arise in the development of the thermodynamic excess functions from the Lennard-Jones model.

If we now write the formal expression for the second virial coefficient with the Lennard-Jones potential in reduced coordinates, we obtain

$$B^*(T^*) = 3\int_0^\infty r^{*2}\{1 - \exp\,[4T^{*-1}(r^{*-6} - r^{*-12})]\}\, dr^* \qquad (5.31)$$

Series expansion of the exponential, as done for the Sutherland potential, and term-by-term integration yield

$$B(T) = b_0 B^*(T^*)$$

where $B^*(T^*)$ is a series

$$B^*(T^*) = \sum_{j=0}^{\infty} b^{(j)} T^{*-(2j+1)/4} \qquad (5.32)$$

with the coefficients

$$b^{(j)} = -\frac{2^{j+1/2}}{4j!}\Gamma\left(\frac{2j-1}{4}\right)$$

and similarly for C^*, B_k^*, and C_k^*. Each of these quantities depends solely on T^*. It is convenient to prepare a table of their values, as in Table 5.6.

Each of the series converges well. At low temperatures, where kT is of the order of the depth of the potential well, the colliding molecules "see" this attraction; the actual pressure must then be lowered from the ideal gas value of zero attraction, and the second virial coefficient is then negative. At high temperatures, the thermal energy of kT is large compared to well depth; the molecules do not then "see" the well, and then the pressure must go up just as it did for the rigid sphere

TABLE 5.6 Reduced Second and Third Virial Coefficients and Their First Derivative with Respect to the Reduced Temperature T^* as a Function of T^* for the Lennard-Jones Potential Function

T^*	B^*	B_1^*	C^*	C_1^*
0.30	- 27.8806	76.6073		
0.35	- 18.7549	45.2477		
0.40	- 13.7988	30.2671		
0.45	- 10.7550	21.9895		
0.50	- 8.7202	16.9237		
0.55	- 7.2741	13.5822		
0.60	- 6.1980	11.2488		
0.65	- 5.3682	9.5455		
0.70	- 4.7100	8.2571	- 3.3766	28.68
0.75	- 4.1759	7.2540	- 1.7920	18.05
0.80			- 0.8495	11.60
0.85	- 3.3631	5.8034	- 0.2766	7.561
0.90			0.0765	4.953
0.95	- 2.7749	4.8128	0.2951	3.234
1.00			0.4297	2.078
1.05	- 2.3302	4.0977	0.5108	1.292
1.10			0.5576	0.7507
1.15			0.5822	0.3760
1.20	- 1.8359	3.3375	0.5924	0.1159
1.25			0.5933	- 0.0646
1.30			0.5882	- 0.1889
1.35	- 1.4753	2.8058	0.5793	- 0.2731
1.40			0.5683	- 0.3288
1.45			0.5561	- 0.3641
1.50	- 1.2009	2.4141	0.5434	- 0.3845
1.60			0.5180	- 0.3963
1.70	- 0.9236	2.0293		
1.80			0.4728	- 0.3643
1.90	- 0.7141	1.7454		
2.20	- 0.4817	1.4366	0.4100	- 0.2588
2.40			- 0.3636	1.2819
2.60	- 0.2661	1.1552	0.3738	- 0.1777
2.80	- 0.1845	1.0495		
3.00	- 0.1152	0.9600	0.3523	- 0.1247
3.10	- 0.0844	0.9202		
3.20	- 0.0558	0.8833		
3.30	- 0.0291	0.8489		
3.40	- 0.0043	0.8168	0.3389	- 0.0913
3.50	0.0190	0.7867		
3.60	0.0407	0.7585		
3.70	0.0611	0.7321		
3.80			0.3300	- 0.0702
3.90	0.0984	0.6837		
4.20	0.1467	0.6206	0.3237	- 0.0571
4.60	0.1999	0.5505	0.3189	- 0.0487
5.0	0.2433	0.4926	0.3151	- 0.0436
6.0	0.3229	0.3840	0.3077	- 0.0389
7.0	0.3761	0.3083	0.3017	- 0.0399
8.0	0.4134	0.2525		
10.0	0.4609	0.1759	0.2861	- 0.0483
20.0	0.5254	0.0287	0.2464	- 0.0644

SOURCE: Abstracted from a more detailed table in J.O. Hirschfelder, C. F. Curtiss, and R. B. Bird, p. 1114 ff. Linear Interpolation in this table will yield the omitted numbers from the original table to within ±1%.

model that was described earlier. The second virial coefficient is then positive. A comparison of the experimental and theoretical variation of B^* with T^* appears in Fig. 5.8, where we note that the second virial coefficient is independent of temperature above T^* of about 10. The agreement is excellent. Figure 5.8 reveals from still another perspective the molecular basis of the principle of corresponding states. Here data for all four species fall on a single curve.

Note that the rigid-sphere model predicts a constant $B^* = 1$ independent of T^* and thus completely fails to represent experimental pVT data.

Now imagine yourself as an equation-of-state builder. What analytic form will you fit to experimental data? Something "virial-like" sounds sensible at low to moderate densities. And we have in Fig. 5.8 a prediction of the form of the second virial coefficient that agrees well with experiment for several simple gases. If we replot B^* versus T^{*-1}, we obtain almost a straight line, so perhaps it would be reasonable to attempt to fit real second virial data by an expression like A_0 +

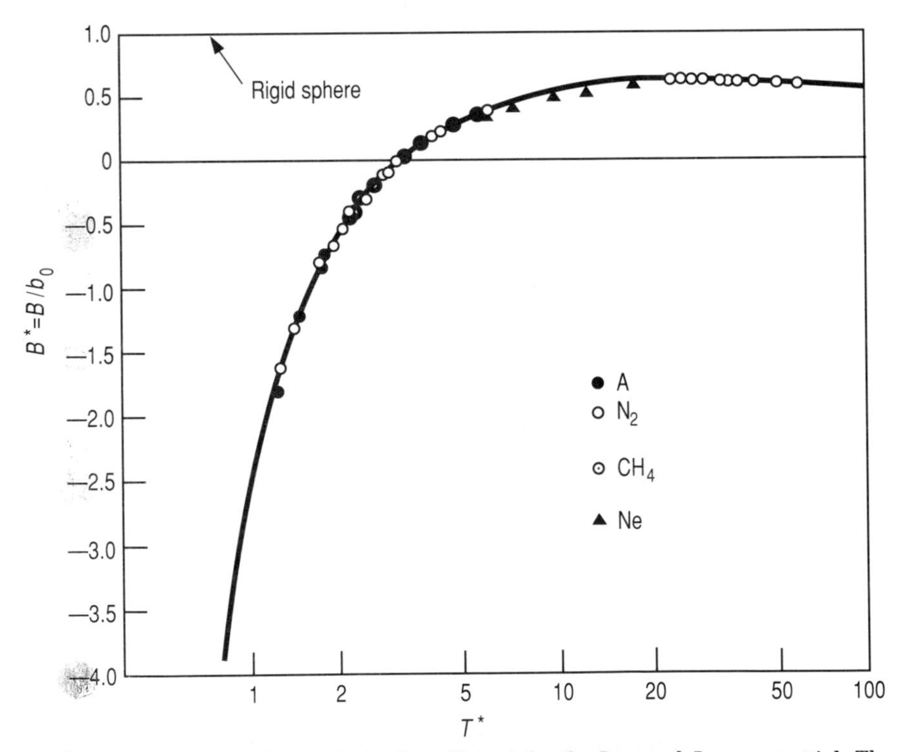

Figure 5.8 The reduced second virial coefficient for the Lennard-Jones potential. The calculated curve of $B^*(T^*)$ is shown with the experimental points for several simple gases. *(From J. O. Hirschfelder, C. F. Curtiss, and R. B. Bird.[4] Used by permission.)*

$B_0/T + C_0/T^2$. This is almost the form used by Benedict, Webb, and Rubin in developing their complex, but accurate, equation of state. They found a T^{-3} term to fit data on hydrocarbons better than the above polynomial with a T^{-2} term. Similarly, the data on the third virial coefficient were reasonably well fit by a function linear in T^{-1}. This form, however, clearly does not predict the maximum in C^* versus T^* that is an artifact of the Lennard-Jones potential that is evident in Fig. 5.9.

The point is that the theory is not good enough for accurate calculations, but it can well suggest the form to use and what should be plotted against what for best results.

Some typical values of the Lennard-Jones parameters are summarized in Table 5.7.

The values of σ and ε/k that appear in this table have been obtained by selecting those values that will best reproduce experimental second virial coefficient data. So a few pVT data permit the evaluation of the Lennard-Jones parameters, which can then be used to interpolate and extrapolate these few pVT data into regions not yet experimentally studied. The agreement with experiment is poor for cigar-shaped or pill-shaped molecules, that is, for molecules where the spherical sym-

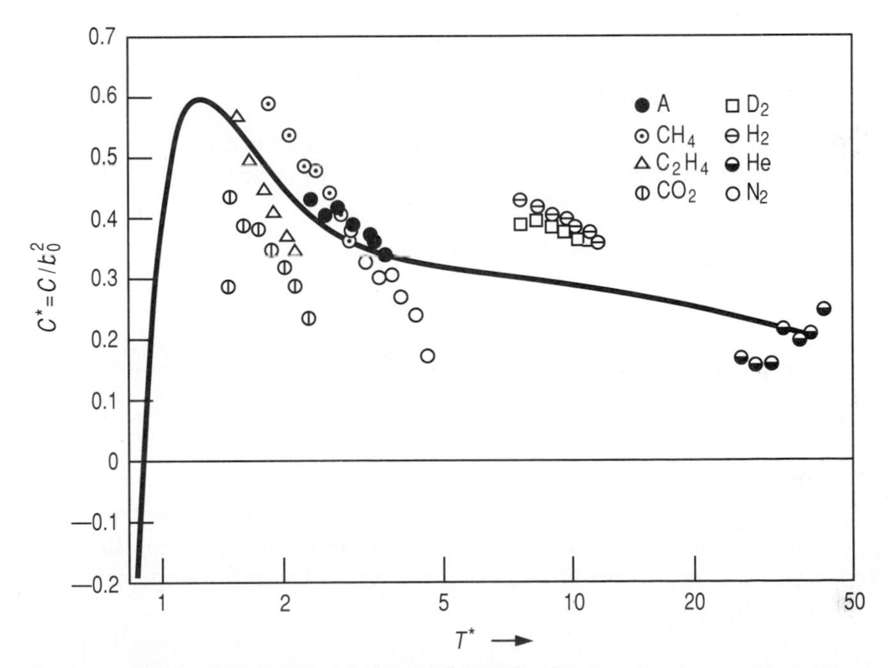

Figure 5.9 The reduced third virial coefficient for the Lennard-Jones potential. The experimental points for several gases are also shown. (*From J. O. Hirschfelder, C. F. Curtiss, and R. B. Bird.*[4] *Used by permission.*)

TABLE 5.7 Constants in the Lennard-Jones Potential as Evaluated from Experimental Second Virial Coefficients for Several Species

Species	σ, nm	ε/k, K
Ne	0.278	34.9
Xe	0.3963	217
CO_2	0.407	205
N_2O	0.459	189
CH_4	0.3817	148.2
SF_6	0.551	200.9
C_2H_4	0.4523	199.2
nC_4H_{10}	0.4971	297

SOURCE: J.O. Hirschfelder, C.F. Curtiss, and R.B. Bird,[4] p. 1110 ff.

metry of the Lennard-Jones approximation is obviously a poor model. The agreement is similarly poor for polar molecules, where the attraction in fact varies as r^{-6} to a first approximation but the Lennard-Jones model assumes spherical symmetry.

It has been empirically observed that $\varepsilon/k \cong 0.77T_c$ and $\frac{2}{3}\pi n_A \sigma^3 = b_0 \cong 0.75V_c$, and thus the Lennard-Jones parameters can be reasonably guessed from knowledge of T_c and V_c.

Second Virial Coefficient for Mixtures

The second virial coefficient for mixtures may be developed in the same manner as was that for pure substances, with the following result:

$$B(T) = \sum_i \sum_j B_{ij}(T) x_i x_j$$

where x_i is the mole fraction of species i in the mixture. For a binary, this will become

$$B(T)_{\text{binary}} = B_{11}x_1^2 + 2B_{12}x_1x_2 + B_{22}x_2^2$$

Clearly B_{11} and B_{22} may be gotten from experiments with pure i and pure j, but what about B_{12}? Measurements of diffusivity provide a good "handle" for study of these bimolecular interaction coefficients. However, as a first approximation we may use the following empirical combining laws:

$$\sigma_{12} = \tfrac{1}{2}(\sigma_{11} + \sigma_{22})$$

and

$$\varepsilon_{12} = (\varepsilon_1 \varepsilon_2)^{1/2}$$

This σ_{12} is correct for rigid-sphere molecules, and the ε_{12} relation has some basis in theory as well.

Example Problem Calculate the second virial coefficient at 50°C for a mixture of 25 mol % N_2 and 75% CH_4.

	N_2	CH_4
b_0, cm^3/mol	63.78	70.16
σ, nm	0.3698	0.3817
ε/k, K	95.05	148.2

$$\text{Then,}\quad \left(\frac{\varepsilon}{k}\right)_{12} = \left[\left(\frac{\varepsilon}{k}\right)_{11}\left(\frac{\varepsilon}{k}\right)_{22}\right]^{1/2} = 118.7\ \text{K}$$

$$\sigma_{12} = \tfrac{1}{2}(\sigma_{11} + \sigma_{22}) = 0.3758$$

$$\text{and}\quad b_0 = 67\ \text{cm}^3/\text{mol}$$

N_2	$T^* = 3.400$	$B^* = -0.0043$	$B = -0.2743$
CH_4	$T^* = 2.180$	$B^* = -0.4942$	$B = -34.67$
$N_2 - CH_4$	$T^* = 2.722$	$B^* = -0.2143$	$B = -14.34$

For the binary mixture, then,

$$B(T)_{\text{mix}} = B_{11}x_1^2 + 2B_{12}x_1x_2 + B_{22}x_2^2$$

$$\text{or}\quad B(T)_{\text{mix}} = -24.90\ \text{cm}^3/\text{mol}$$

We had defined a reduced third virial coefficient $C^*(T^*)$ as

$$C(T) = b_0^2C^*(T^*) \tag{5.33}$$

and with the Lennard-Jones potential, $C^*(T^*)$ may be evaluated again in series form as

$$C^*(T^*) = \sum_{j=0}^{\infty} c^{(j)}T^{*-(j+1)/2}$$

where the $c^{(j)}$ coefficients are complex integrals rather than the gamma functions that appeared in $b^{(j)}$. A comparison of theoretical and experimental third virial coefficients appears in Fig. 5.9, wherein the parameters in the Lennard-Jones potential have been determined from second virial coefficient data. The poor agreement surely reflects the inadequacy of the simple Lennard-Jones potential.

The second virial coefficient from the Lennard-Jones model and from the square-well model will be in agreement with each other if we take $R = 1.8$, if σ for the square well is taken to be the same as σ for the Lennard-Jones model, and if the depth of the square well is taken to be 0.56 times that of the Lennard-Jones gas. The third virial coefficients, on the other hand, will differ by a factor of 2. This means that the second virial coefficient is just not very sensitive to the shape of the potential function, whereas the third virial coefficient is much more sensitive.

The transport properties also depend on this same intermolecular interaction potential (see Chap. 9). To a first approximation, the viscosity of a pure gas is given by

$$\eta = \frac{2.6693 \times 10^{-8}\sqrt{MT}}{\sigma^2 \Omega^{(2,2)*}(T^*)} \tag{5.34}$$

Where η = viscosity, kg/m · s
M = molecular weight
T = temperature, K
σ = collision diameter, nm

The quantity $\Omega^{(2,2)*}(T^*)$ is a complex integral that depends on the reduced temperature, T^* or $T/\varepsilon/k$, deduced from the characteristic energy ε/k of the intermolecular potential function. The omega integral (and other similar integrals) have been evaluated and tabulated, but the form of $\phi(r)$ must be

$$\phi(r) = \varepsilon f\left(\frac{\sigma}{r}\right)$$

With two parameters in the Lennard-Jones potential, two experimental measurements of viscosity at two temperatures will serve to evaluate the two Lennard-Jones parameters. The ratio can be formed

$$\left[\frac{\eta(T_2)}{\eta(T_1)}\right]_{\text{exp}} = \left(\frac{T_2}{T_1}\right)^{1/2} \frac{\Omega^{(2,2)*}(T_1^*)}{\Omega^{(2,2)*}(T_2^*)} \tag{5.35}$$

Then by trial and error one would select a value of ε/k, calculate the right-hand side of Eq. (5.35), and compare it with the left-hand side until the correct ε/k is identified. Then σ is obtained from Eq. (5.34) at either T_1 or T_2. The quality of fit is evident from Fig. 5.10.

Similar arguments may be employed with experimental measurements of thermal conductivity or diffusivity that will also allow the determination of σ and ε/k in the Lennard-Jones potential or similarly for other potentials.

These parameters of ε/k and σ then allow us to calculate second virial coefficients which may be compared with those from experimental pVT data. Interaction potentials such as that of Lennard-Jones then provide a sort of bridge that links transport and volumetric properties. Comparisons with experimental data are, however, not particularly accurate, as is evident for three very simple gases in Table 5.8.

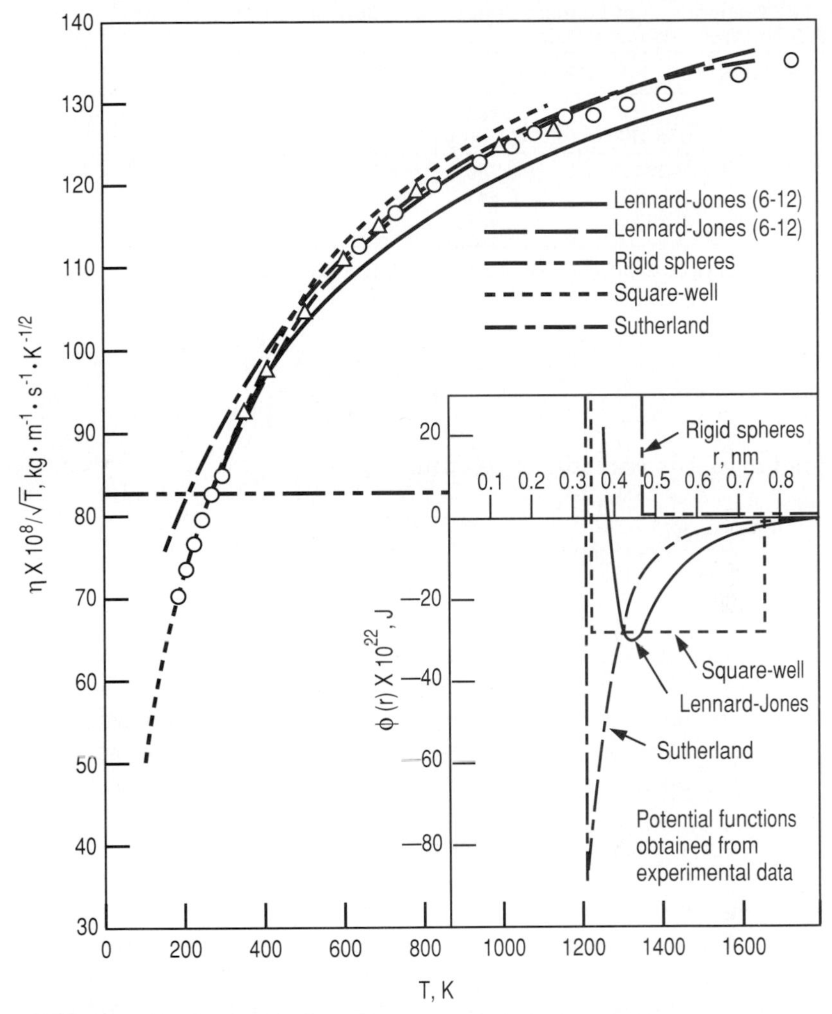

Figure 5.10 Viscosity versus temperature for CO_2 as calculated by using several spherically symmetric molecular models. (*Adapted from J. O. Hirschfelder, C. F. Curtiss, and R. B. Bird.*[4] *Used by permission.*)

The lack of good agreement arises from the lack of realism of the Lennard-Jones potential. These sorts of interrelationships between volumetric and transport properties are pleasing, even though the detailed theory does not allow useful calculations.

The dimensionless enthalpy excess function

$$H - H^0 = \int_V^\infty \left[p - T\left(\frac{\partial p}{\partial T}\right)_V\right] dV + pV - RT$$

TABLE 5.8 Comparison of Lennard-Jones Force Constants as Calculated from $B(T)$ and from Viscosity Data

	σ, nm		ε/k, K	
Gas	From $B(T)$	From viscosity	From $B(T)$	From viscosity
Ne	0.274	0.280	35.7	35.7
A	0.3405	0.3418	119.75	124.0
N_2	0.3698	0.3681	95.05	91.46

SOURCE: From J. O. Hirschfelder, C.F. Curtis, and R.B. Bird,[4] p. 209.

becomes
$$\frac{H - H^0}{RT} = \frac{B^* - B_1^*}{V^*} + \frac{C^* - \frac{1}{2}C_1^*}{V^{*2}} + \cdots \tag{5.36}$$

and, similarly, the entropy excess function is

$$\frac{S - S^0}{R} = -\ln p - \left[\frac{B_1^*}{V^*} + \frac{B^{*2} - C^* + C_1^*}{2V^{*2}}\right] + \cdots \tag{5.37}$$

The first derivatives of B^* and C^* that appear in Table 5.6 can then be used to evaluate these excess functions. These are shown here for completeness to illustrate the very important point that, in principle, real-gas properties may be developed from an intermolecular potential function. Again, this is the only lesson to be learned, for these functions from the Lennard-Jones potential are too inaccurate for practical utility.

This entire formalism is also immediately applicable to mixtures, but the complications regarding the additional knowledge that is required are obvious, for we must now use the appropriate potential for 1,1, 2,2, and 1,2 interactions if we have a binary mixture. Data on the character of the interaction between unlike molecules is difficult to obtain. And, of course, the required unlike molecular interactions increase markedly for multicomponent mixtures.

Several more complex and more physically reasonable models may be manipulated through this same formalism to evaluate virial coefficients and excess functions. These models allow for asymmetrically shaped molecules, imbedded dipoles, and the like. For example, most calculations for polar molecules have been based on the Stockmayer potential, which is a Lennard-Jones interaction plus the polar attraction from Eq. (5.7):

$$\phi(r, \theta_1, \theta_2, \phi_2 - \phi_1) = 4\varepsilon\left[\left(\frac{\sigma}{r}\right)^{12} - \left(\frac{\sigma}{r}\right)^{6}\right] - \left(\frac{\mu^2}{4\pi\varepsilon_0 D r^3}\right) f(\theta_1, \theta_2, \phi_2 - \phi_1) \tag{5.38}$$

The reduced second virial coefficient from the Stockmayer potential appears in Fig. 5.11 as a function of T^* and t^* (defined as $\mu^{*2}/2\sqrt{2}$, where μ^* is the reduced dipole moment, $\mu/\varepsilon^{1/2}\sigma^{3/2}$). For example, a modestly polar species like N_2O with a dipole of $0.17D$ has $t^* = 0.004$ and a B^* of $-$ 0.6 at a T^* of 2. This is the same value that would be obtained from the Lennard-Jones model.

Just as we earlier imagined molecules to be rigid inpenetrable spheres, calculations may also be accomplished with molecules modeled as rigid, inpenetrable pill-shaped or cigar-shaped, or what-have-you-shaped molecules. Calculations and comparisons with experimental data for a number of spherical, nonspherical, and polar models have appeared.[4]

There are at least two more, rather different, sorts of techniques to calculate the macroscopic properties of matter from microscopic prop-

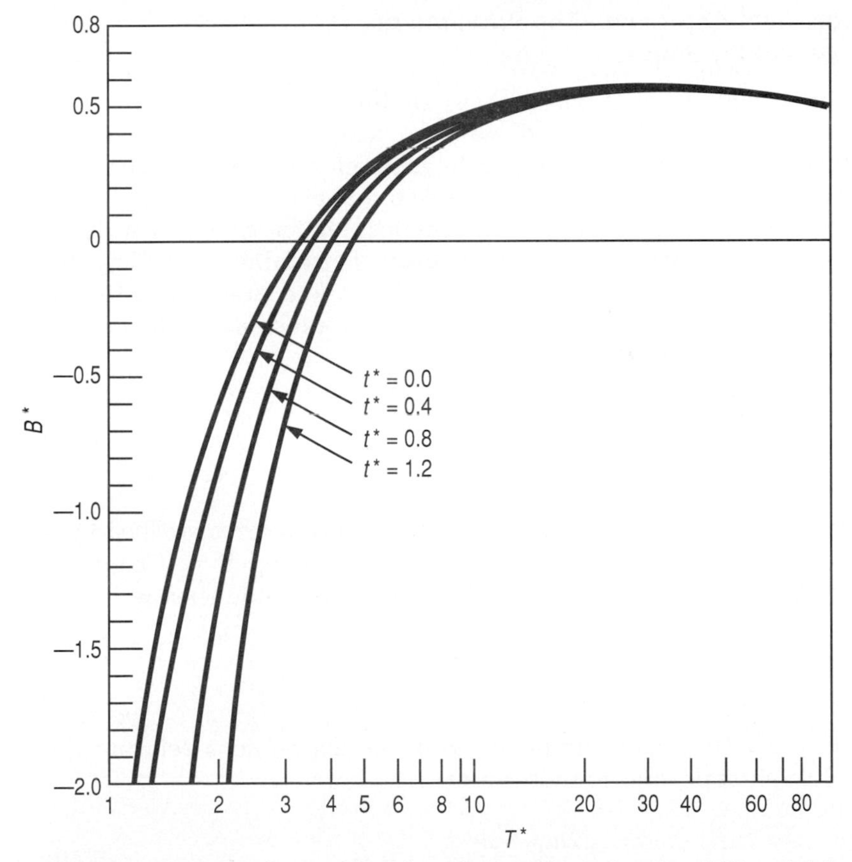

Figure 5.11 The reduced second virial coefficient B^* as a function of T^* and t^* for the Stockmayer potential. *(From J. O. Hirschfelder, C. F. Curtiss, and R. B. Bird.[4] Used by permission.)*

erties. These involve a direct simulation of the molecules on a large computer. The methods still require a potential of interaction (Lennard-Jones or some other) and the results are not yet of practical significance. It is, however, interesting that a simulation using the so-called molecular dynamics approach with only 216 rigid water molecules at a density of 1 g/cm^3 demonstrated that averages over the individual properties of this small number of molecules give reasonable predictions of bulk properties.[5] Another simulation technique based upon a so-called Monte Carlo method[6] is interesting. Here, by random chance, initial circumstances are selected, and subsequent motions are calculated from newtonian mechanics. With sufficient numbers of such choices, one can predict the expected actual behavior. Monte Carlo schemes have also been used in calculations of specific reaction rate coefficients (see later).

An Empirical Approach—The Principle of Corresponding States

Clearly, the theoretical approach to real-gas properties is an enormously complex problem, and we then are not surprised that the greatest successes in describing real-gas phenomena have arisen from semiempirical approaches where we use the above sort of theory to obtain insights into the nature of intermolecular forces which we then develop phenomenologically to the point of usefulness. In fact, this very relationship between molecular insight and practical utility is typical of the use of molecular engineering to understand the properties of matter.

For example, the pVT properties of a real gas may, to a first approximation, be formally written as

$$f(T, p, V, \sigma, \varepsilon) = 0$$

if we imagine only that the intermolecular forces are described by a Lennard-Jones type of two-parameter potential function. Using the standard techniques of dimensional analysis, we could then write

$$f\left(\frac{V}{n_A\sigma^3}, \frac{kT}{\varepsilon}, \frac{p\sigma^3}{\varepsilon}\right) = 0 \tag{5.39}$$

which is exactly analogous to the familiar expressions relating the Reynolds, Nusselt, and Prandtl numbers,

$$f\left[\frac{Du\rho}{\mu}, \frac{C_p\mu}{k}, \frac{hD}{k}\right] = 0 \tag{5.40}$$

Numerous fits of experimental data to Eq. (5.40) allow the estimation of convective film coefficients. Equation (5.39) merely applies the same dimensional analysis to molecules. If the molecule has a dipole moment μ, the corresponding potential function might have three adjustable parameters, ε, σ, and μ, and we could write

$$f\left(\frac{V}{n_A\sigma^3}, \frac{kT}{\varepsilon}, \frac{p\sigma^3}{\varepsilon}, \frac{\mu}{4\pi\varepsilon_0\varepsilon^{1/2}\sigma^{3/2}}\right) = 0 \qquad (5.41)$$

and so on. We may not know the exact form of the potential function, and, even if we knew the form, we would probably be unable to manipulate the potential through the formalism to evaluate the virial coefficients. Nevertheless, this last relationship confirms that a universal function of the four reduced variables exists. The existence of this (and similar) functions is called the principle of corresponding states, and we see that many such principles exist, depending only on our choice of the intermolecular potential function. We also infer that such universal functions exist for dimensionless groups involving any parameters whose numerical values might reflect the character of the intermolecular forces. With exactly this idea in mind, one group[7] has prepared extensive tables based on a "principle of corresponding states" using p_r, T_r, z, and z_c as dimensionless groups; that is, one obtains

$$z = z(p_r, T_r, z_c)$$

Such a correlation has been very successful, and the results are presented in Table 5.9. The correlation and all of the numbers in Table 5.9 are from fitting experimental pVT data on many substances. But the existence of a correlation among these dimensionless groups is a prediction from molecular theory. The theory provides the impetus to pursue development of equations of state like these.

These tables have been prepared for the dimensionless group $z_c = 0.27$, which is a value characteristic of many hydrocarbons. For values of z_c above 0.27, use D_a from the table and the relationship

$$z = z(0.27) + D_a(z_c - 0.27)$$

and similarly for z_c less than 0.27, except now use D_b from the table in this same correction relationship. This tabular equation of state can be manipulated through the rigorous thermodynamic relationship for, say, the fugacity coefficient, to obtain f/p. This quantity may be tabulated as well, as is displayed in Table 5.10. Note that, as was true for z, at higher reduced pressures the effect of z_c is forgettable. All properties depending on the equation of state may be similarly evaluated and tabulated. Many examples of this sort of approach to predicting real-gas behavior have appeared.

TABLE 5.9 Compressibility Factors of Pure Gases and Liquids, z†

	p_r = 0.01			p_r = 0.05			p_r = 0.10			p_r = 0.2			p_r = 0.3		
T_{rs} / z sat. gas / z sat. liquid		0.59 / 0.985 / 0.002			0.690 / 0.942 / 0.009			0.740 / 0.898 / 0.015			0.804 / 0.833 / 0.030			0.847 / 0.783 / 0.045	
T_r	D_b	z	D_a	D_b	z	D_a	D_b	z	D_a	D_b	z	D_a	D_b	z	D_a
0.50	0.01	0.002	0.01	0.05	0.009	0.07	0.11	0.0184	0.14	0.22	0.0367	0.27	0.35	0.0551	0.40
0.60	0.20	0.990	0.02	0.05	0.008	0.07	0.10	0.0164	0.12	0.20	0.0328	0.25	0.31	0.0491	0.37
0.70	0.07	0.992	0.02	0.33	0.943	0.35	0.09	0.0152	0.12	0.19	0.0304	0.23	0.29	0.0456	0.34
0.80	0.01	0.993	0.02	0.13	0.960	0.18	0.28	0.920	0.40	0.18	0.0295	0.20	0.28	0.0441	0.31
0.90	0.01	0.994	0.02	0.07	0.973	0.10	0.14	0.947	0.20	0.28	0.899	0.36	0.44	0.825	0.50
0.92	0.01	0.995	0.02	0.07	0.975	0.10	0.13	0.951	0.19	0.26	0.900	0.34	0.40	0.840	0.47
0.94	0.01	0.995	0.02	0.06	0.977	0.10	0.12	0.954	0.18	0.24	0.908	0.33	0.37	0.854	0.44
0.96	0.01	0.995	0.02	0.05	0.978	0.09	0.11	0.958	0.17	0.22	0.915	0.30	0.33	0.868	0.40
0.98	0.01	0.996	0.02	0.05	0.980	0.09	0.10	0.961	0.16	0.21	0.921	0.28	0.31	0.879	0.37
1.00	0.01	0.996	0.02	0.04	0.982	0.09	0.10	0.964	0.15	0.21	0.927	0.24	0.28	0.889	0.34
1.01	0.01	0.996	0.02	0.04	0.983	0.08	0.10	0.966	0.15	0.20	0.930	0.24	0.26	0.894	0.33
1.02	0.01	0.996	0.02	0.04	0.983	0.08	0.10	0.967	0.15	0.19	0.933	0.23	0.25	0.897	0.34
1.03	0.01	0.996	0.02	0.04	0.984	0.08	0.09	0.968	0.14	0.18	0.935	0.22	0.24	0.902	0.32
1.04	0.01	0.996	0.02	0.04	0.985	0.08	0.09	0.970	0.14	0.18	0.938	0.21	0.24	0.905	0.29
1.05	0.00	0.996	0.02	0.04	0.985	0.08	0.08	0.971	0.14	0.17	0.940	0.20	0.23	0.909	0.28
1.06	0.00	0.996	0.02	0.04	0.986	0.08	0.08	0.972	0.14	0.17	0.942	0.20	0.22	0.913	0.26
1.07	0.00	0.996	0.02	0.04	0.986	0.07	0.08	0.973	0.14	0.16	0.944	0.19	0.21	0.916	0.25
1.08	0.00	0.996	0.02	0.04	0.987	0.07	0.08	0.974	0.13	0.16	0.946	0.18	0.20	0.918	0.24
1.09	0.00	0.997	0.01	0.04	0.987	0.07	0.07	0.975	0.12	0.15	0.948	0.17	0.19	0.922	0.24
1.10	0.00	0.997	0.01	0.04	0.988	0.07	0.07	0.976	0.12	0.14	0.950	0.17	0.18	0.924	0.21
1.12	0.00	0.997	0.01	0.04	0.988	0.06	0.06	0.977	0.12	0.13	0.953	0.16	0.17	0.928	0.20
1.14	0.00	0.997	0.01	0.03	0.989	0.06	0.06	0.979	0.11	0.12	0.956	0.14	0.16	0.933	0.19
1.16	0.00	0.997	0.01	0.03	0.990	0.05	0.06	0.980	0.09	0.12	0.960	0.13	0.14	0.937	0.16
1.18	0.00	0.997	0.01	0.03	0.991	0.04	0.06	0.982	0.09	0.12	0.962	0.12	0.12	0.942	0.15
1.20	0.00	0.998	0.01	0.03	0.991	0.03	0.06	0.983	0.07	0.09	0.965	0.10	0.11	0.945	0.13
1.30	0.00	0.998	0.01	0.03	0.993	0.02	0.04	0.987	0.05	0.07	0.974	0.08	0.07	0.960	0.10
1.40	0.00	0.998	0.00	0.02	0.995	0.01	0.03	0.990	0.03	0.05	0.982	0.05	0.06	0.971	0.07
1.50	0.00	0.999	0.00	0.01	0.995	0.01	0.01	0.991	0.02	0.03	0.986	0.03	0.02	0.980	0.04
1.60	0.00	0.999	0.00	0.01	0.996	0.00	0.00	0.992	0.00	0.01	0.988	0.02	0.01	0.986	0.02
1.70	0.00	0.999	0.00	0.00	0.996	0.00	0.00	0.992	0.00	0.00	0.989	0.01	0.00	0.989	0.01
1.80	0.00	0.999	0.00	0.00	0.996	0.00	0.00	0.993	0.00	0.00	0.991	0.01	0.00	0.991	0.01
1.90	0.00	1.000	0.00	0.00	0.996	0.00	0.00	0.993	0.00	0.00	0.992	0.00	0.00	0.993	0.00
2.00		1.000			0.997	0.00	0.00	0.994	0.00	0.00	0.994	0.00	0.00	0.995	

T_{rs} / z sat. gas / z sat. liquid	$p_r = 0.4$			$p_r = 0.5$			$p_r = 0.6$			$p_r = 0.7$			$p_r = 0.8$		
		0.879 0.738 0.060			0.909 0.693 0.077			0.929 0.641 0.096			0.950 0.583 0.114			0.967 0.519 0.136	
T_r	D_b	z	D_a	D_b	z	D_a	D_b	z	D_a	D_b	z	D_a	D_b	z	D_a
0.50	0.46	0.0734	0.53	0.57	0.0918	0.66	0.70	0.110	0.81	0.81	0.128	0.95	0.93	0.147	1.07
0.60	0.41	0.0654	0.49	0.52	0.0817	0.60	0.63	0.0980	0.71	0.74	0.113	0.82	0.84	0.130	0.95
0.70	0.39	0.0605	0.45	0.49	0.0758	0.55	0.59	0.0906	0.65	0.69	0.106	0.77	0.79	0.121	0.88
0.80	0.37	0.0588	0.40	0.47	0.0735	0.52	0.57	0.0879	0.62	0.66	0.102	0.73	0.76	0.116	0.85
0.90	0.73	0.763	0.63	0.45	0.0761	0.50	0.55	0.0908	0.60	0.64	0.105	0.71	0.74	0.120	0.82
0.92	0.60	0.783	0.59	0.81	0.710	0.70	0.55	0.0929	0.60	0.65	0.108	0.70	0.74	0.122	0.82
0.94	0.50	0.800	0.55	0.63	0.735	0.64	0.77	0.660	0.73	0.65	0.111	0.70	0.74	0.126	0.82
0.96	0.44	0.817	0.51	0.53	0.760	0.59	0.65	0.700	0.67	0.75	0.613	0.76	0.76	0.133	0.82
0.98	0.39	0.832	0.47	0.46	0.781	0.54	0.54	0.729	0.62	0.62	0.665	0.68	0.70	0.580	0.76
1.00	0.34	0.845	0.42	0.41	0.800	0.48	0.47	0.755	0.54	0.52	0.704	0.60	0.60	0.636	0.65
1.01	0.33	0.852	0.42	0.38	0.809	0.47	0.44	0.765	0.51	0.50	0.718	0.56	0.55	0.659	0.61
1.02	0.30	0.858	0.39	0.36	0.817	0.44	0.41	0.775	0.48	0.45	0.732	0.52	0.50	0.678	0.56
1.03	0.29	0.863	0.37	0.34	0.825	0.42	0.38	0.786	0.46	0.42	0.745	0.50	0.46	0.696	0.54
1.04	0.28	0.869	0.34	0.32	0.832	0.38	0.35	0.794	0.40	0.39	0.755	0.44	0.43	0.710	0.46
1.05	0.27	0.873	0.30	0.30	0.838	0.33	0.33	0.802	0.35	0.36	0.765	0.38	0.39	0.723	0.39
1.06	0.26	0.878	0.29	0.29	0.845	0.32	0.31	0.810	0.33	0.34	0.773	0.35	0.35	0.735	0.36
1.07	0.25	0.883	0.27	0.27	0.850	0.28	0.29	0.817	0.30	0.32	0.781	0.31	0.33	0.745	0.33
1.08	0.24	0.886	0.26	0.26	0.856	0.27	0.28	0.824	0.28	0.30	0.790	0.28	0.31	0.755	0.29
1.09	0.22	0.890	0.25	0.24	0.862	0.25	0.26	0.830	0.25	0.23	0.798	0.25	0.28	0.764	0.25
1.10	0.21	0.894	0.22	0.23	0.867	0.22	0.24	0.836	0.22	0.25	0.805	0.22	0.25	0.773	0.23
1.12	0.19	0.900	0.20	0.21	0.876	0.20	0.21	0.848	0.20	0.22	0.818	0.20	0.22	0.789	0.20
1.14	0.18	0.907	0.20	0.18	0.884	0.20	0.19	0.859	0.20	0.19	0.830	0.20	0.19	0.803	0.20
1.16	0.15	0.913	0.19	0.16	0.891	0.20	0.17	0.868	0.20	0.17	0.842	0.20	0.17	0.816	0.18
1.18	0.13	0.918	0.16	0.14	0.898	0.17	0.15	0.877	0.18	0.15	0.852	0.18	0.15	0.830	0.17
1.20	0.12	0.924	0.15	0.13	0.905	0.15	0.13	0.885	0.15	0.14	0.862	0.14	0.14	0.841	0.15
1.30	0.09	0.944	0.11	0.10	0.931	0.11	0.10	0.916	0.11	0.10	0.900	0.11	0.10	0.888	0.12
1.40	0.06	0.959	0.08	0.07	0.949	0.08	0.07	0.937	0.09	0.07	0.928	0.09	0.07	0.920	0.09
1.50	0.04	0.970	0.05	0.04	0.963	0.06	0.05	0.952	0.07	0.05	0.948	0.07	0.05	0.945	0.07
1.60	0.02	0.978	0.03	0.02	0.973	0.04	0.03	0.965	0.05	0.03	0.964	0.06	0.03	0.960	0.06
1.70	0.01	0.983	0.02	0.01	0.980	0.03	0.02	0.974	0.03	0.02	0.974	0.03	0.02	0.970	0.04
1.80	0.00	0.987	0.02	0.00	0.985	0.02	0.01	0.982	0.02	0.01	0.982	0.02	0.01	0.980	0.02
1.90	0.00	0.991	0.01	0.00	0.989	0.01	0.00	0.987	0.01	0.00	0.987	0.01	0.00	0.987	0.02
2.00	0.00	0.994	0.00	0.00	0.993	0.00	0.00	0.992	0.01	0.00	0.992	0.01	0.00	0.989	0.02

†Values of z are recorded for $z_c = 0.27$. At other values of z_c, $z' = z + D(z_c - 0.27)$.

TABLE 5.9 Compressibility Factors of Pure Gases and Liquids, z† (*Contd.*)

	p_r = 0.9			p_r = 1.0			p_r = 1.05			p_r = 1.1			p_r = 1.2		
T_{rs}		0.984			1.000										
z sat. gas		0.443			0.270										
z sat. liquid		0.164			0.270										
	D_b	z	D_a	D_b	z	D_a	D_b	z	D_a	D_b	z	D_a	D_b	z	D_a
T_r															
0.50	1.05	0.165	1.20	1.17	0.183	1.35	1.22	0.192	1.40	1.28	0.201	1.48	1.40	0.220	1.62
0.60	0.95	0.147	1.05	1.05	0.163	1.17	1.11	0.171	1.23	1.16	0.179	1.28	1.27	0.195	1.39
0.70	0.90	0.136	0.99	1.00	0.151	1.10	1.05	0.158	1.15	1.10	0.165	1.20	1.20	0.180	1.30
0.80	0.86	0.131	0.95	0.95	0.145	1.05	1.00	0.152	1.10	1.05	0.159	1.15	1.15	0.173	1.25
0.90	0.83	0.134	0.92	0.92	0.148	1.02	0.97	0.155	1.07	1.01	0.162	1.11	1.10	0.176	1.20
0.92	0.83	0.137	0.92	0.92	0.151	1.02	0.97	0.158	1.06	1.01	0.165	1.11	1.10	0.179	1.19
0.94	0.83	0.141	0.92	0.93	0.155	1.01	0.98	0.162	1.06	1.01	0.169	1.10	1.10	0.183	1.18
0.96	0.85	0.147	0.92	0.94	0.161	1.01	0.99	0.169	1.05	1.03	0.176	1.09	1.13	0.189	1.17
0.98	0.87	0.161	0.92	0.97	0.174	1.00	1.02	0.182	1.05	1.07	0.189	1.09	1.16	0.202	1.16
1.00	0.70	0.520	0.82	1.00	0.270	1.00	1.06	0.230	1.05	1.14	0.224	1.09	1.20	0.220	1.15
1.01	0.60	0.568	0.68	0.65	0.424	0.75	0.67	0.365	0.79	0.68	0.256	0.83	0.70	0.242	0.88
1.02	0.55	0.600	0.62	0.58	0.509	0.67	0.59	0.447	0.70	0.60	0.374	0.73	0.60	0.295	0.77
1.03	0.49	0.627	0.57	0.51	0.555	0.60	0.52	0.505	0.62	0.52	0.461	0.63	0.52	0.369	0.66
1.04	0.44	0.642	0.49	0.45	0.585	0.51	0.45	0.546	0.53	0.45	0.505	0.54	0.45	0.422	0.56
1.05	0.40	0.670	0.41	0.41	0.611	0.43	0.41	0.577	0.44	0.41	0.541	0.45	0.41	0.478	0.46
1.06	0.37	0.687	0.38	0.38	0.633	0.39	0.38	0.603	0.40	0.38	0.568	0.40	0.38	0.517	0.40
1.07	0.34	0.700	0.34	0.35	0.654	0.35	0.35	0.627	0.35	0.35	0.594	0.35	0.35	0.548	0.36
1.08	0.32	0.715	0.29	0.32	0.671	0.30	0.32	0.647	0.30	0.32	0.616	0.30	0.32	0.573	0.30
1.09	0.28	0.726	0.25	0.28	0.686	0.25	0.28	0.662	0.25	0.28	0.637	0.25	0.28	0.600	0.25
1.10	0.25	0.738	0.23	0.25	0.700	0.23	0.25	0.678	0.23	0.25	0.655	0.23	0.25	0.620	0.23
1.12	0.22	0.756	0.21	0.22	0.723	0.21	0.22	0.704	0.21	0.22	0.686	0.21	0.22	0.654	0.21
1.14	0.19	0.773	0.20	0.19	0.745	0.20	0.19	0.731	0.20	0.19	0.712	0.20	0.19	0.683	0.20
1.16	0.18	0.790	0.18	0.18	0.764	0.18	0.18	0.750	0.18	0.18	0.735	0.18	0.18	0.707	0.20
1.18	0.17	0.805	0.17	0.17	0.780	0.17	0.17	0.771	0.17	0.17	0.756	0.17	0.17	0.730	0.18
1.20	0.15	0.818	0.15	0.15	0.795	0.15	0.15	0.787	0.15	0.15	0.775	0.15	0.15	0.751	0.15
1.30	0.11	0.874	0.12	0.11	0.857	0.12	0.11	0.849	0.12	0.11	0.841	0.13	0.11	0.827	0.13
1.40	0.07	0.912	0.10	0.07	0.899	0.10	0.07	0.890	0.10	0.07	0.888	0.10	0.07	0.875	0.10
1.50	0.05	0.938	0.08	0.05	0.927	0.08	0.05	0.922	0.08	0.05	0.918	0.08	0.05	0.911	0.08
1.60	0.03	0.955	0.06	0.03	0.948	0.07	0.03	0.944	0.07	0.03	0.940	0.08	0.03	0.935	0.08
1.70	0.01	0.968	0.04	0.02	0.964	0.05	0.02	0.958	0.05	0.03	0.956	0.06	0.03	0.951	0.07
1.80	0.00	0.976	0.03	0.00	0.974	0.03	0.00	0.968	0.04	0.00	0.968	0.05	0.02	0.963	0.06
1.90	0.00	0.985	0.02	0.00	0.983	0.02	0.00	0.978	0.02	0.00	0.978	0.03	0.01	0.974	0.05
2.00		0.990	0.02	0.00	0.988	0.02	0.00	0.986	0.02	0.00	0.984	0.03	0.01	0.981	0.03

T_r	p_r = 1.4 z	p_r = 1.6 z	p_r = 1.8 z	p_r = 2.0 z	p_r = 4.0 z	p_r = 6.0 z	p_r = 8.0 z	p_r = 10.0 z	p_r = 20.0 z	p_r = 30.0 z
0.50	0.256	0.293	0.329	0.365	0.726	1.083	1.439	1.791	3.551	5.28
0.60	0.227	0.259	0.291	0.323	0.640	0.952	1.262	1.568	3.098	4.59
0.70	0.210	0.239	0.268	0.297	0.584	0.862	1.139	1.413	2.769	4.08
0.80	0.201	0.229	0.257	0.284	0.549	0.804	1.056	1.305	2.525	3.70
0.90	0.203	0.230	0.257	0.283	0.532	0.768	1.005	1.233	2.341	3.40
0.92	0.206	0.233	0.259	0.284	0.530	0.763	0.997	1.222	2.310	3.35
0.94	0.210	0.237	0.262	0.287	0.530	0.760	0.991	1.201	2.278	3.30
0.96	0.217	0.242	0.267	0.291	0.531	0.757	0.985	1.202	2.250	3.25
0.98	0.228	0.253	0.276	0.298	0.532	0.755	0.980	1.195	2.224	3.20
1.00	0.234	0.254	0.279	0.306	0.536	0.756	0.975	1.193	2.200	3.15
1.01	0.246	0.262	0.287	0.312	0.538	0.757	0.974	1.188	2.188	3.14
1.02	0.264	0.276	0.296	0.318	0.540	0.758	0.973	1.184	2.175	3.11
1.03	0.288	0.289	0.307	0.326	0.543	0.759	0.972	1.181	2.164	3.08
1.04	0.323	0.305	0.317	0.333	0.546	0.760	0.972	1.177	2.153	3.06
1.05	0.366	0.323	0.332	0.341	0.548	0.761	0.972	1.174	2.142	3.04
1.06	0.403	0.347	0.347	0.351	0.552	0.762	0.971	1.171	2.130	3.02
1.07	0.438	0.370	0.365	0.361	0.554	0.763	0.970	1.168	2.119	3.00
1.08	0.472	0.396	0.380	0.372	0.558	0.764	0.970	1.165	2.109	2.96
1.09	0.507	0.424	0.398	0.386	0.562	0.766	0.970	1.162	2.098	2.95
1.10	0.534	0.455	0.416	0.400	0.565	0.768	0.970	1.160	2.088	2.93
1.12	0.577	0.505	0.454	0.432	0.572	0.772	0.970	1.156	2.068	2.89
1.14	0.615	0.549	0.494	0.466	0.581	0.776	0.970	1.153	2.049	2.85
1.16	0.647	0.588	0.540	0.503	0.589	0.780	0.972	1.151	2.030	2.81
1.18	0.677	0.622	0.583	0.542	0.599	0.786	0.973	1.150	2.013	2.78
1.20	0.705	0.653	0.620	0.573	0.609	0.792	0.975	1.148	1.995	2.74
1.30	0.795	0.768	0.742	0.716	0.687	0.824	0.984	1.144	1.921	2.63
1.40	0.855	0.837	0.819	0.801	0.763	0.863	0.996	1.144	1.862	2.56
1.50	0.894	0.882	0.869	0.852	0.813	0.893	1.012	1.146	1.818	2.49
1.60	0.923	0.914	0.904	0.888	0.852	0.918	1.028	1.150	1.790	2.44
1.70	0.945	0.934	0.929	0.915	0.883	0.940	1.041	1.154	1.767	2.39
1.80	0.960	0.950	0.946	0.935	0.909	0.960	1.052	1.156	1.744	2.33
1.90	0.972	0.965	0.960	0.952	0.932	0.977	1.061	1.158	1.714	2.29
2.00	0.979	0.974	0.971	0.966	0.952	0.993	1.070	1.159	1.691	2.24
3.00	1.000	0.997	0.995	0.986	0.990	1.008	1.068	1.130	1.500	1.84
4.00	1.000	1.000	0.997	0.992	1.000	1.014	1.065	1.120	1.400	1.66
6.00	1.004	1.003	1.000	1.000	1.013	1.024	1.064	1.100	1.300	1.50
8.00	1.008	1.008	1.005	1.005	1.016	1.030	1.063	1.085	1.250	1.40
10.00	1.010	1.010	1.008	1.010	1.020	1.035	1.062	1.080	1.185	1.30
15.00	1.020	1.020	1.020	1.020	1.030	1.045	1.061	1.070	1.140	1.20

†Values of z are recorded for $z_c = 0.27$. At other values of z_c, $z' = z + D(z_c - 0.27)$.
SOURCE: From Lydersen et al.[7]

TABLE 5.10 Fugacity Coefficients of Pure Gases and Liquids, f/p†

T_{rs} sat. gas and liquid	p_r = 0.01			p_r = 0.05			p_r = 0.10			p_r = 0.20			p_r = 0.30		
		0.59 0.993			0.690 0.972			0.740 0.946			0.804 0.887			0.847 0.840	
	D_b	f/p	D_a	D_b	f/p	D_a	D_b	f/p	D_a	D_b	f/p	D_a	D_b	f/p	D_a
T_r															
0.50		0.021			0.015			0.0102			0.00518			0.00351	
0.60	0.06	0.994	0.09	7.9	0.17	12.50	8.40	0.111	12.80	11.05	0.0563	13.50	11.0	0.0381	14.00
0.70	0.02	0.995	0.03	0.06	0.975	0.13	3.30	0.640	6.40	4.20	0.325	6.80	4.5	0.220	7.20
0.80	0.00	0.996	0.01	0.02	0.980	0.06	0.04	0.961	0.18	2.85	0.848	2.90	2.53	0.574	3.20
0.90	0.00	0.997	0.01	0.01	0.986	0.03	0.03	0.972	0.11	0.07	0.922	0.23	0.19	0.871	0.32
0.92	0.00	0.997	0.01	0.01	0.987	0.00	0.01	0.975	0.10	0.06	0.929	0.21	0.12	0.882	0.28
0.94	0.00	0.998	0.01	0.01	0.988	0.03	0.01	0.977	0.09	0.05	0.934	0.19	0.10	0.892	0.25
0.96	0.00	0.998	0.01	0.01	0.989	0.03	0.01	0.979	0.08	0.05	0.939	0.17	0.09	0.900	0.22
0.98	0.00	0.998	0.00	0.00	0.990	0.02	0.01	0.981	0.07	0.05	0.943	0.15	0.08	0.907	0.20
1.00	0.00	0.998	0.00	0.00	0.991	0.02	0.02	0.983	0.06	0.05	0.947	0.14	0.08	0.916	0.19
1.01	0.00	0.998	0.00	0.00	0.991	0.02	0.02	0.983	0.06	0.04	0.949	0.13	0.08	0.917	0.17
1.02	0.00	0.998	0.00	0.00	0.991	0.02	0.01	0.983	0.05	0.04	0.951	0.12	0.07	0.919	0.16
1.03	0.00	0.998	0.00	0.00	0.992	0.02	0.01	0.984	0.05	0.04	0.953	0.11	0.07	0.922	0.16
1.04	0.00	0.998	0.00	0.00	0.992	0.02	0.01	0.985	0.04	0.04	0.955	0.10	0.07	0.926	0.15
1.05	0.00	0.998	0.00	0.00	0.992	0.01	0.01	0.985	0.04	0.03	0.956	0.09	0.06	0.928	0.14
1.06	0.00	0.999	0.00	0.00	0.993	0.01	0.01	0.986	0.03	0.03	0.958	0.08	0.06	0.930	0.13
1.07	0.00	0.999	0.00	0.00	0.993	0.01	0.01	0.986	0.03	0.03	0.959	0.08	0.05	0.933	0.13
1.08	0.00	0.999	0.00	0.00	0.993	0.01	0.00	0.986	0.03	0.03	0.962	0.07	0.05	0.935	0.12
1.09	0.00	0.999	0.00	0.00	0.993	0.01	0.01	0.986	0.03	0.02	0.962	0.07	0.04	0.938	0.11
1.10	0.00	0.999	0.00	0.00	0.993	0.01	0.01	0.987	0.02	0.02	0.964	0.06	0.04	0.942	0.10
1.12	0.00	0.999	0.00	0.00	0.994	0.01	0.01	0.988	0.02	0.02	0.967	0.05	0.03	0.943	0.09
1.14	0.00	0.999	0.00	0.00	0.994	0.01	0.01	0.989	0.01	0.02	0.969	0.04	0.03	0.947	0.08
1.16	0.00	0.999	0.00	0.00	0.995	0.00	0.01	0.990	0.01	0.02	0.970	0.04	0.02	0.950	0.07
1.18	0.00	0.999	0.00	0.00	0.995	0.01	0.01	0.991	0.01	0.02	0.973	0.03	0.02	0.954	0.06
1.20	0.00	0.999	0.00	0.00	0.995	0.01	0.01	0.991	0.01	0.01	0.974	0.03	0.02	0.958	0.05
1.30	0.00	0.999	0.00	0.00	0.996	0.01	0.01	0.993	0.01	0.01	0.980	0.02	0.01	0.968	0.03
1.40	0.00	0.999	0.00	0.00	0.997	0.01	0.00	0.995	0.01	0.0	0.985	0.02	0.00	0.976	0.02
1.50	0.00	1.000	0.00	0.00	0.998	0.01	− 0.01	0.996	0.01	− 0.01	0.987	0.02	− 0.02	0.980	0.04
1.60	0.00	1.000	0.00	0.00	0.998	0.01	− 0.01	0.997	0.01	− 0.02	0.989	0.04	− 0.02	0.983	0.05
1.70	0.00	1.000	0.00	0.00	0.998	0.01	− 0.01	0.997	0.02	− 0.02	0.989	0.05	− 0.03	0.984	0.08
1.80	0.00	1.000	0.00	0.00	0.998	0.01	− 0.01	0.997	0.03	− 0.02	0.990	0.08	− 0.04	0.986	0.10
1.90	0.00	1.000	0.00	0.00	0.999	0.01	− 0.01	0.998	0.04	− 0.03	0.990	0.10	− 0.05	0.987	0.13
2.00		1.000	0.00	0.00	0.999	0.01		0.999	0.06		0.992	0.13		0.989	0.15

T_{rs} sat. gas and liquid	$p_r = 0.40$			$p_r = 0.50$			$p_r = 0.60$			$p_r = 0.70$			$p_r = 0.80$		
		0.879 0.805			0.907 0.777			0.929 0.752			0.950 0.730			0.967 0.710	
	D_b	f/p	D_a	D_b	f/p	D_a	D_b	f/p	D_a	D_b	f/p	D_a	D_b	f/p	D_a
T_r															
0.50		0.0027			0.0022			0.00186			0.00162			0.00144	
0.60	10.90	0.0291	14.40	10.70	0.0236	14.70	10.60	0.0200	15.00	10.50	0.0174	15.20	10.18	0.0155	15.40
0.70	4.70	0.168	7.50	4.75	0.136	7.80	4.80	0.115	8.00	5.00	0.100	8.00	5.03	0.0890	8.00
0.80	2.41	0.437	3.50	2.40	0.354	3.70	2.45	0.300	3.83	2.50	0.261	4.00	2.60	0.231	4.22
0.90	0.21	0.820	0.38	1.32	0.739	1.50	1.20	0.642	1.63	1.11	0.559	1.71	1.15	0.496	1.77
0.92	0.19	0.836	0.34	0.24	0.789	0.40	1.04	0.722	1.12	0.94	0.629	1.40	0.92	0.559	1.43
0.94	0.17	0.849	0.31	0.22	0.806	0.36	0.25	0.762	0.42	0.69	0.699	0.97	0.71	0.622	1.09
0.96	0.15	0.860	0.28	0.21	0.820	0.33	0.28	0.781	0.38	0.35	0.741	0.43	0.60	0.688	0.68
0.98	0.13	0.870	0.25	0.18	0.834	0.31	0.23	0.798	0.35	0.29	0.762	0.39	0.33	0.725	0.43
1.00	0.11	0.879	0.23	0.15	0.846	0.28	0.19	0.812	0.32	0.23	0.779	0.35	0.27	0.745	0.38
1.01	0.11	0.884	0.22	0.14	0.852	0.25	0.18	0.818	0.29	0.22	0.786	0.32	0.25	0.755	0.34
1.02	0.11	0.887	0.21	0.13	0.856	0.24	0.17	0.826	0.28	0.21	0.794	0.31	0.24	0.764	0.33
1.03	0.10	0.891	0.20	0.15	0.860	0.24	0.16	0.831	0.27	0.19	0.802	0.30	0.22	0.772	0.32
1.04	0.10	0.896	0.19	0.12	0.868	0.23	0.15	0.838	0.26	0.18	0.810	0.28	0.20	0.781	0.30
1.05	0.09	0.899	0.18	0.11	0.871	0.22	0.14	0.842	0.25	0.17	0.816	0.27	0.18	0.788	0.29
1.06	0.08	0.903	0.17	0.10	0.876	0.21	0.13	0.848	0.24	0.15	0.822	0.26	0.17	0.795	0.28
1.07	0.08	0.906	0.16	0.09	0.879	0.20	0.13	0.853	0.23	0.14	0.826	0.25	0.16	0.800	0.27
1.08	0.08	0.910	0.15	0.09	0.883	0.18	0.12	0.857	0.21	0.13	0.832	0.23	0.15	0.807	0.25
1.09	0.07	0.913	0.14	0.08	0.887	0.17	0.11	0.863	0.19	0.12	0.838	0.21	0.14	0.813	0.23
1.10	0.06	0.916	0.13	0.08	0.892	0.15	0.10	0.867	0.18	0.12	0.843	0.19	0.13	0.820	0.21
1.12	0.06	0.920	0.12	0.07	0.898	0.14	0.09	0.876	0.17	0.10	0.853	0.18	0.12	0.831	0.19
1.14	0.06	0.926	0.10	0.06	0.905	0.12	0.08	0.884	0.14	0.09	0.863	0.15	0.11	0.843	0.16
1.16	0.05	0.931	0.09	0.06	0.910	0.11	0.07	0.890	0.12	0.08	0.871	0.13	0.10	0.852	0.14
1.18	0.05	0.936	0.08	0.05	0.917	0.09	0.07	0.898	0.10	0.08	0.880	0.11	0.09	0.862	0.12
1.20	0.04	0.939	0.07	0.05	0.920	0.08	0.07	0.906	0.09	0.08	0.889	0.10	0.08	0.872	0.10
1.30	0.02	0.955	0.05	0.02	0.942	0.06	0.03	0.929	0.07	0.04	0.917	0.07	0.05	0.903	0.07
1.40	0.00	0.967	0.03	0.00	0.956	0.04	0.00	0.947	0.04	0.01	0.938	0.05	0.02	0.928	0.05
1.50	− 0.02	0.973	0.05	− 0.02	0.966	0.06	0.02	0.958	0.06	0.01	0.950	0.06	0.01	0.944	0.06
1.60	− 0.03	0.978	0.07	− 0.03	0.972	0.07	0.04	0.962	0.07	0.03	0.961	0.07	0.03	0.956	0.07
1.70	− 0.04	0.980	0.09	− 0.04	0.976	0.09	0.05	0.972	0.09	0.05	0.968	0.09	0.05	0.965	0.09
1.80	− 0.06	0.983	0.11	− 0.06	0.980	0.11	0.07	0.977	0.11	0.07	0.974	0.11	0.07	0.971	0.11
1.90	− 0.07	0.985	0.13	− 0.08	0.983	0.13	0.08	0.981	0.13	0.08	0.979	0.13	0.08	0.977	0.13
2.00		0.987	0.15		0.985	0.15		0.983	0.15		0.982	0.15		0.981	0.15

†Values of f/p are recorded for $z_c = 0.27$. At other values of z_c, $(f/p)' = (f/p)(10^{D(z_c - 0.27)})$.

TABLE 5.10 Fugacity Coefficients of Pure Gases and Liquids, f/p† (*Contd.*)

T_{rs} sat. gas and liquid	$p_r = 0.90$			$p_r = 1.00$			$p_r = 1.05$			$p_r = 1.10$			$p_r = 1.20$		
		0.984 0.688			1.000 0.665										
	D_b	f/p	D_a	D_b	f/p	D_a	D_b	f/p	D_a	D_b	f/p	D_a	D_b	f/p	D_a
T_r															
0.50		0.00131			0.00120			0.00115			0.00111			0.00104	
0.60	10.24	0.0150	15.60	10.14	0.0134	15.70	10.10	0.0123	15.70	10.00	0.0119	15.80	9.81	0.0110	15.80
0.70	5.20	0.0803	8.00	5.30	0.0734	8.00	5.35	0.0704	8.00	5.40	0.0677	8.00	5.50	0.0619	8.00
0.80	2.60	0.209	4.30	2.70	0.191	4.40	2.70	0.183	4.40	2.75	0.176	4.45	2.80	0.163	4.60
0.90	1.19	0.448	1.77	1.20	0.409	1.88	1.20	0.392	1.91	1.25	0.377	1.98	1.25	0.351	2.12
0.92	0.92	0.504	1.42	0.98	0.461	1.54	1.00	0.442	1.56	1.05	0.425	1.67	1.13	0.396	1.69
0.94	0.71	0.561	1.15	0.78	0.513	1.15	0.81	0.493	1.16	0.84	0.474	1.20	0.90	0.441	1.29
0.96	0.57	0.622	0.86	0.57	0.569	0.82	0.65	0.546	0.93	0.67	0.526	0.97	0.72	0.489	1.04
0.98	0.44	0.678	0.57	0.45	0.621	0.62	0.51	0.606	0.67	0.54	0.574	0.71	0.58	0.535	0.76
1.00	0.32	0.710	0.41	0.36	0.665	0.42	0.39	0.641	0.47	0.41	0.618	0.49	0.46	0.578	0.52
1.01	0.29	0.721	0.37	0.37	0.685	0.40	0.35	0.665	0.43	0.38	0.643	0.45	0.41	0.602	0.48
1.02	0.27	0.733	0.36	0.30	0.699	0.38	0.32	0.681	0.40	0.34	0.664	0.42	0.37	0.620	0.44
1.03	0.25	0.742	0.35	0.28	0.710	0.37	0.29	0.695	0.38	0.31	0.678	0.39	0.33	0.645	0.41
1.04	0.23	0.752	0.33	0.26	0.722	0.34	0.27	0.706	0.35	0.28	0.692	0.36	0.30	0.660	0.38
1.05	0.21	0.760	0.31	0.24	0.731	0.32	0.25	0.717	0.33	0.26	0.703	0.34	0.28	0.674	0.35
1.06	0.19	0.768	0.29	0.22	0.741	0.30	0.23	0.727	0.31	0.24	0.713	0.32	0.26	0.685	0.33
1.07	0.18	0.776	0.28	0.20	0.750	0.29	0.21	0.736	0.29	0.23	0.723	0.30	0.24	0.697	0.31
1.08	0.17	0.783	0.26	0.19	0.759	0.27	0.20	0.746	0.27	0.21	0.733	0.28	0.23	0.706	0.29
1.09	0.16	0.789	0.24	0.18	0.766	0.25	0.19	0.753	0.26	0.20	0.741	0.26	0.21	0.716	0.27
1.10	0.15	0.796	0.22	0.17	0.772	0.22	0.18	0.762	0.23	0.19	0.750	0.23	0.20	0.726	0.24
1.12	0.13	0.809	0.20	0.15	0.787	0.20	0.16	0.776	0.21	0.16	0.766	0.21	0.18	0.745	0.21
1.14	0.12	0.822	0.17	0.13	0.802	0.17	0.14	0.792	0.18	0.14	0.781	0.18	0.16	0.761	0.18
1.16	0.11	0.832	0.14	0.11	0.812	0.14	0.12	0.802	0.15	0.12	0.793	0.15	0.13	0.774	0.16
1.18	0.10	0.844	0.12	0.10	0.825	0.13	0.10	0.816	0.14	0.11	0.807	0.14	0.11	0.790	0.14
1.20	0.09	0.855	0.11	0.09	0.838	0.12	0.09	0.829	0.13	0.10	0.821	0.13	0.10	0.805	0.13
1.30	0.05	0.890	0.07	0.05	0.878	0.07	0.05	0.872	0.08	0.05	0.866	0.08	0.05	0.854	0.08
1.40	0.02	0.920	0.05	0.02	0.910	0.05	0.02	0.905	0.05	0.02	0.903	0.05	0.02	0.892	0.05
1.50	− 0.01	0.937	0.06	− 0.01	0.931	0.06	− 0.01	0.927	0.06	− 0.01	0.923	0.06	− 0.01	0.916	0.06
1.60	− 0.03	0.951	0.07	− 0.03	0.946	0.07	− 0.03	0.944	0.07	− 0.03	0.941	0.07	− 0.03	0.935	0.07
1.70	− 0.05	0.961	0.09	− 0.05	0.957	0.09	− 0.05	0.955	0.09	− 0.05	0.953	0.09	− 0.05	0.949	0.09
1.80	− 0.07	0.969	0.11	− 0.07	0.966	0.11	− 0.07	0.965	0.11	− 0.07	0.964	0.11	− 0.07	0.961	0.11
1.90	− 0.08	0.975	0.13	− 0.08	0.973	0.13	− 0.08	0.972	0.13	− 0.08	0.971	0.13	− 0.08	0.970	0.13
2.00		0.980	0.15		0.979	0.15		0.978	0.15		0.977	0.15		0.975	0.15

T_r	$p_r = 1.40$	$p_r = 1.60$	$p_r = 1.80$	$p_r = 2.0$	$p_r = 4.0$	$p_r = 6.0$	$p_r = 8.0$	$p_r = 10.0$	$p_r = 20.0$	$p_r = 30.0$
	f/p	f/p	f/p	f/p	f/p	f/p	f/p	f/p	f/p	f/p
0.50	0.00092	0.00084	0.00077	0.00072	0.09052	0.00050	0.00053	0.00061	0.00182	0.00709
0.60	0.00978	0.00884	0.00811	0.00798	0.00520	0.00477	0.00511	0.0058	0.0127	0.0396
0.70	0.0556	0.0502	0.0459	0.0426	0.0286	0.0255	0.0254	0.0270	0.0547	0.144
0.80	0.144	0.130	0.119	0.110	0.0727	0.0636	0.062	0.0648	0.121	0.271
0.90	0.310	0.279	0.255	0.236	0.155	0.135	0.131	0.134	0.222	0.469
0.92	0.349	0.315	0.288	0.267	0.175	0.152	0.146	0.150	0.245	0.508
0.94	0.390	0.351	0.321	0.298	0.196	0.169	0.163	0.167	0.269	0.550
0.96	0.433	0.391	0.358	0.332	0.219	0.189	0.182	0.186	0.296	0.595
0.98	0.474	0.429	0.393	0.365	0.241	0.209	0.201	0.204	0.322	0.638
1.00	0.513	0.464	0.426	0.396	0.263	0.228	0.219	0.223	0.348	0.679
1.01	0.540	0.485	0.446	0.414	0.275	0.238	0.228	0.232	0.358	0.697
1.02	0.559	0.507	0.465	0.433	0.288	0.248	0.238	0.242	0.372	0.721
1.03	0.580	0.529	0.485	0.451	0.303	0.261	0.250	0.255	0.392	0.750
1.04	0.599	0.546	0.505	0.470	0.316	0.272	0.262	0.265	0.403	0.769
1.05	0.615	0.563	0.520	0.485	0.327	0.282	0.272	0.276	0.418	0.786
1.06	0.630	0.580	0.536	0.499	0.339	0.294	0.282	0.286	0.430	0.815
1.07	0.644	0.594	0.552	0.515	0.351	0.304	0.292	0.296	0.444	0.835
1.08	0.658	0.610	0.566	0.531	0.362	0.314	0.302	0.306	0.457	0.855
1.09	0.668	0.622	0.581	0.545	0.374	0.324	0.312	0.315	0.470	0.875
1.10	0.680	0.636	0.594	0.555	0.385	0.334	0.322	0.325	0.484	0.891
1.12	0.701	0.660	0.620	0.585	0.408	0.356	0.342	0.346	0.509	0.934
1.14	0.721	0.682	0.645	0.611	0.431	0.376	0.362	0.366	0.536	0.966
1.16	0.737	0.700	0.664	0.632	0.453	0.396	0.382	0.386	0.561	1.004
1.18	0.756	0.721	0.689	0.658	0.481	0.422	0.408	0.412	0.591	1.051
1.20	0.773	0.741	0.710	0.681	0.504	0.445	0.429	0.434	0.620	1.088
1.30	0.830	0.806	0.783	0.762	0.609	0.548	0.533	0.540	0.756	1.248
1.40	0.874	0.857	0.840	0.824	0.700	0.647	0.634	0.643	0.881	1.420
1.50	0.901	0.888	0.876	0.863	0.768	0.722	0.712	0.724	0.976	1.545
1.60	0.924	0.913	0.903	0.893	0.813	0.766	0.753	0.768	1.030	1.595
1.70	0.942	0.933	0.925	0.918	0.852	0.820	0.817	0.834	1.106	1.688
1.80	0.955	0.949	0.943	0.937	0.886	0.862	0.862	0.882	1.164	1.753
1.90	0.966	0.962	0.958	0.954	0.912	0.893	0.897	0.919	1.201	1.775
2.00	0.971	0.968	0.964	0.960	0.929	0.916	0.923	0.947	1.224	1.781
3.00	1.000	1.000	1.000	0.999	0.989	0.986	0.994	1.016	1.242	1.621
4.00	1.000	1.000	1.000	1.000	0.991	0.989	0.999	1.019	1.204	1.476
6.00	1.000	1.000	1.000	1.000	0.996	0.995	1.005	1.024	1.171	1.372
8.00	1.000	1.000	1.000	1.001	1.001	1.001	1.010	1.027	1.148	1.306
10.00	1.000	1.000	1.003	1.003	1.003	1.004	1.015	1.034	1.146	1.224
15.00	1.000	1.000	1.006	1.008	1.025	1.042	1.057	1.071	1.145	1.220

†Values of f/p are recorded for $z_c = 0.27$. At other values of z_c, $(f/p)' = (f/p)(10^{D(z_c - 0.27)})$.
SOURCE: From Lydersen et al.[7]

Certainly the vapor pressure is also a measure of the intermolecular forces, and another popular formulation of the principle of corresponding states defines an arbitrary dimensionless factor, called the acentric factor, in terms of the vapor pressure at $T_r = 0.7$, which is generally near the normal boiling point of many nonpolar liquids:

$$\omega \equiv -\log \frac{p_s}{p_c} - 1.00$$

where p_s is the vapor pressure at $T_r = 0.7$. The above definition will make $\omega = 0$ for inert gases and small spherical molecules. Other molecules will have small positive values of ω (see Table 5.11). The "dimensionless group", acentric factor, was adopted as an indication of the deviation of the real intermolecular potential function from the spherically symmetric functions, such as those of Lennard-Jones or Sutherland, which we have discussed. We can then write $z = f(p_r, T_r, \omega)$. The functional form of $f(p_r, T_r, \omega)$ is, as is always the case in dimensional analysis, dictated by the pVT data themselves. The compressibility may be adequately presented as

$$z = z^{(0)}(p_r, T_r) + \omega z^{(1)}(p_r, T_r)$$

where $z^{(0)}$ and $z^{(1)}$ are best presented in tabular form (see Tables 5.12 to 5.17).

TABLE 5.11 Accentric Factors of Several Compounds

Compound	ω	Compound	ω
CH_4	0.008	$C_6H_5CH_3$	0.257
nC_6H_{14}	0.296	Ar	0
C_2H_4	0.085	CCl_4	0.194
$(CH_3)_2CO$	0.309	NO	0.607
C_6H_6	0.212	H_2O	0.344

Example Problem Calculate the specific volume of steam at 1500 lb/in^2 (absolute) and 800°F. Then

$$T_r = \frac{800 + 460}{1.8(647.3)} = 1.08$$

$$p_r = \frac{1500}{14.696/217.6} = 0.47$$

and $\omega = 0.348$

Then $z = 0.88 + 0.348(0) = 0.88$

or $$V = \frac{zRT}{18p} = \frac{0.88(10.73)(1260)}{(18)(1514.7)} = 0.44 \text{ ft}^3/\text{lb}$$

which compares well with a value of 0.43 ft^3/lb read from the steam table.

TABLE 5.12 Values of $z^{(0)}$ for Compressibility Factor Calculations

T_r	P_r																								
	0.2	0.4	0.6	0.8	1.0	1.2	1.4	1.6	1.8	2.0	2.2	2.4	2.6	2.8	3.0	3.2	3.4	3.6	3.8	4.0	4.5	5.0	6.0	7.0	8.0	9.0
0.80	0.851	0.066	0.100	0.133	0.164	0.192	0.225	0.258	0.287	0.318	0.347	0.376	0.405	0.433	0.461	0.490	0.519	0.547	0.576	0.605	0.675	0.746	0.883	1.017	1.15	1.28
0.85	0.882	0.067	0.101	0.134	0.165	0.194	0.226	0.258	0.287	0.316	0.345	0.374	0.403	0.431	0.459	0.487	0.515	0.542	0.569	0.597	0.663	0.730	0.861	0.990	1.115	1.24
0.90	0.904	0.778	0.102	0.135	0.167	0.198	0.229	0.258	0.288	0.316	0.345	0.373	0.402	0.430	0.458	0.485	0.512	0.538	0.565	0.591	0.655	0.718	0.842	0.966	1.089	1.21
0.95	0.920	0.819	0.697	0.145	0.176	0.205	0.235	0.262	0.292	0.321	0.347	0.375	0.403	0.430	0.457	0.484	0.510	0.536	0.561	0.587	0.647	0.709	0.828	0.947	1.066	1.185
1.00	0.932	0.849	0.756	0.638	0.291	0.231	0.250	0.278	0.304	0.329	0.356	0.381	0.407	0.433	0.458	0.484	0.509	0.534	0.557	0.582	0.642	0.702	0.819	0.932	1.048	1.166
1.05	0.942	0.874	0.800	0.714	0.609	0.470	0.341	0.320	0.332	0.350	0.372	0.393	0.417	0.441	0.466	0.489	0.512	0.535	0.557	0.580	0.639	0.700	0.814	0.923	1.032	1.147
1.10	0.950	0.893	0.833	0.767	0.691	0.607	0.512	0.442	0.408	0.402	0.405	0.420	0.440	0.462	0.484	0.504	0.525	0.547	0.567	0.589	0.643	0.699	0.810	0.916	1.019	1.129
1.15	0.958	0.908	0.858	0.805	0.746	0.684	0.620	0.562	0.514	0.484	0.477	0.478	0.485	0.498	0.513	0.529	0.546	0.563	0.581	0.600	0.651	0.705	0.809	0.911	1.008	1.113
1.20	0.963	0.921	0.879	0.835	0.788	0.737	0.690	0.640	0.598	0.568	0.553	0.545	0.544	0.548	0.554	0.563	0.574	0.587	0.601	0.618	0.664	0.714	0.810	0.907	1.000	1.100
1.25	0.968	0.930	0.896	0.858	0.820	0.778	0.740	0.702	0.664	0.636	0.618	0.606	0.599	0.597	0.598	0.602	0.609	0.618	0.629	0.643	0.682	0.726	0.816	0.907	0.994	1.088
1.30	0.971	0.940	0.909	0.878	0.846	0.811	0.780	0.749	0.718	0.691	0.671	0.657	0.649	0.644	0.642	0.642	0.645	0.651	0.659	0.668	0.701	0.740	0.824	0.910	0.992	1.078
1.4	0.977	0.952	0.929	0.908	0.883	0.859	0.838	0.817	0.795	0.777	0.759	0.745	0.734	0.725	0.720	0.718	0.718	0.722	0.727	0.734	0.754	0.781	0.844	0.921	0.994	1.071
1.5	0.982	0.963	0.945	0.927	0.909	0.892	0.875	0.859	0.844	0.831	0.819	0.808	0.800	0.794	0.790	0.785	0.784	0.784	0.786	0.790	0.805	0.826	0.877	0.934	1.000	1.070
1.6	0.985	0.971	0.957	0.944	0.930	0.917	0.904	0.893	0.882	0.872	0.863	0.855	0.848	0.843	0.840	0.836	0.834	0.833	0.834	0.835	0.844	0.860	0.904	0.953	1.010	1.075
1.7	0.988	0.977	0.966	0.956	0.946	0.936	0.926	0.919	0.911	0.903	0.896	0.889	0.883	0.879	0.875	0.873	0.872	0.872	0.873	0.874	0.882	0.895	0.930	0.972	1.023	1.082
1.8	0.991	0.982	0.974	0.966	0.958	0.950	0.944	0.937	0.931	0.926	0.921	0.916	0.913	0.910	0.908	0.907	0.906	0.906	0.907	0.908	0.914	0.925	0.955	0.993	1.039	1.091
1.9	0.993	0.986	0.980	0.974	0.968	0.962	0.958	0.952	0.948	0.944	0.940	0.936	0.933	0.931	0.930	0.929	0.929	0.930	0.932	0.934	0.941	0.950	0.976	1.010	1.051	1.097
2.0	0.995	0.989	0.984	0.979	0.975	0.971	0.968	0.964	0.961	0.959	0.956	0.954	0.953	0.953	0.952	0.952	0.953	0.954	0.954	0.956	0.962	0.972	0.996	1.027	1.064	1.106
2.5	1.000	0.999	0.999	0.998	0.998	0.998	0.998	0.997	0.999	1.000	1.001	1.001	1.002	1.004	1.006	1.008	1.009	1.012	1.014	1.018	1.026	1.035	1.055	1.079	1.105	1.136
3.0	1.001	1.002	1.003	1.004	1.005	1.007	1.008	1.010	1.012	1.014	1.016	1.019	1.022	1.025	1.028	1.030	1.033	1.036	1.038	1.041	1.049	1.058	1.077	1.10	1.124	1.150
3.5	1.002	1.004	1.006	1.008	1.011	1.013	1.015	1.018	1.020	1.022	1.024	1.027	1.030	1.033	1.036	1.039	1.042	1.045	1.048	1.051	1.058	1.067	1.086	1.105	1.126	1.148
4.0	1.003	1.005	1.008	1.010	1.013	1.015	1.017	1.020	1.022	1.024	1.026	1.029	1.032	1.035	1.038	1.041	1.044	1.047	1.050	1.053	1.060	1.068	1.086	1.104	1.124	1.143

SOURCE: K. S. Pitzer et al., *J. Am. Chem. Soc.*, *77*, 3427 (1955). See Tables 5.13 and 5.14 for additional data near the two-phase region.

TABLE 5.13 Values of $z^{(0)}$ near the Two-Phase Region

T_r	P_r						
	0.4	0.5	0.6	0.7	0.8	0.9	1.0
0.90	0.778	0.701	0.102	0.118	0.135	0.151	0.167
0.91	0.787	0.715	0.104	0.120	0.136	0.152	0.168
0.92	0.796	0.728	0.650	0.122	0.138	0.153	0.169
0.93	0.805	0.740	0.666	0.124	0.140	0.155	0.170
0.94	0.812	0.751	0.681	0.125	0.142	0.157	0.173
0.95	0.819	0.762	0.697	0.612	0.145	0.160	0.176
0.96	0.826	0.772	0.711	0.632	0.149	0.164	0.180
0.97	0.832	0.782	0.724	0.652	0.56	0.170	0.186
0.98	0.838	0.791	0.735	0.669	0.591	0.177	0.193
0.99	0.844	0.800	0.746	0.685	0.616	0.514	0.205
1.00	0.849	0.807	0.757	0.699	0.638	0.554	0.291
1.01	0.854	0.813	0.767	0.713	0.654	0.583	0.476
1.02	0.860	0.820	0.776	0.726	0.672	0.608	0.525
1.03	0.865	0.826	0.784	0.737	0.687	0.630	0.558
1.04	0.870	0.833	0.793	0.748	0.701	0.648	0.586
1.05	0.874	0.838	0.800	0.758	0.714	0.665	0.609

SOURCE: K. S. Pitzer et al., *J. Am. Chem. Soc.*, *77*, 3427 (1955).

TABLE 5.14 Values of $z^{(0)}$ in the Critical Region

T_r	P_r										
	1.0	1.1	1.2	1.3	1.4	1.5	1.6	1.7	1.8	1.9	2.0
0.98	0.193	0.204	0.217	0.230	0.244	0.257	0.270	0.284	0.298	0.313	0.324
0.99	0.205	0.210	0.223	0.235	0.247	0.260	0.273	0.287	0.301	0.315	0.327
1.00	0.291	0.220	0.231	0.241	0.250	0.265	0.278	0.290	0.304	0.317	0.329
1.01	0.476	0.283	0.243	0.248	0.259	0.271	0.283	0.294	0.307	0.319	0.332
1.02	0.525	0.402	0.273	0.260	0.270	0.278	0.291	0.300	0.311	0.323	0.335
1.03	0.558	0.466	0.34	0.29	0.283	0.288	0.297	0.306	0.316	0.328	0.339
1.04	0.586	0.509	0.41	0.33	0.307	0.302	0.307	0.314	0.324	0.334	0.343
1.05	0.609	0.543	0.470	0.375	0.341	0.324	0.320	0.323	0.332	0.341	0.350
1.06	0.628	0.572	0.505	0.423	0.370	0.349	0.336	0.333	0.343	0.348	0.358
1.07	0.645	0.597	0.534	0.468	0.408	0.379	0.358	0.349	0.356	0.358	0.367
1.08	0.663	0.618	0.562	0.504	0.445	0.412	0.385	0.373	0.370	0.369	0.375
1.09	0.677	0.636	0.587	0.535	0.480	0.443	0.412	0.396	0.387	0.383	0.387
1.10	0.691	0.652	0.607	0.561	0.512	0.473	0.442	0.422	0.408	0.400	0.402
1.11	0.703	0.667	0.625	0.584	0.538	0.502	0.469	0.448	0.428	0.418	0.417
1.13	0.726	0.693	0.658	0.621	0.584	0.549	0.520	0.494	0.472	0.456	0.450
1.15	0.746	0.715	0.684	0.652	0.620	0.589	0.562	0.536	0.514	0.495	0.484

SOURCE: K. S. Pitzer et al., *J. Am. Chem. Soc.*, *77*, 3427 (1955).

TABLE 5.15 Values of $z^{(1)}$ for Compressibility Factor Calculations

T_r	P_r																				
	0.2	0.4	0.6	0.8	1.0	1.2	1.4	1.6	1.8	2.0	2.2	2.4	2.6	2.8	3.0	4.0	5.0	6.0	7.0	8.0	9.0
0.80	-0.095	-0.028	-0.044	-0.058	-0.07	-0.08	-0.10	-0.11	-0.12	-0.13	-0.14	-0.15	-0.16	-0.17	-0.18	-0.23	-0.26	-0.29	-0.32	-0.35	-0.37
0.85	-0.067	-0.031	-0.049	-0.064	-0.08	-0.09	-0.11	-0.12	-0.13	-0.14	-0.15	-0.16	-0.17	-0.18	-0.18	-0.22	-0.25	-0.28	-0.31	-0.34	-0.36
0.90	-0.042	-0.09	-0.053	-0.068	-0.085	-0.10	-0.11	-0.12	-0.13	-0.14	-0.15	-0.16	-0.17	-0.17	-0.18	-0.21	-0.24	-0.27	-0.30	-0.32	-0.35
0.95	-0.025	-0.050	-0.10	-0.072	-0.091	-0.10	-0.11	-0.12	-0.12	-0.13	-0.14	-0.15	-0.15	-0.16	-0.17	-0.20	-0.22	-0.25	-0.28	-0.31	-0.34
1.00	-0.012	-0.016	-0.020	-0.05	-0.080	-0.090	-0.099	-0.108	-0.115	-0.123	-0.13	-0.13	-0.14	-0.14	-0.15	-0.17	-0.20	-0.23	-0.26	-0.30	-0.33
1.05	0.000	+0.001	+0.005	+0.015	+0.02	+0.01	-0.01	-0.04	-0.06	-0.07	-0.08	-0.09	-0.10	-0.10	-0.11	-0.14	-0.17	-0.20	-0.24	-0.28	-0.31
1.10	+0.002	0.008	0.016	0.030	0.055	0.082	+0.11	+0.082	+0.035	0.000	-0.02	-0.03	-0.05	-0.06	-0.07	-0.10	-0.13	-0.16	-0.21	-0.25	-0.28
1.15	0.004	0.012	0.012	0.040	0.064	0.093	0.12	0.140	0.136	+0.100	+0.07	+0.04	+0.02	0.00	-0.01	-0.04	-0.08	-0.12	-0.16	-0.20	-0.24
1.20	0.009	0.018	0.028	0.044	0.069	0.10	0.13	0.16	0.17	0.17	0.16	0.14	0.12	+0.09	+0.07	0.00	-0.04	-0.08	-0.12	-0.16	-0.19
1.25	0.011	0.023	0.036	0.050	0.069	0.10	0.13	0.16	0.18	0.19	0.19	0.18	0.16	0.14	0.12	+0.05	0.00	-0.03	-0.07	-0.11	-0.13
1.30	0.013	0.027	0.041	0.055	0.072	0.10	0.13	0.16	0.18	0.20	0.20	0.20	0.20	0.19	0.18	0.10	+0.04	0.00	-0.04	-0.07	-0.09
1.4	0.016	0.032	0.049	0.065	0.082	0.10	0.13	0.16	0.18	0.19	0.20	0.21	0.21	0.21	0.20	0.15	0.11	+0.07	+0.04	+0.01	-0.01
1.5	0.017	0.035	0.052	0.070	0.088	0.10	0.13	0.15	0.17	0.18	0.20	0.20	0.21	0.21	0.21	0.20	0.17	0.14	0.11	0.09	+0.07
1.6	0.018	0.036	0.054	0.07	0.08	0.10	0.12	0.14	0.16	0.17	0.18	0.19	0.20	0.20	0.21	0.22	0.21	0.19	0.17	0.15	0.14
1.7	0.018	0.036	0.054	0.07	0.09	0.10	0.11	0.13	0.15	0.16	0.17	0.18	0.19	0.20	0.21	0.24	0.25	0.26	0.25	0.24	0.22
1.8	0.018	0.036	0.054	0.07	0.09	0.10	0.11	0.13	0.15	0.16	0.17	0.18	0.19	0.20	0.21	0.26	0.29	0.31	0.32	0.32	0.30
1.9	0.018	0.035	0.05	0.07	0.09	0.10	0.11	0.13	0.15	0.16	0.17	0.18	0.19	0.20	0.21	0.26	0.30	0.35	0.38	0.40	0.40
2.0	0.016	0.031	0.05	0.07	0.08	0.10	0.11	0.13	0.14	0.15	0.16	0.17	0.19	0.20	0.21	0.26	0.30	0.35	0.40	0.43	0.45
2.5	0.01	0.02	0.04	0.05	0.07	0.08	0.10	0.11	0.12	0.13	0.15	0.16	0.18	0.19	0.20	0.25	0.30	0.35	0.40	0.45	0.50
3.0	0.01	0.02	0.03	0.05	0.06	0.07	0.08	0.09	0.10	0.11	0.13	0.14	0.15	0.16	0.17	0.23	0.28	0.34	0.38	0.45	0.50
3.5	0.01	0.02	0.03	0.04	0.05	0.06	0.07	0.08	0.08	0.09	0.10	0.11	0.12	0.13	0.14	0.19	0.24	0.28	0.33	0.38	0.42
4.0	0.01	0.02	0.02	0.03	0.04	0.05	0.06	0.06	0.07	0.08	0.09	0.10	0.10	0.11	0.12	0.16	0.20	0.23	0.27	0.31	0.35

SOURCE: K. S. Pitzer et al., *J. Am. Chem. Soc.*, **77**, 3427 (1955). See Tables 5.16 and 5.17 for additional data in the regions enclosed by dashed lines.

TABLE 5.16 Values of $z^{(1)}$ near the Two-Phase Region

T_r	P_r			
	0.4	0.6	0.8	1.0
0.90	−0.09	−0.053	−0.068	−0.085
0.91	−0.08	−0.053	−0.069	−0.087
0.92	−0.072	−0.18	−0.070	−0.089
0.93	−0.066	−0.15	−0.071	−0.090
0.94	−0.058	−0.12	−0.072	−0.091
0.95	−0.050	−0.10	−0.072	−0.091
0.96	−0.042	−0.08	−0.072	−0.091
0.97	−0.035	−0.065	−0.14	−0.091
0.98	−0.027	−0.050	−0.11	−0.090
0.99	−0.021	−0.033	−0.08	−0.087
1.00	−0.016	−0.020	−0.05	−0.080
1.01	−0.012	−0.012	−0.02	−0.02
1.02	−0.008	−0.006	0.00	−0.01
1.03	−0.005	−0.001	+0.005	0.00
1.04	−0.002	+0.002	+0.010	+0.01
1.05	+0.001	+0.005	+0.015	+0.02

SOURCE: K. S. Pitzer et al., *J. Am. Chem. Soc.*, *77*, 3427 (1955).

TABLE 5.17 Values of $z^{(1)}$ in the Critical Region

T_r	P_r					
	1.0	1.2	1.4	1.6	1.8	2.0
0.98	− 0.090	− 0.099	− 0.109	− 0.118	− 0.125	− 0.130
0.99	− 0.087	− 0.095	− 0.104	− 0.114	− 0.121	− 0.127
1.00	− 0.080	− 0.090	− 0.099	− 0.108	− 0.115	− 0.123
1.01	− 0.02	− 0.080	− 0.091	− 0.102	− 0.10	− 0.100
1.02	− 0.01	− 0.065	− 0.082	− 0.095	− 0.09	− 0.09
1.03	0.00	− 0.047	− 0.068	− 0.085	− 0.08	− 0.09
1.04	+ 0.01	− 0.025	− 0.050	− 0.073	− 0.07	− 0.08
1.05	+ 0.02	+ 0.01	− 0.01	− 0.04	− 0.06	− 0.07
1.06	+ 0.03	+ 0.06	+ 0.07	− 0.02	− 0.05	− 0.073
1.07	+ 0.04	+ 0.08	+ 0.09	0.000	− 0.038	− 0.059
1.08	+ 0.047	+ 0.08	+ 0.10	+ 0.30	− 0.015	− 0.041
1.09	+ 0.050	+ 0.08	+ 0.11	+ 0.056	+ 0.012	− 0.022
1.10	+ 0.055	+ 0.082	+ 0.11	+ 0.082	+ 0.035	0.000
1.11	+ 0.057	+ 0.085	+ 0.12	+ 0.099	+ 0.062	+ 0.020
1.13	+ 0.062	+ 0.089	+ 0.12	+ 0.123	+ 0.105	+ 0.060
1.15	+ 0.064	+ 0.093	+ 0.122	+ 0.140	+ 0.136	+ 0.100

SOURCE: K. S. Pitzer et al., *J. Am. Chem. Soc.*, *77*, 3427 (1955).

All quantities that depend on the equation of state may be evaluated and tabulated for convenient use with the principle of corresponding states based on the dimensionless groups of p_r, T_r, z_c, and ω. One of the most frequently needed derived properties from the equation of state of a gas is the fugacity coefficient of that gas at some pressure and temperature. The rigorous thermodynamic relationship is

$$\ln\left(\frac{f}{p}\right) = \int_0^{p_r}(z - 1)\, d \ln p_r$$

The expression for z linear in ω is reflected in the fugacity coefficient, which, after conversion to base 10 logarithms, is

$$\log\left(\frac{f}{p}\right) = \log\left(\frac{f}{p}\right)^{(0)} + \omega \log\left(\frac{f}{p}\right)^{(1)}$$

where the quantities may be tabulated as displayed in Tables 5.18 and 5.19.

Properties other than pVT data also yield to correlation in terms of reduced variables. Riedel[8] wrote for the vapor pressure

$$f(p_r, T_r, \alpha) = 0$$

and obtained a successful vapor pressure equation.

$$\ln p_r = \alpha \ln T_r - 0.0838(\alpha - 3.75)\left(\frac{36}{T_r} - 35 - T_r^6 + 42 \ln T_r\right) \tag{5.42}$$

Here the single parameter α is evaluated from any one point on the vapor pressure curve, say the normal boiling point. Knowledge of this one point permits the prediction of the entire vapor pressure curve.

In fact, the corresponding-states idea lends itself, in principle, for correlations of all equilibrium properties, for all are manifestations of innumerable molecular events, each of which depends on the intermolecular forces. For example, the surface tension has been rather well correlated by using the same principle of corresponding states as proposed by Riedel in successfully correlating vapor pressure. The corresponding-states relationship that was obtained,

$$\gamma = p_c^{2/3}\, T_c^{1/3}\, (0.0133\alpha - 0.0281)\, (1 - T_r)^{11/9} \tag{5.43}$$

reproduced data on surface tension on over 80 compounds to within an average deviation of about 3 percent.[9] Some representative data appear in Table 5.20. In Eq. (5.43) above, p_c = critical pressure, atm, T_c = critical temperature, K, and γ = surface tension, N/m.

TABLE 5.18 Values of $[\log (f/P)]^{(0)}$

T_r	P_r												
	0.2	0.4	0.6	0.8	1.0	1.2	1.4	1.6	1.8	2.0	2.2	2.4	2.6
0.80	− 0.060	− 0.262	− 0.425	− 0.535	− 0.618	− 0.683	− 0.736	− 0.780	− 0.817	− 0.849	− 0.877	− 0.901	− 0.922
0.85	− 0.046	− 0.120	− 0.281	− 0.392	− 0.474	− 0.539	− 0.592	− 0.636	− 0.673	− 0.705	− 0.733	− 0.757	− 0.779
0.90	− 0.042	− 0.087	− 0.163	− 0.273	− 0.356	− 0.421	− 0.474	− 0.517	− 0.554	− 0.587	− 0.614	− 0.639	− 0.680
0.95	− 0.033	− 0.070	− 0.112	− 0.173	− 0.255	− 0.319	− 0.372	− 0.415	− 0.452	− 0.483	− 0.511	− 0.535	− 0.557
1.00	− 0.028	− 0.059	− 0.094	− 0.131	− 0.175	− 0.237	− 0.287	− 0.330	− 0.367	− 0.398	− 0.425	− 0.449	− 0.470
1.05	− 0.024	− 0.051	− 0.079	− 0.109	− 0.142	− 0.178	− 0.218	− 0.257	− 0.292	− 0.322	− 0.349	− 0.372	− 0.393
1.10	− 0.021	− 0.044	− 0.067	− 0.093	− 0.120	− 0.147	− 0.177	− 0.207	− 0.237	− 0.264	− 0.289	− 0.311	− 0.331
1.15	− 0.018	− 0.037	− 0.058	− 0.079	− 0.101	− 0.123	− 0.146	− 0.170	− 0.194	− 0.217	− 0.238	− 0.258	− 0.276
1.20	− 0.016	− 0.032	− 0.050	− 0.067	− 0.086	− 0.104	− 0.124	− 0.143	− 0.163	− 0.182	− 0.200	− 0.217	− 0.233
1.25	− 0.014	− 0.029	− 0.044	− 0.059	− 0.075	− 0.091	− 0.107	− 0.123	− 0.139	− 0.155	− 0.171	− 0.186	− 0.199
1.3	− 0.012	− 0.025	− 0.038	− 0.051	− 0.065	− 0.078	− 0.092	− 0.106	− 0.119	− 0.133	− 0.146	− 0.159	− 0.171
1.4	− 0.010	− 0.021	− 0.031	− 0.041	− 0.052	− 0.062	− 0.072	− 0.082	− 0.092	− 0.102	− 0.111	− 0.120	− 0.130
1.5	− 0.008	− 0.016	− 0.024	− 0.032	− 0.040	− 0.047	− 0.055	− 0.063	− 0.070	− 0.078	− 0.085	− 0.092	− 0.099
1.6	− 0.007	− 0.013	− 0.019	− 0.026	− 0.032	− 0.038	− 0.044	− 0.050	− 0.056	− 0.062	− 0.067	− 0.072	− 0.077
1.7	− 0.005	− 0.010	− 0.015	− 0.020	− 0.025	− 0.030	− 0.034	− 0.039	− 0.043	− 0.047	− 0.051	− 0.056	− 0.059
1.8	− 0.004	− 0.008	− 0.012	− 0.015	− 0.019	− 0.022	− 0.026	− 0.030	− 0.033	− 0.036	− 0.039	− 0.042	− 0.045
1.9	− 0.003	− 0.006	− 0.009	− 0.012	− 0.015	− 0.018	− 0.020	− 0.023	− 0.025	− 0.028	− 0.030	− 0.033	− 0.035
2.0	− 0.002	− 0.004	− 0.007	− 0.009	− 0.011	− 0.013	− 0.015	− 0.017	− 0.019	− 0.021	− 0.023	− 0.025	− 0.026
2.5	0.000	0.000	0.000	0.000	− 0.001	− 0.001	− 0.001	− 0.001	− 0.001	− 0.001	− 0.001	− 0.001	− 0.001
3.0	0.000	+ 0.001	+ 0.001	+ 0.002	+ 0.002	+ 0.003	+ 0.003	+ 0.004	+ 0.004	+ 0.005	+ 0.005	+ 0.006	+ 0.007
3.5	+ 0.001	0.002	0.003	0.003	0.004	0.005	0.006	0.007	0.008	0.009	0.010	0.011	0.012
4.0	0.001	0.002	0.003	0.005	0.006	0.007	0.008	0.009	0.010	0.011	0.012	0.013	0.014

T_r	P_r												
	2.8	3.0	3.2	3.4	3.6	3.8	4.0	4.5	5.0	6.0	7.0	8.0	9.0
0.80	− 0.941	− 0.957	− 0.972	− 0.985	− 0.997	− 1.007	− 1.016	− 1.035	− 1.048	− 1.064	− 1.067	− 1.063	− 1.052
0.85	− 0.797	− 0.814	− 0.829	− 0.842	− 0.854	− 0.864	− 0.874	− 0.893	− 0.907	− 0.924	− 0.929	− 0.926	− 0.917
0.90	− 0.679	− 0.696	− 0.710	− 0.724	− 0.736	− 0.746	− 0.756	− 0.775	− 0.789	− 0.807	− 0.814	− 0.813	− 0.805
0.95	− 0.575	− 0.592	− 0.607	− 0.621	− 0.632	− 0.643	− 0.652	− 0.672	− 0.687	− 0.706	− 0.713	− 0.713	− 0.707
1.00	− 0.489	− 0.505	− 0.520	− 0.534	− 0.545	− 0.556	− 0.566	− 0.586	− 0.601	− 0.620	− 0.629	− 0.630	− 0.624
1.05	− 0.411	− 0.428	− 0.442	− 0.455	− 0.467	− 0.478	− 0.488	− 0.508	− 0.523	− 0.543	− 0.552	− 0.553	− 0.549
1.10	− 0.348	− 0.364	− 0.378	− 0.391	− 0.403	− 0.413	− 0.422	− 0.442	− 0.457	− 0.477	− 0.487	− 0.489	− 0.486
1.15	− 0.293	− 0.307	− 0.321	− 0.333	− 0.344	− 0.354	− 0.363	− 0.383	− 0.397	− 0.417	− 0.427	− 0.429	− 0.426
1.20	− 0.247	− 0.261	− 0.273	− 0.285	− 0.295	− 0.305	− 0.314	− 0.332	− 0.346	− 0.366	− 0.375	− 0.378	− 0.376
1.25	− 0.212	− 0.224	− 0.236	− 0.246	− 0.256	− 0.264	− 0.273	− 0.290	− 0.304	− 0.322	− 0.331	− 0.334	− 0.332
1.3	− 0.182	− 0.193	− 0.203	− 0.212	− 0.221	− 0.229	− 0.237	− 0.253	− 0.266	− 0.283	− 0.292	− 0.295	− 0.294
1.4	− 0.138	− 0.146	− 0.154	− 0.162	− 0.169	− 0.175	− 0.181	− 0.194	− 0.205	− 0.220	− 0.228	− 0.231	− 0.229
1.5	− 0.104	− 0.112	− 0.117	− 0.124	− 0.129	− 0.134	− 0.139	− 0.149	− 0.158	− 0.170	− 0.176	− 0.178	− 0.176
1.6	− 0.082	− 0.087	− 0.092	− 0.096	− 0.100	− 0.104	− 0.108	− 0.116	− 0.123	− 0.132	− 0.137	− 0.138	− 0.136
1.7	− 0.063	− 0.067	− 0.071	− 0.074	− 0.077	− 0.080	− 0.083	− 0.089	− 0.094	− 0.101	− 0.105	− 0.105	− 0.102
1.8	− 0.048	− 0.051	− 0.053	− 0.056	− 0.058	− 0.060	− 0.063	− 0.067	− 0.071	− 0.076	− 0.078	− 0.077	− 0.074
1.9	− 0.037	− 0.039	− 0.041	− 0.043	− 0.045	− 0.046	− 0.048	− 0.051	− 0.054	− 0.057	− 0.057	− 0.055	− 0.051
2.0	− 0.028	− 0.029	− 0.031	− 0.032	− 0.033	− 0.034	− 0.035	− 0.037	− 0.039	− 0.040	− 0.039	− 0.038	− 0.034
2.5	− 0.001	− 0.001	− 0.001	− 0.001	0.000	0.000	0.000	+ 0.001	+ 0.003	+ 0.006	+ 0.011	+ 0.016	+ 0.022
3.0	+ 0.007	+ 0.008	+ 0.009	+ 0.010	+ 0.011	+ 0.012	+ 0.012	0.015	0.017	0.023	0.028	0.035	0.042
3.5	0.013	0.014	0.015	0.016	0.017	0.018	0.020	0.022	0.025	0.031	0.038	0.044	0.051
4.0	0.015	0.016	0.017	0.019	0.020	0.021	0.022	0.025	0.028	0.034	0.040	0.047	0.054

SOURCE: K. S. Pitzer et al., *J. Am. Chem. Soc.* **77**, 3427 (1955).

TABLE 5.19 Values of [log (*f*/*P*)][1]

T_r	P_r																				
	0.2	0.4	0.6	0.8	1.0	1.2	1.4	1.6	1.8	2.0	2.2	2.4	2.6	2.8	3.0	4.0	5.0	6.0	7.0	8.0	9.0
0.80	− 0.04	− 0.47	− 0.48	− 0.48	− 0.48	− 0.49	− 0.50	− 0.50	− 0.51	− 0.51	− 0.52	− 0.52	− 0.53	− 0.53	− 0.54	− 0.56	− 0.59	− 0.61	− 0.63	− 0.65	− 0.67
0.85	− 0.03	− 0.31	− 0.31	− 0.32	− 0.33	− 0.33	− 0.34	− 0.35	− 0.35	− 0.36	− 0.37	− 0.37	− 0.38	− 0.38	− 0.39	− 0.41	− 0.44	− 0.46	− 0.48	− 0.50	− 0.51
0.90	− 0.02	− 0.04	− 0.18	− 0.20	− 0.20	− 0.21	− 0.21	− 0.22	− 0.23	− 0.23	− 0.24	− 0.24	− 0.25	− 0.26	− 0.26	− 0.29	− 0.31	− 0.33	− 0.35	− 0.36	− 0.38
0.95	− 0.01	− 0.02	− 0.03	− 0.09	− 0.10	− 0.11	− 0.12	− 0.12	− 0.13	− 0.13	− 0.14	− 0.15	− 0.15	− 0.16	− 0.16	− 0.18	− 0.20	− 0.22	− 0.24	− 0.26	− 0.27
1.00	− 0.01	− 0.01	− 0.01	− 0.02	− 0.03	− 0.03	− 0.04	− 0.05	− 0.05	− 0.06	− 0.06	− 0.07	− 0.07	− 0.08	− 0.08	− 0.10	− 0.12	− 0.13	− 0.15	− 0.17	− 0.18
1.05	0.00	0.00	0.00	0.00	+ 0.01	+ 0.01	+ 0.01	+ 0.01	0.00	0.00	0.00	0.00	− 0.01	− 0.01	− 0.01	− 0.03	− 0.05	− 0.06	− 0.07	− 0.09	− 0.11
1.10	0.00	0.00	0.00	+ 0.01	0.01	0.02	0.02	0.03	+ 0.03	+ 0.03	+ 0.03	+ 0.03	+ 0.03	+ 0.03	+ 0.03	+ 0.02	0.00	− 0.01	− 0.02	− 0.03	− 0.05
1.15	0.00	0.00	0.00	0.01	0.02	0.02	0.03	0.04	0.04	0.05	0.05	0.05	0.06	0.06	0.06	0.05	+ 0.05	+ 0.04	+ 0.02	+ 0.01	0.00
1.20	0.00	+ 0.01	+ 0.01	0.01	0.02	0.03	0.04	0.05	0.05	0.06	0.07	0.07	0.08	0.08	0.08	0.09	0.09	0.08	0.07	0.07	+ 0.06
1.25	0.00	0.01	0.01	0.02	0.03	0.03	0.04	0.05	0.06	0.07	0.07	0.08	0.09	0.09	0.10	0.11	0.11	0.11	0.10	0.10	0.09
1.3	+ 0.01	0.01	0.02	0.02	0.03	0.04	0.04	0.05	0.06	0.07	0.08	0.08	0.09	0.10	0.10	0.12	0.13	0.13	0.13	0.12	0.12
1.4	0.01	0.01	0.02	0.03	0.04	0.04	0.05	0.06	0.07	0.08	0.08	0.09	0.10	0.11	0.11	0.13	0.15	0.15	0.16	0.16	0.16
1.5	0.01	0.02	0.02	0.03	0.04	0.05	0.05	0.06	0.06	0.07	0.08	0.08	0.09	0.10	0.11	0.13	0.15	0.16	0.17	0.17	0.18
1.6	0.01	0.02	0.02	0.03	0.04	0.05	0.05	0.06	0.06	0.07	0.08	0.08	0.09	0.10	0.11	0.14	0.16	0.18	0.19	0.20	0.21
1.7	0.01	0.02	0.02	0.03	0.04	0.05	0.05	0.06	0.06	0.07	0.08	0.08	0.09	0.10	0.11	0.14	0.16	0.18	0.20	0.21	0.23
1.8	0.01	0.02	0.02	0.03	0.04	0.05	0.05	0.06	0.06	0.07	0.08	0.08	0.09	0.10	0.11	0.14	0.16	0.19	0.21	0.23	0.24
1.9	0.01	0.02	0.02	0.03	0.04	0.05	0.05	0.06	0.06	0.07	0.08	0.08	0.09	0.10	0.11	0.14	0.16	0.19	0.21	0.23	0.25
2.0	0.01	0.01	0.02	0.03	0.04	0.05	0.05	0.06	0.07	0.07	0.08	0.08	0.09	0.09	0.10	0.13	0.16	0.19	0.21	0.23	0.26
2.5	0.01	0.01	0.02	0.02	0.03	0.04	0.04	0.05	0.06	0.06	0.07	0.07	0.08	0.08	0.09	0.12	0.14	0.17	0.19	0.22	0.24
3.0	0.00	0.01	0.01	0.02	0.02	0.03	0.04	0.04	0.05	0.05	0.05	0.06	0.06	0.07	0.07	0.10	0.12	0.15	0.17	0.20	0.22
3.5	0.00	0.01	0.01	0.02	0.02	0.02	0.03	0.03	0.04	0.04	0.04	0.05	0.05	0.06	0.06	0.08	0.10	0.13	0.15	0.17	0.19
4.0	0.00	0.01	0.01	0.02	0.02	0.02	0.02	0.03	0.03	0.03	0.04	0.04	0.04	0.05	0.05	0.07	0.09	0.10	0.12	0.14	0.15

SOURCE: K. S. Pitzer et al., *J. Am. Chem. Soc.* **77**, 3427 (1955)

TABLE 5.20 Correlation and Prediction of Surface Tension of Pure Liquids by a Principle of Corresponding States

Compound	T, °C	γ(exp), N/m	Percent error†
Aniline	20	4.267	11
	80	3.615	8.0
Benzene	20	2.888	− 2.0
	80	2.120	− 1.9
CCl_4	35	2.521	− 5.1
	95	1.786	− 4.4
$(C_2H_5)_2O$	30	1.620	− 2.5
C_2H_5Br	30	2.304	13
n-C_7	20	2.014	0.3
	80	1.426	0.8
n-propyl	20	2.998	− 1.1
Benzene	100	2.038	1.9

†Percent error = [(calculated − experimental)/(experimental)] × 100.

Not surprisingly, the transport coefficients, i.e., the nonequilibrium properties, also lend themselves to corresponding-states-type correlations, as we shall see.

An analytic equation of state based on still another principle of corresponding states utilizes five rather than four dimensionless groups. This is the formulation of Hirschfelder et al.,[10,11] who developed the equation of state and then derived all of the thermodynamic excess functions, $H - H^0$, $S - S^0$, $C_p - C_p^0$, $\ln f/p$, etc. These workers divided the pVT surface into three regions and developed equations of state that well fit pVT data on all substances. Their equations also joined in a thermodynamically consistent manner at the borders between the three regions. These workers wrote

$$p_r = f(T_r, V_r, \alpha, \beta)$$

where α is the Riedel α and β is simply related to z_c:

$$z_c = \frac{3\beta - 1}{(1 + \beta)^3}$$

We expect this equation to fit data better than either of the two previous tabular presentations because we are here using two correlating parameters, α and β, for the intermolecular forces, rather than only one, z_c or ω, as was used earlier. This equation of state also has an enormous advantage in that it can be readily coded to allow extensive computer calculations. All of this has been done, and the equation is the most accurate (and the most useful) of any of the generalized formulations that are now available. The formal equation in region I of Hirschfelder, et al.[10] has been cast in the form

$$\frac{p_r}{T_r} = -\omega_1(T_r)\rho^2 - \omega_2(T_r)\rho^3 + g(\rho)$$

where $\omega_1(T_r) = k_0 T_r^{-1} + (\beta - k_0)T_r^{-2}$ (5.44)

$$\omega_2(T_r) = \tfrac{1}{2}[1 - k_0 - \alpha + 2\beta](1 - T_r^{-2})$$

$$g(\rho) = \frac{(1 + \beta)^3\rho}{\beta(3\beta - 1) - (3\beta^2 - 6\beta - 1)\rho + \beta(\beta - 3)\rho^2}$$

$$k_0 = 5.5$$

Region I encompasses all of the superheat region. The boundaries are reduced densities of less than 1 above the critical temperature and densities less than saturation at all temperatures less than critical. Their equation may also be cast in the form

$$\left[p + \frac{a(T)}{V^2} + \frac{a^1(T)}{V^3}\right]\left[V - b + \frac{b^1}{V}\right] = RT \qquad (5.45)$$

which is reminiscent of the van der Waals equation.

We recall again that each of these macroscopic principles of corresponding states is just a scale change from that using parameters from the intermolecular potential. The practical development of these expressions has been discussed in detail elsewhere.[10–13] The important thing here from the viewpoint of statistical thermodynamics is that the root of these sorts of expressions lies in the evaluation of the thermodynamic properties in terms of some intermolecular potential function. We can readily appreciate the structure of these theoretical arguments, and we can well see why the semitheoretical approaches have been so very successful as compared to the purely theoretical approaches.

The equation of state and the intermolecular potential are intimately linked. While it is impossible in practice to obtain a potential function by manipulating an equation of state like Eq. (5.45) through the formalism that we have developed, it is nonetheless quite possible to do so in principle. And that is an important point in molecular engineering.

We have seen these same sorts of generalizations at a wholly empirical level before. The idea was first proposed by van der Waals, who said that all substances obey the same equation of state in reduced variables, $p_r = p/p_c$, $V_r = V/V_c$, and $T_r = T/T_c$. And using the conditions at the critical point, $(\partial p/\partial V)_T = 0$ and $(\partial^2 p/\partial V^2)_T = 0$, van der Waals used his two-constant equation to derive a universal equation of state:

$$\left(p_r + \frac{3}{V_r^2}\right)(3V_r - 1) = 8T_r$$

We note that similar universal equations of state can be immediately derived from any two-parameter equation of state. Any property dependent on the equation of state can then be expressed as a universal function of the reduced variables. For example, the compressibility is

$$z = f(p_r, T_r)$$

and the fugacity coefficient is

$$\phi = g(p_r, T_r)$$

This suggests that at the critical point where p_r and T_r are both equal to one, z will have some fixed value for all substances. For the Redlich-Kwong equation, this fixed value is 0.333, which is very near the experimental range of 0.23 to 0.30. The value of z_c for the van der Waals equation is 0.375, for the Berthelot equation, 0.281, and for the Dieterici equation, 0.271. The idea of improving the fit of $f(p_r, T_r)$ to z by adding a third parameter, as has been so successfully done by Pitzer et al. and Hougan et al., is nonetheless an old idea. It was first proposed by Nernst in 1907. The idea is general in scope and very successful in application.

References

1. J. M. Prausnitz, *Science, 205*, 759 (1979).
2. A. W. Rosenbluth, *J. Chem. Phys., 21*, 1087 (1953).
3. G. Mie, *Ann. Physik, 11*, 657 (1903).
4. J. O. Hirschfelder, C. F. Curtiss, and R. B. Bird, *Molecular Theory of Gases and Liquids*, Wiley, New York, 1954.
5. A. Rahmanad and F. H. Stillinger, *J. Chem. Phys., 55*, 3336 (1971).
6. N. Metropolis et al., *J. Chem. Phys., 21*, 1087 (1953).
7. A. L. Lydersen, R. A. Greenkorn, and O. A. Hougen, *Univ. Wisconsin Eng. Exp. Sta. Rept. 4*, October 1955.
8. L. Riedel, *Chemie Ing. Tech., 26*, 83 (1954).
9. J. R. Brock and R. B. Bird, *AIChE Journal, 1*, 174 (1955).
10. J. O. Hirschfelder, R. J. Buehler, H. A. McGee, Jr., and J. R. Sutton, *Ind. Eng. Chem., 50*, 375 (1958).
11. J. O. Hirschfelder, R. J. Buehler, H. A. McGee, Jr., and J. R. Sutton, *Ind. Eng. Chem., 50*, 386 (1958).
12. K. S. Pitzer et al., *J. Am. Chem. Soc., 77*, 3433 (1955).
13. K. S. Pitzer et al., *Ind. Eng. Chem., 50*, 265 (1958).

Further Reading

1. J. O. Hirschfelder, C. F. Curtiss, and R. B. Bird, *Molecular Theory of Gases and Liquids*, Wiley, New York, 1954.
2. J. M. Prausnitz, R. N. Lichtenthaler, and E. G. deAzevedo, *Molecular Theory of Fluid Phase Equilibria*, 2d ed., Prentice-Hall, Englewood Cliffs, N.J., 1986.
3. D. A. McQuarrie, *Statistical Thermodynamics*, Harper and Row, New York, 1973.

Chapter

6

The Properties of Materials

The demand continues to grow for materials that exhibit properties far beyond known limits of performance. Harder, lower thermal expansion, higher-temperature service, greater inertness are examples from a long litany of requirements arising from projected applications. The ability to design new materials to address such increasing requirements rests on understanding of materials at a fundamental—at an atomic and molecular—level.

As an introduction to this understanding, consider the thermodynamic properties of materials. As early as 1819, Dulong and Petit[1] noted that the heat capacity was the same and was equal to 6 cal/g · atom · K for all of the solid elements. As more data were collected, Kopp and Neumann[2] recognized that this was not uniquely the case, and they proposed that the heat capacity of each element was different and that the heat capacity for a compound was just the sum of the values for its constituent elements. Thus, the elements are as shown in Table 6.1, where we can well-recognize how each value was obtained from observed heat capacities on many solids. If it had not been measured, we might predict the heat capacity of $KClO_3$ to be $C_p = 6 + 5.4 + 3(4) = 23.4$ cal/g · mole · K.

TABLE 6.1 Heat Capacities of Some Solid Elements at Room Temperature

Element	Heat capacity	Element	Heat capacity
Cu	6	B	2.7
Pb	6	S	3.8
Zn	6	O	4.0
Fe	6	F	5.0
C	1.8	P	5.4
H	2.3	I	5.4

From a theoretical perspective, and prior to the advent of quantum mechanics, it was possible to imagine the atoms of, say, a crystal of copper to be vibrating about their equilibrium lattice sites. We could also assume any crystal to be a perfect array of its component atoms, ions, or molecules; we take the crystal as having no imperfections such as dislocations or holes; and we assume the constituent units to perform only harmonic oscillations about their equilibrium lattice sites. If we imagine the crystal then to be rather like a giant molecule, it will have $(3n_A - 6)$ of these harmonic frequencies, and if n_A is Avogadro's number, i.e., if we are considering a mole of the solid, then any theory will have to describe the effect on the macroscopic properties of these essentially $3n_A$ frequencies. As we will see, both Einstein and Debye proposed such theories. The Debye theory is used today, and it describes (1) atomic solids where the units are held together by covalent bonds (e.g., diamond), (2) ionic solids held together by coulombic forces (e.g., NaCl), (3) molecular solids held together by van der Waals forces (e.g., benzene), and (4) metallic solids held together by the free electron gas that is the conduction electrons which interpenetrate the three-dimensional array of positive ions of the metal. These solids span about 2 orders of magnitude in binding energy, but as long as the oscillations may be reasonably considered to be harmonic, the Debye theory will be useful, as we shall soon see. It is a remarkably successful theory.

Classical mechanics treats each of the $3n_A$ oscillators as having a kinetic energy of $\frac{1}{2}mv^2$ and a harmonic restoring potential energy proportional to the square of the displacement, or $\frac{1}{2}Kq^2$. Then by the principle of equipartitioning of energy (see later), the total energy of each oscillation will be $\frac{1}{2}kT + \frac{1}{2}kT$, or kT, and for $3n_A$ oscillations this becomes a total internal energy of $3n_AkT$ or $3RT$. The heat capacity C_V is then $3R$, the Dulong and Petit value, and independent of temperature. This prediction of classical mechanics was totally wrong, for the Dulong and Petit value is not correct for all atomic solids and certainly C_V falls to zero with decreasing temperature for everything.

Although the heat capacities of most metals and heavy elements was constant at about 6.2 cal/mol · K, Dewar (1907) had shown that there was no such regularity at all in the heat capacities of species at very low temperatures. So the description of this then-striking low-temperature behavior was a theoretical problem of some concern at the time. It then remained for quantum mechanics to provide an acceptable and accurate explanation of this variation of heat capacity of atomic solids with temperature.

But before proceeding with the development of the appropriate partition function, it is well to recall the underlying statistical arguments. One of the expressions for w that we derived in Chap. 2 was

$w = n!/\Pi n_i!$, and this statistical result required distinguishability of the molecules. This is impossible, of course, for like molecules are indistinguishable, but if we limit our attention to solids, that is to systems of rigidly localized molecules, it is possible to label not the molecules, but the fixed sites where the molecules are localized. We will apply this statistical expression for w then, but we will imagine that it is the sites that are labeled, not the molecules. A moment's reflection reveals that this substitution is sensible, and the statistics based on the above expression for w will be applied to solids.

Einstein Model of a Solid

For the moment, let us take the molecules to be structureless and let us assume, with Einstein, that the displacement of a molecule from its equilibrium position in, for example, the x direction in the solid lattice is governed by Hooke's law. The motion will then be harmonic, and, according to quantum mechanics, the energy levels will be given by the expression

$$\varepsilon = (n + \tfrac{1}{2})h\nu$$

where n is a quantum number having values of 0, 1, 2,.... As in the earlier discussion of gases, we have already seen that the partition function for this one-dimensional harmonic motion is

$$f = e^{-h\nu/2kT} + e^{-3h\nu/2kT} + e^{-5h\nu/2kT} + \cdots$$

We neglect the zero-point vibrational energy by setting the zero of energy at $n = 0$. This is of no consequence, for this zero-point energy merely leads to an additive constant in the energy and enthalpy, and it cancels out in the entropy. These same arguments were made in our earlier discussion of vibrations of gaseous molecules. The partition function becomes

$$f = 1 + e^{-h\nu/kT} + e^{-2h\nu/kT} + \cdots$$

If one forms $fe^{-h\nu/kT}$,

$$fe^{-h\nu/kT} = e^{-h\nu/kT} + e^{-2h\nu/kT} + e^{-3h\nu/kT} + \cdots$$

and subtracts, one obtains

$$f(1 - e^{-h\nu/kT}) = 1$$

or

$$f = (1 - e^{-h\nu/kT})^{-1}$$

This is the same relationship, of course, as that developed earlier to describe intramolecular vibration.

The displacement of an atom in its solid lattice is not one-dimensional as assumed here. Rather its actual displacement may be resolved into three dimensions with the above quantization and resulting partition function in each. Then the total partition function for one unit (atom, molecule, ion, etc.) at its lattice site is

$$\ln f = -3 \ln (1 - e^{-h\nu/kT}) \tag{6.1}$$

where the vibration in each direction occurs at the same frequency ν. It is convenient to define an Einstein characteristic temperature, θ_E, as $h\nu_E/k$. We can then rewrite $\ln f$ as

$$\ln f = -3 \ln (1 - e^{-\theta_E/T})$$

This is now all we require for an experimental test of the theory, since all of the thermodynamic properties have been expressed in terms of $\ln f$ and its derivatives.

Let us calculate the heat capacity as a function of temperature to compare with measured values. First, recall

$$U = nkT^2 \left(\frac{\partial \ln f}{\partial T}\right)_V$$

and, with the above partition function, and taking all $3n_A$ vibration frequencies to be the same, we obtain for the internal energy

$$U - U_0 = 3R\theta_E(e^{\theta_E/T} - 1)^{-1} \tag{6.2}$$

Differentiating again, we obtain the so-called Einstein heat capacity function,

$$C_V(\text{lattice}) = \frac{3R(\theta_E/T)^2 e^{\theta_E/T}}{(e^{\theta_E/T} - 1)^2} = \frac{3Rx_E^2\, e^{x_E}}{(e^{x_E} - 1)^2} \tag{6.3}$$

where x_E is defined as θ_E/T. If we have a measured heat capacity at some one temperature, this will serve to evaluate the characteristic Einstein temperature θ_E, from which the heat capacity may be predicted at all temperatures. So the Einstein heat capacity of a solid fits another principle of corresponding states where a dimensionless heat capacity C_V/R is a universal function of a dimensionless temperature θ_E/T. Unfortunately, the theory does not often exist to allow such a neat calculation of the relationship between the dimensionless groups, and we are left with empirical correlations. Compare, for example, correlations of the Nusselt number versus the Reynolds and Prandtl numbers to obtain values of film coefficients in convective heat transfer. The heat capacity as a function of temperature as given by Eq. (6.3) appears as shown in Fig. 6.1, where C_V is 0 at 0 K and $3R$ J/mol · K at temperatures of the order of θ_E and above. Both the boundary values and the general shape in between are correct, and this agreement was a powerful argument in the early 1900s that was instrumental in leading to the acceptance of quantum mechanics. In the expression for heat capacity, C_V becomes 0/0 for large T, but successive differentiation of Eq. (6.3) reveals the high-temperature limiting value of C_V to be $3R$. Similarly, at low temperatures C_V tends to zero. This limiting high-temperature value of 25 J/mol · K is in keeping with the old Dulong and Petit (1819) and Kopp and Neumann (1869) rules, and with the very low values of C_V that were observed at low temperatures which were heretofore inexplicable. The variation of C_V with temperature as predicted by Einstein was qualitatively correct, and it was a huge improvement over classical mechanical arguments which predicted a constant C_V of $3R$. The Debye theory for solids which we will now develop is superior to the Einstein theory, but nevertheless, the Einstein theory is of far more than mere historical interest and value, for it provides an excellent description of the properties that arise from the internal vibrational modes of molecules, as we have seen. That is, for the intramolecular vibration frequencies, the vibration frequency is indeed fixed, even though this single fixed

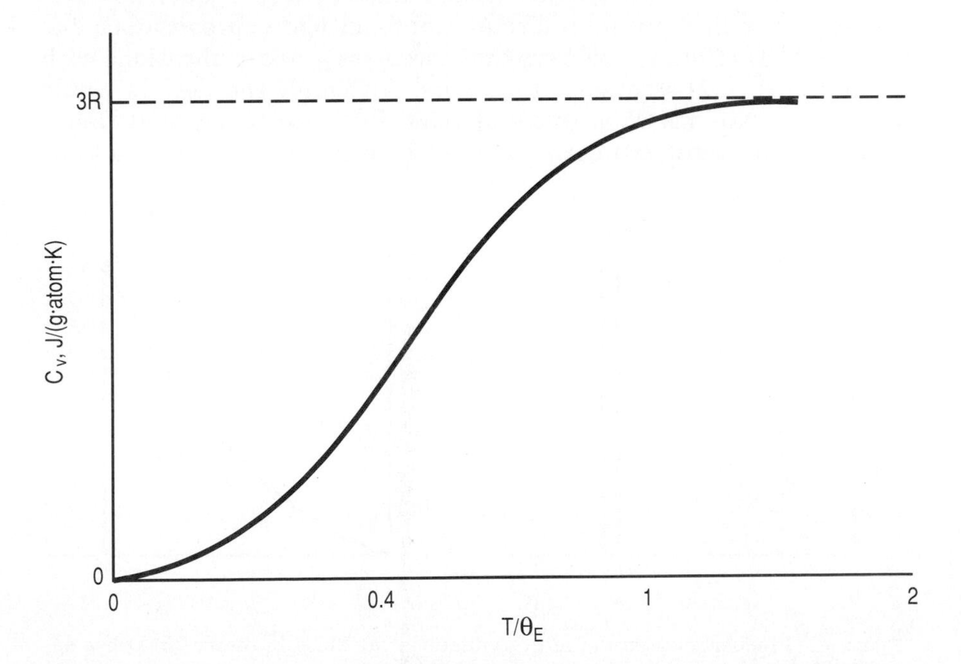

Figure 6.1 Heat capacity of solids as a function of temperature from Einstein theory.

frequency was an incorrect assumption in the description of lattice vibrations.

Finally, for the entropy of an Einstein solid, we write

$$S = R \ln f + RT \left(\frac{\partial \ln f}{\partial T} \right)_V$$

$$= 3R \left[\frac{\theta_E/T}{e^{\theta_E/T} - 1} - \ln \left(1 - e^{-\theta_E/T}\right) \right]$$

and similarly for all of the other thermodynamic functions.

Debye Model of a Solid

The difficulty then with the Einstein model is its assumption of a constant vibration frequency for each atom about its lattice site. The Debye model pictures these frequencies as varying quadratically with ν from 0 up to a maximum value. The difference between the two assumptions is evident in Fig. 6.2. In other words, the Debye model pictures a crystalline solid as one big molecule not too unlike a protein or a polymer. With Debye we may visualize that the mole of atoms n_A are vibrating about their equilibrium lattice sites with $3n_A$ frequencies just as with the Einstein model. We do not, however, know exactly how the atoms are distributed among these frequencies. But suppose we merely write a formal mathematical expression, or distribution function, which says that there are $g(\nu)\, d\nu$ vibrations with frequencies in the range ν to $\nu + d\nu$. Although the details would take us too far into theoretical physics, suffice it to say that Debye developed a useful expression for $g(\nu)$ by considering the solid to be

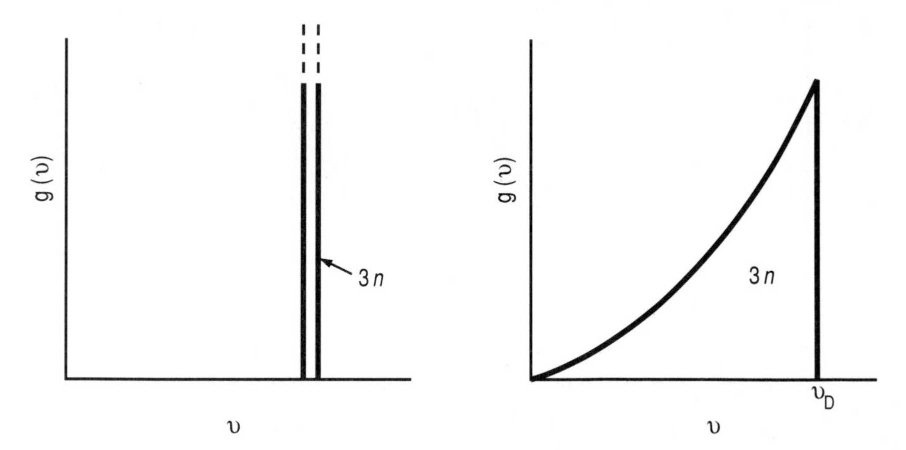

Figure 6.2 Frequency spectra of lattice vibrations. (*a*) Einstein theory; (*b*) Debye approximation.

a continuous (not atomic) elastic body with $g(\nu)$ proportional to ν^2. We take the different frequencies of $g(\nu)$ to be continuous, so their summation by integration is reasonable. If the partition function for a single vibrator in one dimension is $(1 - e^{-h\nu/kT})^{-1}$, as we have seen, then for a vibrator in three dimensions, it is $(1 - e^{-h\nu_x/kT})^{-1}(1 - e^{-h\nu_y/kT})^{-1}(1 - e^{-h\nu_z/kT})^{-1}$. This is true because any energy level for vibration in the x direction may combine with any other level in either the y or z directions. Then for $3n_A$ such vibrators, the partition function is $(1 - e^{-h\nu_i/kT})^{-3n_A}$.

In the Debye approximation, the partition function for all $3n_A$ vibrators is again a continued product of terms $(1 - e^{-h\nu_i/kT})^{-1}$, but there are multiple numbers of vibrators at each frequency, so each such term at each such frequency is raised to a power $ng(\nu)$. That is, each of the product terms is raised to a power that is the number of vibrators with that frequency ν. As with Einstein's single frequency, $\int_0^{\nu_{max}} ng(\nu)\, d\nu$ must equal the total number of vibrators, $3n_A$. The frequencies are continuous, and then

$$\ln f = -n\int_0^{\nu_{max}} g(\nu) \ln(1 - e^{-h\nu/kT})\, d\nu$$

There are $3n_A$ vibrators altogether, so

$$\int_0^{\nu_{max}} ng(\nu)\, d\nu = 3n_A$$

Before we can proceed further, we must make some sort of assumption about the behavior of $g(\nu)$ as a function of ν. When Einstein was first concerned about the heat capacity of solids, he merely assumed that $g(\nu) = 0$ except for a single value at ν_E,

$$\ln f = -\ln[1 - e^{-h\nu_E/kT}]\int_0^{\nu_{max}} ng(\nu)\, d\nu \tag{6.4}$$

The integral is just equal to $3n_A$, and hence the Einstein partition function for all vibrators is

$$\ln f = -3n_A \ln[1 - e^{-h\nu_E/kT}] \tag{6.5}$$

which is just Eq. (6.1) that we had previously derived from an initial assumption of a single fixed frequency. With the form of the partition function of Eq. (6.5) already multiplied by $3n_A$ for the $3n_A$ vibrators, the internal energy is

$$U - U_0 = kT^2\left[\frac{\partial \ln f}{\partial T}\right]_V$$

Debye assumed another and more realistic distribution function which is able to well-describe the low-temperature heat capacity of solids. He took the distribution as being quadratic in ν, i.e.,

$$g(\nu) = \alpha\nu^2$$

where α is some constant, and this quadratic distribution contained frequencies from zero up to some maximum value ν_D, which may, however, be readily evaluated since there can be no more than $3n_A$ frequencies altogether; that is,

$$n\int_0^{\nu_D} \alpha\nu^2\, d\nu = 3n_A$$

or

$$\frac{\alpha\nu_D^3}{3} = 3$$

or

$$\alpha = \frac{9}{\nu_D^3}$$

And the complete description of the Debye assumptions is

$$g(\nu) = \begin{cases} \dfrac{9\nu^2}{\nu_D^3} & \text{for } \nu < \nu_D \\ 0 & \text{for } \nu > \nu_D \end{cases}$$

The Debye theory is then a curious mixture of continuous frequencies from classical mechanics, with each frequency treated as a quantum-mechanical oscillator.

With these assumptions, the partition function, Eq. (6.4), becomes,

$$\ln f = -\frac{9n}{\nu_D^3}\int_0^{\nu_D} \nu^2 \ln\left(1 - e^{-h\nu/kT}\right) d\nu \tag{6.6}$$

As with the Einstein model, we may define a so-called Debye characteristic temperature, $\theta_D = h\nu_D/k$, and, letting $x = h\nu/kT$, we can write the above partition function in the more usual manner as follows:

$$\ln f = -9n\left(\frac{1}{x_D}\right)^3 \int_0^{x_D} x^2 \ln\left(1 - e^{-x}\right) dx$$

It is easy to manipulate Eq. (6.6) through the general relationship between, say, the internal energy and the partition function to obtain

$$U - U_0 = 9RT\left[\frac{1}{x_D^3}\int_0^{x_D} \frac{x^3}{e^x - 1}\, dx\right] \tag{6.7}$$

which may be compared with the Einstein internal energy of Eq. (6.2). The integral is a function only of the upper limit x_D, where $x_D = h\nu_D/kT$ or θ_D/T and where we have again set the zero of energy at 0 K; i.e., we have neglected the zero-point vibrational energy. The heat capacity $C_V = (\partial U/\partial T)_V$ may be written

$$C_V(\text{lattice}) = 9R\left(\frac{1}{x_D^3}\right)\int_0^{x_D} \frac{x^4 e^x}{(e^x - 1)^2}\,dx \tag{6.8}$$

which may be compared with the Einstein expression of Eq. (6.3). The Debye model quantitatively describes the heat capacity of solids composed of structureless particles at each lattice site, as may be seen in Fig. 6.3, where data on a number of solids have been well-correlated by the reduced temperature T/θ_D. This is another example of a principle of corresponding states.

Properties of Monatomic Solids

The above arguments of Einstein and Debye have assumed structureless molecules at each lattice site, and both models have described U and C_V. To further describe the properties of the most simple solids, i.e., atomic solids, it is necessary only to additionally account for the ground state multiplicity of the atom which is just a multiplicative

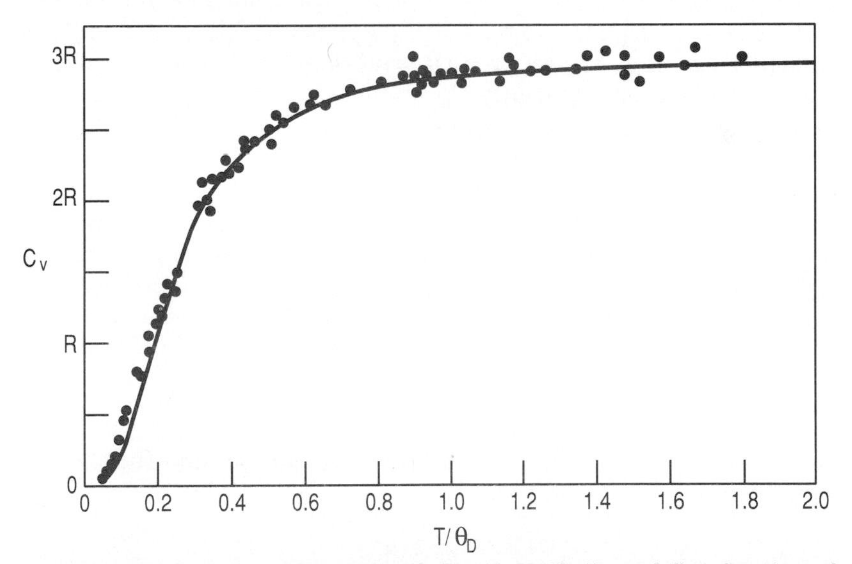

Figure 6.3 The heat capacities of Pb, Hg, Zn, Al, NaCl, Cu, and C as functions of T/θ_D form a single curve. *(From H. S. Taylor and S. Glasstone, Treatise on Physical Chemistry, Van Nostrand, New York, 1942. Used by permission.)*

constant on f(lattice) [see development of Eq. (6.1).] With the Debye partition function, the Helmholtz free energy becomes

$$A - U_0 = 9RT\left(\frac{1}{x_D}\right)^3 \int_0^{x_D} \ln(1 - e^{-x})x^2\,dx - RT\ln g_0 \qquad (6.9)$$

and with both internal energy and the Helmholtz free energy, we can determine the entropy by simple combination. That is,

$$S = -\frac{A - U_0}{T} + \frac{U - U_0}{T} = \frac{9R}{x_D^3}\int_0^{x_D}\left[-\ln(1 - e^{-x}) + \frac{x}{e^x - 1}\right]x^2\,dx + R\ln g_0 \qquad (6.10)$$

Since $(A - U_0)$ and S depend on $\ln f$ as well as its derivative, we have included the electronic contribution to $\ln f$ of $\ln g_0$ in Eqs. (6.9) and (6.10). The electronic partition function is just the ground-state multiplicity. One could ignore g_0 in calculations of energy and related properties, since we can safely ignore the possibility of electronic excitation, but the ground-state multiplicity is not necessarily conserved in either phase changes or in chemical reactions. It is then included here. There are a few otherwise normal molecular solids that have other than unit ground-state multiplicities. For example O_2 has a triplet ground state, i.e., $g_0 = 3$, and NO is a doublet, $g_0 = 2$.

Although the functions of Eqs. (6.9) and (6.10) are troublesome and must be evaluated numerically, convenient tables have been prepared and values of $(U - U_0)$, C_V, and $(A - U_0)$ appear in Tables 6.2 and 6.3. For condensed phases, the pV product is so small that A is very nearly equal to G, and C_V is essentially C_p.

Low-temperature heat capacity measurements reveal that below 10 or 12 K, the heat capacity seems to decrease as T^3. Interestingly, the Debye function predicts exactly that behavior. As T gets small, θ_D/T or x_D gets large, and the upper limit of the integral of Eq. (6.7) may be replaced by ∞. The integral that appears in the expression for internal energy looks very similar to a standard integral in statistical mechanics that is called the *Riemann zeta function*, defined as

$$\zeta(s) = \frac{1}{\Gamma(s)}\int_0^{\infty} \frac{x^{s-1}}{e^x - 1}\,dx$$

where $\Gamma(s)$ is the gamma function. Recall that the gamma function is a generalized factorial, and that

$$\Gamma(n + 1) = n!$$

The value of the $\zeta(s)$ for $s = 4$ may be shown to be $\pi^4/90$. Then at low temperatures the internal energy becomes

TABLE 6.2 Debye Heat Capacity and Internal Energy as a Function of θ_D/T†

θ_D/T	$C_V/3R$	$(U - U_0)/3RT$	θ_D/T	$C_V/3R$	$(U - U_0)/3RT$
0	1.000	1.000	8.5	0.1182	0.03084
0.5	0.9876	0.8250	9.0	0.1015	0.02620
1.0	0.9517	0.6744	9.5	0.08751	0.02241
1.5	0.8960	0.5471	10.0	0.07582	0.01930
2.0	0.8254	0.4411	10.5	0.06600	0.01672
2.5	0.7459	0.3541	11.0	0.05773	0.01457
3.0	0.6628	0.2836	11.5	0.05073	0.01277
3.5	0.5807	0.2269	12.0	0.04478	0.01125
4.0	0.5031	0.1817	12.5	0.03970	0.00996
4.5	0.4320	0.1459	13.0	0.03535	0.00886
5.0	0.3686	0.1176	13.5	0.03160	0.00791
5.5	0.3133	0.09524	14.0	0.02835	0.00710
6.0	0.2656	0.07758	14.5	0.02553	0.00639
6.5	0.2251	0.06360	15.0	0.02307	0.00577
7.0	0.1909	0.05251	15.5	0.02092	0.00523
7.5	0.1622	0.04366	16.0	0.01902	0.00476
8.0	0.1382	0.03656			

†At temperatures less than $\theta_D/T > 16$, $C_V/3R = 77.927(T/\theta_D)^3$ and $(U - U_0)/3RT = 19.482(T/\theta_D)^3$. Linear interpolation in this table gives values to within 1% of those from Eqs. (6.7) and (6.8).

SOURCE: From K. S. Pitzer, *Quantum Chemistry*, Prentice-Hall, New York, 1953.

TABLE 6.3 Debye Helmholtz Free Energy $(A - U_0)/3RT$ as a Function of θ_D/T†

θ_D/T	0.0	0.2	0.4	0.6	0.8	1.0
0	∞	2.0168	1.3956	1.0602	0.8405	0.6835
1.0	0.6835	0.5653	0.4734	0.4003	0.3410	0.2925
2.0	0.2925	0.2522	0.2185	0.1901	0.1661	0.1456
3.0	0.1456	0.1281	0.1130	0.1000	0.08882	0.07906
4.0	0.07906	0.07057	0.06316	0.05667	0.05097	0.04596
5.0	0.04596	0.04154	0.03763	0.03416	0.03108	0.02834
6.0	0.02834	0.02590	0.02371	0.02175	0.02000	0.01841
7.0	0.01841	0.01699	0.01570	0.01454	0.01348	0.01252
8.0	0.01252	0.01165	0.01085	0.01013	0.00946	0.00886
9.0	0.00886	0.00830	0.00779	0.00731	0.00688	0.00648
10.0	0.00648	0.00611	0.00576	0.00544	0.00515	0.00487
11.0	0.00487	0.00462	0.00438	0.00416	0.00395	0.00376
12.0	0.00376	0.00357	0.00340	0.00325	0.00310	0.00296
13.0	0.00296	0.00282	0.00270	0.00258	0.00247	0.00237
14.0	0.00237	0.00227	0.00217	0.00209	0.00200	0.00192
15.0	0.00192	0.00185	0.00178	0.00171	0.00165	0.00159

†The contribution arising from the ground state multiplicity is not included in this table. At $\theta_D/T = 0.1$, $-(A - U_0)/3RT = 2.6732$. For $\theta_D/T > 16$, use $-(A - U_0)/3RT = 6.494(T/\theta_D)^3$. Linear interpolation in this table gives values to within 1% of those from Eq. (6.9).

SOURCE: From K. S. Pitzer, *Quantum Chemistry*, Prentice-Hall, New York, 1953.

$$U - U_0 = 9RT\left(\frac{1}{x_D}\right)^3\left(\frac{\pi^4}{90}\right)(3!) = \frac{3RT^4\pi^4}{5\theta_D^3}$$

and the heat capacity becomes

$$C_V = \frac{12}{5}\pi^4 R\left(\frac{T}{\theta_D}\right)^3 = 233.78\,R\left(\frac{T}{\theta_D}\right)^3 \tag{6.11}$$

This is the famous Debye T^3 relationship that is always used to extrapolate heat capacity measurements from the lowest temperatures at which calorimetric measurements are reasonable to 0 K. Calorimetric measurements rarely go below about 12 K, the lowest temperature conveniently obtained with solid hydrogen by merely evacuating the vapor space over the boiling liquid. Lower temperatures require liquid helium. The experimental determination of all absolute entropies that appear in standard tables has involved the Debye T^3 law for the evaluation of the very low temperature contribution to the entropy. The exact form of the low-temperature extrapolation is, however, relatively unimportant since C_V is so small and its contribution to the entropy is small. A comparison of the heat capacities for silver predicted by the two theories appears in Table 6.4.

Conversely, at very high temperatures, x is small, e^x is essentially $1 + x$, and the expression for the internal energy becomes

$$U - U_0 = 9RT\left(\frac{1}{x_D}\right)^3 \int_0^{x_D} x^2\,dx = 3RT$$

or

$$C_V = 3R \tag{6.12}$$

which is, of course, just the expected Dulong and Petit value. The parameter θ_D may be determined from one heat capacity measurement at some one temperature, and then the theory allows the calculation of the heat capacity at any temperature. C_V varies from 0 at 0 K to $3R$ as the temperature rises. The spectrum of lattice vibrational frequen-

TABLE 6.4 Heat Capacity of Silver at Different Temperatures from Einstein and Debye Theories

		C_V (theoretical)	
T	C_V(exp)	Einstein	Debye
7	0.0151	1.30×10^{-7}	0.0172
10	0.0475	1.27×10^{-4}	0.0502
20	0.3995	0.0945	0.394
47.09	2.582	2.272	2.60
103.14	4.797	4.795	4.86
205.30	5.605	5.633	5.66

SOURCE: From C. Kittel, *Solid State Physics*, 2d ed., Wiley, New York, 1956.

cies that appears in the Debye heat capacity theory came from arguments based on the elastic properties of the solid. It is possible then to determine θ_D from velocity-of-sound measurements in the solid and to then use that θ_D to calculate the heat capacity of that solid at all temperatures. Thus we unearth still another unexpected interrelationship—this time between acoustical or velocity-of-sound measurements and calorimetric measurements.

Table 6.5 shows a comparison of Debye temperatures determined from calorimetric and from acoustical measurements, and we see that the agreement is very good.

Lead will reach its asymptotic value of $3R$ at much lower temperatures than will aluminum, and we expect the C_V of lead to be significantly greater than that of aluminum at all temperatures below T/θ_D of about one, or below 400 K in this case. At very low temperatures, the heat capacity of the free-electron gas may represent a significant part of the heat capacity solely on the basis of lattice arguments such as these. We will discuss this issue subsequently.

We can also reason qualitatively when comparing one substance to another. Silver is heavier than potassium which is heavier than sodium, and since the atomic sizes are also in this same order, the atoms of the lattices will be farther apart in this same order. Heavier masses and greater distances imply weaker forces and then lower frequencies. Lower frequencies imply a lower θ_D and then a higher heat capacity at a corresponding temperature. All of this is obeyed, for example, in the series AgCl, KCl, and NaCl.

Diamond and beryllium in particular do not fit the Dulong and Petit law, for there C_V = 5.65 and 14.6 J/mol · K, respectively, at room temperature (not 25 J/mol · K). The characteristic temperature θ_D is anomamously high for these species, where for diamond θ_D = 1860 K and for Be θ_D = 1440 K. Here light masses and strong bonding be-

TABLE 6.5 Debye Characteristic Temperatures from Calorimetric and Acoustical Measurements

Substance	Elastic θ_D	Calorimetric θ_D
Al	399	396
Cu	329	313
Ag	212	220
Au	166	186
Cd	168	164
Sn	185	165
Pb	72	86
Bi	111	111
Pt	226	220

SOURCE: From D. A. McQuarrie, *Statistical Thermodynamics*, Harper and Row, New York, 1973, p. 206.

tween atoms make for high frequencies of oscillation and then lower heat capacities than would be expected for substances with more ordinary values of θ_D.

The hardness of a solid is a measure of the difficulty of its deformation. Strong interatomic forces mean difficult deformation and high θ_D, as we have seen. As a rule then, the harder the solid, the higher is its θ_D.

Upon cooling below 13°C, tin may undergo a transformation from tetragonal white tin to cubic gray tin. Heat is evolved to accompany this white to gray transformation. We expect then that the atoms will be more tightly bonded in the gray form, the frequencies will be higher, θ_D will be higher for gray than for white tin, and the entropy will be lower than for the less strongly bonded white tin. So why does the process reverse, for white tin is readily formed from gray? To do so ΔG, or G(white) − G(gray), must be negative. Now $\Delta G = \Delta H - T\Delta S$, and from the above argument, ΔS for conversion of gray to white tin must be positive. Also ΔH, or H(white) − H(gray), is positive. At high temperatures, $T\Delta S$ is large and ΔG is negative. At low temperatures, $T\Delta S$ is small and ΔG is positive. The ΔS then determines stability, and allotropes stable at high temperatures will always have lower lattice vibrational frequencies than do modifications occurring at lower temperatures.

A severe test of the Debye theory is to calculate θ_D at every measured C_V. The values that one determines do not define a single value as would be expected. Experiment reveals that the real frequency distribution is not the smooth monatonically increasing function assumed by Debye, but nevertheless use of the real rather than the Debye frequency distribution does not significantly affect the calculated thermodynamic properties of the solid. The oscillations about the lattice sites are also not harmonic as the temperature rises, and the Debye theory becomes less reliable. Finally this sort of restoring force disappears altogether and we say the solid has melted.

More accurate spectra of vibration frequencies than that of Debye may be determined and these may be manipulated through the same statistical formulas to calculate the thermodynamic properties. But these will not be discussed here, for from a practical perspective the increased accuracy is not worth the effort.

Free-Electron Gas

We may well imagine a mass of electrically conducting metal, m, to be a box of volume V, or m/ρ, that contains a free-electron gas that interpenetrates the lattice of positive ions. It is the drift of these conducting electrons caused by a voltage difference that appears as the flow of an electric current. Electrons have, of course, two spin states, and there can be no more than two electrons in each allowed energy level, unlike the molecules of a gas where any number may occupy the same translational energy level. The electrons then are piled into the energy levels two at a time, and it is interesting to ask what is the energy of the electrons at the top of the pile. This is called the Fermi energy, and it is readily developed. As was the case for the translational motion of the molecules of an ideal gas, the energy levels available to the electrons moving in a cubical box of side L are

$$\varepsilon = \frac{h^2}{8mL^2}(n_x^2 + n_y^2 + n_z^2)$$

Each triplet of quantum numbers, n_x, n_y, n_z, refers to a single quantum state, and those states of energy ε lie on the surface of a sphere of radius r in a cartesian space of n_x, n_y, n_z. Then with

$$r^2 = n_x^2 + n_y^2 + n_z^2$$

the quantum states of energy between ε and $\varepsilon + d\varepsilon$ lie between spheres of radius r and $r + dr$. The volume of such a spherical shell is $4\pi r^2\,dr$. We have

$$r^2 = \frac{8mL^2}{h^2}\varepsilon$$

and, differentiating,

$$2r\,dr = \frac{8mL^2}{h^2}\,d\varepsilon$$

The physically meaningful quantum states lie within the positive octet (1/8) of this sphere, for only positive quantum numbers are allowed. If a volume of metal V m^3 (or L^3 m^3) contains n free electrons, at 0 K they will pile up into and fill the $n/2$ lowest energy levels, or

$$\frac{n}{2} = \frac{1}{8}\int_0^{r_{\max}} 4\pi r^2\,dr = \frac{1}{8}\int_0^{\varepsilon_F} 4\pi\left(\frac{8mL^2}{h^2}\varepsilon\right)\left(\frac{1}{2}\right)\left(\frac{8mL^2}{h^2}\right)\left(\frac{h^2}{8mL^2}\right)^{1/2}\left(\frac{1}{\varepsilon^{1/2}}\right)d\varepsilon$$

and $$\frac{n}{V} = \frac{\pi}{2}\left(\frac{8m}{h^2}\right)^{3/2}\int_0^{\varepsilon_F}\varepsilon^{1/2}\,d\varepsilon = \frac{\pi}{3}\left(\frac{8m}{h^2}\right)^{3/2}\varepsilon_F^{3/2}$$

Let us convert the Fermi energy ε_F to its corresponding temperature where $\varepsilon_F = kT_F$. Consider copper with a density of 8.92 or a molar volume of 63.5 g of $V = (1/8.92)(63.5)(10^{-2})^3(10^3) = 7.1 \times 10^{-3}$ m^3/kg · mol. Assuming one free electron for every two copper atoms,

$$\frac{6 \times 10^{23} \times 10^{3} \times 0.5}{7.1 \times 10^{-3}} = \frac{1}{3}(3.14)\left[\frac{8\,(9.1 \times 10^{-31})}{(6.6 \times 10^{-34})^2}\right]^{3/2} \varepsilon_F^{3/2}$$

and
$$\varepsilon_F \cong 7.1 \times 10^{-19}\ \text{J} \tag{6.13}$$

or
$$T_F \cong 50{,}000\ \text{K} \tag{6.14}$$

Thus the highest-energy electrons of the free-electron gas that have piled into the lowest energy levels at 0 K will nonetheless have translational energies comparable to that of an ordinary gas at 50,000 K. As the temperature of the copper is raised, only the highest-energy electrons, i.e., those near ε_F, can move up in energy. The contribution to the heat capacity is then small but nonetheless certainly measurable. The Pauli exclusion principle applies to electrons, and hence only one electron may go into each energy level. Here one obtains a somewhat different-looking expression for w, analogous to $w = n!/\pi n_i!$ that we wrote for Maxwell-Boltzmann statistics in Chap. 2. The electrons rather obey what is called *Fermi-Dirac statistics*. From this alternative statistical picture, the partition function and then the heat capacity due to the free electron gas may be developed. Unlike the complex dependence of the Debye lattice heat capacity on the temperature, the heat capacity due to the free-electron gas is linear in temperature. Typical values of θ_D and γ that characterize the lattice and free-electron contributions to the heat capacity, respectively, appear in Table 6.6. Then

$$C_V = \text{Debye} + \text{free-electron gas}$$

or
$$C_V = \text{Debye (see Table 6.2)} + \gamma T \tag{6.15}$$

Using Eq. (6.15), we calculate for copper at 1200 K, C_V (free elect) = 0.83 J/mol · K, which will then be in excess of the Dulong and Petit value for a fully excited lattice. The value of $C_V = 3R + 0.83$, or 26 J/mol · K, is in good agreement with the experimental values shown in Fig. 6.4. Because of the necessity of Fermi-Dirac statistics for electrons, the heat capacity of the free-electron gas of a conductor

TABLE 6.6 Debye Temperatures and Free-Electron Constants for Several Metals

Metal	θ_D, K	γ, J/mol · K^2	Metal	θ_D	γ, J/mol · K^2
Li	344	1.63	Zr	291	2.8
Be	1440	0.17	Hf	252	2.16
Al	428	1.35	Pt	240	6.8
Ti	420	3.5	Au	165	0.69
Fe	420	3.1	Pb	105	3.0
Cu	343	0.69	U	207	10.0

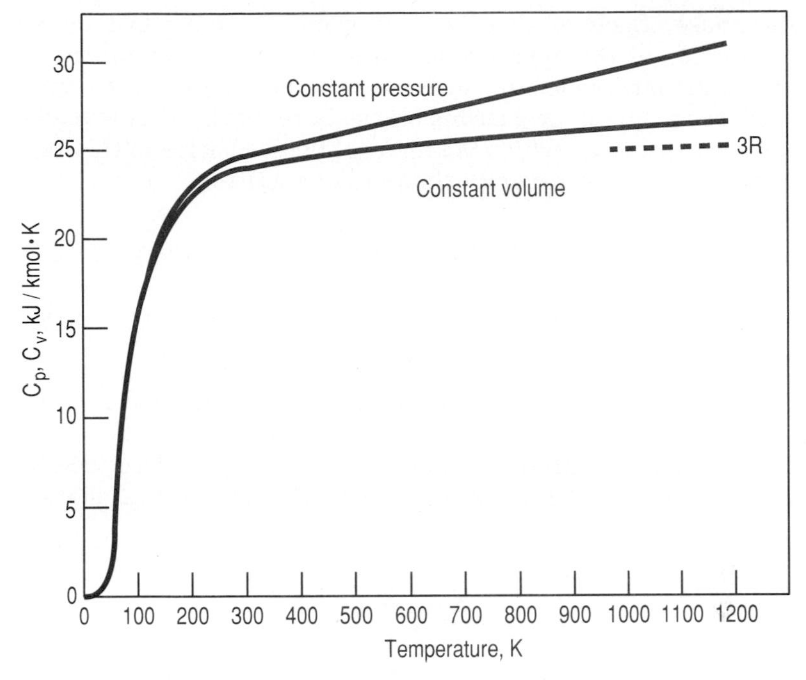

Figure 6.4 Temperature variation of C_p and C_V of copper.

is not nearly as great as the $\frac{3}{2}R$ that would be the expected contribution to C_V according to Maxwell-Boltzmann statistics. It is interesting that the free-electron gas within a metal contributes to the thermal expansion, as well as to the heat capacity, in a manner also linear in temperature,

$$\alpha(\text{free elect}) = \frac{1}{V}\left(\frac{\partial V}{\partial T}\right)_p = aT$$

Molecular Solids

Let us now consider how we might describe the thermodynamic properties of a molecular solid such as benzene. Since a molecular solid has structure, it may take up energy by the rocking of the molecules in three dimensions as well as by the vibration of the center of mass in three dimensions, as was characteristic of the monatomic solid. We might characterize this situation by imagining two Debye functions, one to characterize the oscillations about the lattice sites and a second to characterize the torsional motions about the three principal axis. In addition, the intramolecular vibrational modes would contribute to the thermodynamic properties just as though the molecule were in the ideal gas phase. Each of these modes would be described by an Einstein function at the appropriate value of ω/T. This similarity of treatment of the intramolecular motions, whether the molecule finds itself in the solid, liquid, or gaseous phase, is realistic, as is revealed by the infrared spectrum, which is remarkably insensitive to the state of aggregation. The heat capacity then may be written

$$C_p = C_V(\text{Debye, lattice}) + C_V(\text{Debye, torsion}) + \sum_{i=1}^{3n-6} C_V(\text{vib}) + (C_p - C_V) \quad (6.16)$$

where n is, as before, the number of atoms in the molecule and $(C_p - C_V)$ arises from the work of expanding the lattice and is given by the usual relationship, $C_p - C_V = \alpha^2 VT/\beta$. For simplicity, let us take both the lattice oscillations and torsions as being described by the same θ_D; that is, let us relax our claim of hard physical significance for θ_D. Then

$$C_p = 18R\left(\frac{1}{x_D}\right)^3 \int_0^{x_D} \frac{x^4 e^x}{(e^x - 1)^2}\, dx + \sum_{i=1}^{3n-6} C_V(\text{vib}) + (C_p - C_V) \quad (6.17)$$

or perhaps more poignantly as

$$C_p = C_V(\text{acoustical}) + C_V(\text{optical}) + (C_p - C_V)$$

The Debye heat capacity function was designed to apply to a crystal lattice of mass points having no moment of inertia. The molecule, however, possesses three (or two, if linear) moments of inertia, and if they were freely rotating in the lattice, their contribution to the thermodynamic properties would be the same as their contribution in the gas phase. The molecules are, however, not freely rotating in the lattice, for each must assuredly feel the force field of its nearest neighbors. If the molecule is tilted somewhat, it will return to its equilibrium position, executing then torsional oscillations. As the temperature increases, these torsional oscillations may become free rotations. The solid could then well be thought of as partially "melted" even though the lattice is intact. And in the same sense that one can describe liquid crystals as partially ordered liquids, we might describe a solid with freely rotating molecules as a "crystal liquid." These transitions are evident as sharp peaks in the heat capacity. Some examples of molecules displaying this phenomenon and their transition temperatures appear in Table 6.7.

It is impressive that such a simple combination as Eq. (6.17) leads to a very satisfactory portrayal of the thermodynamic properties of molecular solids. The Debye idea is here used to describe torsions, for which it was certainly not derived, and then averaged with the lattice

TABLE 6.7 Transitions to Free Rotation in Several Solids

Solid	Transition temperature, K	New motion
CO	61.5	Rotation of the molecule
CH_4	20.4	Rotation of the molecule
HI	70	Rotation of the molecule
NH_4Cl	243	Rotation of NH_4^+
KNO_3	401	Rotation of NO_3^-
CH_3OH	159	Hindered rotation of hydroxyl group; methanol melts at 179 K

oscillations to yield a single characteristic θ_D. The internal vibrations of the molecules themselves may absorb energy which leads to the "optical" term. And finally, energy will be absorbed in the work of expanding the lattice itself, and this leads to the final term labeled just $(C_p - C_V)$.

With a value of θ_D and the $3n - 6$ vibration frequencies, we can then expect to predict the heat capacity of any molecular solid at any temperature to within reasonable accuracy. For benzene, with $3n = 3(12) = 36$ degrees of freedom, at 100 K, C_V(Debye) + C_V(Einstein) + $(C_p - C_V)$ is 44.77 + 2.13 + 3.39, respectively, for a total C_p of 50.29 J/mol · K, which compares well with the experimental value of 50.17 J/mol · K. At 200 K, these corresponding numbers are 48.45 + 20.71 + 14.77 = 83.93, which compares well with the experimental value of 83.76.[3]

At temperatures low compared to the characteristic vibration frequencies, it is possible to freeze out these internal vibrations, and we would then expect the heat capacity of the molecular solid to arise solely from the six degrees of freedom of the lattice and to then be not too unlike twice the heat capacity of, say, Cu with its three degrees of freedom, but with, of course, a different value of θ_D. In fact, one experimentally evaluates θ_D by selecting that value that best reproduces experimental heat capacity data at temperatures sufficiently low that all of the vibrational modes have been frozen out.

Similarly, at temperatures above the order of θ_D, the lattice is contributing its asymptotic value of $6R$ and the temperature dependency of C_V arises from the internal vibrational contributions. To obtain the entropy of such a solid, one merely adds the lattice contribution using the appropriate θ_D to the contributions from the $3n - 6$ (or $3n - 5$) molecular vibrations, each with its appropriate ω/T. And similarly for any extensive property.

Solids Composed of Atomic and Molecular Ions

First consider a crystal of atomic ions. In solids such as NaCl there are no single molecules, but it is reasonable to assume (1) each ion pair to be vibrating about its lattice site or center of mass in concert due to the strong coulombic binding forces and (2) for each Na^+/Cl^- pair to be vibrating relative to each other as would any other diatomic species. NaCl will have six degrees of freedom, three of these will be lattice Debye modes of the center of mass. One would be in vibration as in any diatomic species, and the remaining two would then be torsional oscillations as in a molecular solid. Because of the strong Coulomb forces of the ionic solid, the torsional oscillations actually appear in the infrared, and we can best describe all three modes as Einstein-like vibrations. A diatomic species in the gas phase has, of course, one vibrational mode and two rotational modes. In "diatomic" NaCl in the solid phase, all three of these appear as low-frequency vibrations. We thus take the six degrees of freedom to be composed of three lattice or Debye modes and three Einstein modes. The Debye θ_D may be determined from low-temperature heat capacity measurements where the Einstein modes are frozen out, and the intramolecular frequencies may be determined from the infrared spectrum of the crystal.

A solid composed of molecular ions can also be well-approximated by the above sorts of arguments. Consider, for example, a salt like $KClO_3$. Here, the three-dimensional ion pair K^+/ClO_3^- will have three degrees of torsional vibrational freedom in three dimensions about its equilibrium lattice position in just the same way as the linear ion pair Na^+/Cl^- underwent two-dimensional torsion.

The ClO_3^- ion will have six Debye lattice degrees of freedom, as do all molecular solids. The motion of these negative ions will be characterized by its particular value of θ_D. And the ClO_3^- ion, which is pyramidal like ammonia, will have $3n - 6$, or 6, characteristic molecular vibration frequencies, and each of these will contribute to the heat capacity according to its particular ω/T. Thus $KClO_3$ has 3 + 6 + 6 or 15 degrees of freedom, i.e., the required $3n$. This vibrational contribution is unchanged whatever the phase in which it finds itself, as is revealed by spectral evidence where the vibration frequencies are essentially unchanged in the solid, liquid, dissolved, or gaseous phase. Thus the total heat capacity will be given by

$$C_V = 18R\left[\frac{1}{x_D^3}\int_0^{x_D}\frac{x^4e^x}{(e^x-1)^2}\,dx\right] + R\sum_{i=1}^{i=3}\left[\frac{x^2e^x}{(e^x-1)^2}\right] + R\sum_{i=1}^{i=6}\left[\frac{x_i^2e^{x_i}}{(e^{x_i}-1)^2}\right] \tag{6.18}$$

or C_V = (six lattice modes from translation and torsion of ClO_3^-; here translational and torsion have the same θ_D) + (Einstein vibrations from coulombic bond between the ions of the solid) + (intramolecular Einstein vibrations of ClO_3^-)

and similarly for other molecular ionic solids. The Einstein modes arising from the coulombic forces of the lattice we expect to be of low frequency because of the large masses and distances involved. It is usually reasonable to take all three Einstein frequencies to be the same and still adequately describe the "optical" part of the heat capacity due to ionic motion. Then the second term in Eq. (6.18) would be just 3 times a single Einstein contribution. Therefore we expect the infrared absorption spectrum of solid $KClO_3$ to consist of nine frequencies, of which three will be low and six will be significantly higher. That is in fact the case.

The value of θ_D, or x_D, is best determined by selecting that value that will best reproduce calorimetrically measured values of C_V of the solid at such low temperatures that the intramolecular vibrational modes are frozen out. As always, velocity-of-sound measurements also permit a calculation of θ_D.

In general then, we see that the heat capacity of a solid of whatever complexity is rather well reproduced by a sum of an "acoustical" part arising from either three or six Debye modes and the "optical" parts arising from the Einstein intramolecular vibrational modes and from the Einstein vibrational modes of the ion pairs when we are describing an ionic solid.

$C_p - C_V$

In all of these arguments we have calculated C_V because the energy levels of the lattice are functions of volume, and we have then always held volume fixed to write

$$U = RT^2\left(\frac{\partial \ln f}{\partial T}\right)_V$$

and then C_V as $(\partial U/\partial T)_V$. This is unfortunate, since, from a practical point of view, C_p is the more usually required quantity, as well as the quantity that is virtually always measured for a solid. Of course, C_p and C_V are thermodynamically rigorously related:

$$C_p - C_V = \frac{\alpha^2 VT}{\beta}$$

where α is the coefficient of thermal expansion

$$\alpha = \frac{1}{V}\left(\frac{\partial V}{\partial T}\right)_p$$

and β is the coefficient of isothermal compressibility

$$\beta = -\frac{1}{V}\left(\frac{\partial V}{\partial P}\right)_T$$

One should not forget that the C_p of a solid continues to increase with temperature after the lattice is fully excited due to this $(C_p - C_V)$ term and its dependence on the pVT behavior of the solid. The heat capacity difference can be rather well correlated by an empirical equation of the form

$$C_p - C_V = aC_V^2T \tag{6.19}$$

It is this factor (along with the heat capacity of the free-electron gas for metals) that is responsible for the increase in C_p with temperature of solids in which the C_V has already reached its asymptotic Dulong and Petit value of $3R$. When the solid melts, the arrangement of the molecules does not change too much, and we are not surprised that the heat capacities of the solid and liquid phases are about the same at the melting point. The difference $C_p(\text{liq}) - C_p(\text{solid})$ is largely determined by the $(C_p - C_V)$ for each phase. Since liquids are usually more expansive than solids, that is, since α, which is $(1/V)\ (\partial V/\partial T)_p$, is usually greater for liquids, we expect that the heat capacity of the liquid will usually exceed that of the solid at the melting point, i.e., $(C_p - C_V)_{\text{liq}}$ is usually greater than $(C_p - C_V)_{\text{sol}}$, and if $C_V(\text{liq}) = C_V(\text{sol})$, $C_p(\text{liq})$ will be greater than $C_p(\text{sol})$. This is usually the experimental observation.

On the other hand, when a liquid vaporizes, the "lattice" vibrational contribution of $3R$ or $6R$ will become the free translational and rotational values of $3/2R$ and $3/2R$ respectively. Thus the heat capacity of liquids is always greater than that of the coexisting equilibrium vapor.

Anomalies

There are a number of anomalies that occur in the heat capacity of solids, and these give rise to distinctive peaks, or so-called lambda points, in the curve of C_p versus T. These may arise from solid-state phase transitions, from reorientations in the solid, from the onset of free rotation of a molecular ion, and the like. The Debye and Einstein theories, of course, make no allowance for such phenomena; their presence can only be determined experimentally, and such transitions will significantly affect, for example, the absolute entropy. Some transitions associated with several sorts of molecular rotations appear in Table 6.7, but this is only one type of behavior leading to anomalies in the heat capacity of solids.

References

1. P. L. Dulong and A. T. Petit, *Ann. Chim. Phys., 10*, 395 (1819).
2. H. Kopp, *Ann. Chem. Pharm. Suppl., 3*, 1, 289 (1864).
3. R. C. Lord, J. E. Ahlberg, and D. H. Andrews, *J. Chem. Phys., 5*, 649 (1937).

Further Reading

1. F. Seitz, *The Modern Theory of Solids*, McGraw-Hill, New York, 1940.
2. C. Kittel, *Solid State Physics*, Wiley, New York, 1967.
3. A. T. Stewart, *Rev. Mod. Phys., 30*, 250 (1958).
4. S. Blinder, *Adv. Phys. Chem. 1969*, Macmillan, New York, 1969.

Chapter

7

Equilibrium

Physical processes such as the vaporization of a liquid or diffusion of a species across a membrane, as well as the chemical conversion of one species into another, all occur at varying rates and all proceed to an asymptotic state of no further change. Thermodynamics is concerned with that state of no further change, while physical and chemical kinetics are concerned with the rate processes toward that end. Thermodynamics is legislative in character in that it tells what changes in a system are even possible, while both physical and chemical kinetics are executive in character, in that of all allowed changes, they tell us which changes will in fact occur. Both equilibrium processes and rate processes may be rewardingly discussed from the perspective of molecular engineering.

Our quantum and statistical ideas have much to say about equilibrium, but first we need to develop a microscopic picture of the equilibrium phenomenon itself. Equilibrium is a dynamic thing, and if we think of a liquid in equilibrium with its vapor, many molecules are continuously evaporating and an equal many molecules are simultaneously condensing. And so it also is with the sublimation of a solid, or in the transition of one solid phase into another solid phase, or in a chemical conversion wherein atomic groupings are being rearranged, as in an isomerization. In all of these there is a kinetic balance at equilibrium. At a molecular level, single-component equilibrium processes correspond to two sets of energy levels wherein species are free to distribute themselves between them. This is schematically evident in Fig. 7.1, wherein we might think of state A as the energy levels available to the molecules of a solid and state B as those available to the molecules of gas. And we have drawn the density of states as much greater for the gas which is, of course, the case.

Let us interrupt this argument for a moment to consider the rather

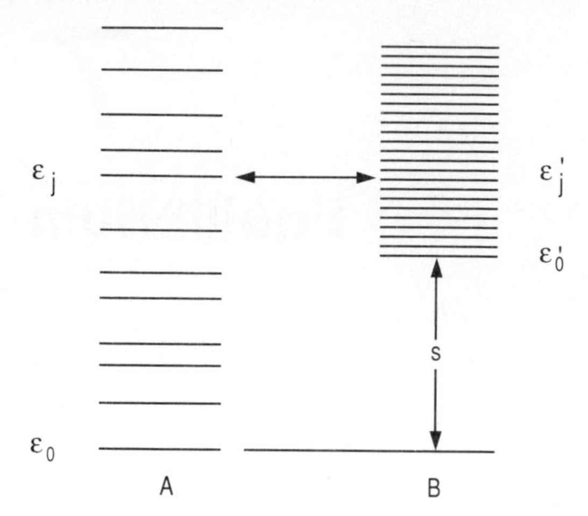

Figure 7.1 Schematic representation of equilibrium wherein species may occupy two sets of energy levels.

simpler case of two different substances in thermal contact with each other. Imagine, for example, a slab of copper in thermal contact with a slab of iron. Here the sets of energy levels are different for the two species and there can be no intercombinations, but heat can flow from one to the other. Imagine each system to be in its most probable exponential distribution according to its particular temperature. Equilibrium will exist when the combined system has achieved that pair of sets of occupation numbers that can occur in the most number of ways, and since each set of occupation numbers in the copper (A) is independent of the arrangement in iron (B), we seek the distribution in A and in B that will together maximize the product of W_A and W_B. That is, in the single system we sought $\delta \ln W = 0$ while in the combined system we seek $\delta \ln W_A W_B = 0$. The product, $W_A W_B$, appears, for clearly any single arrangement of A can coexist with all possible arrangements of B. Recall that for a single system, $\delta \ln W = \delta Q/kT$, so for the combined system, $\delta \ln W_A + \delta \ln W_B = \delta Q(1/kT_A - 1/kT_B)$, or T_A must equal T_B at equilibrium. And we would have been disappointed if the result had been anything else.

Now if we have the same species, but distributed between two alternate sets of energy levels as, for example, when molecules may be in either the solid or the gaseous phase, again, and for the same reasons, equilibrium will exist when

$$\delta \ln W_A W_B = 0 \tag{7.1}$$

Since the statistics, or the way of counting, is different depending on the phase and the nature of the particles, we must make some specification before proceeding with the implications of this fundamental criterion for equilibrium, $\delta \ln W_A W_B = 0$.

Vapor Pressure of a Solid

Let us first consider a pure monatomic solid in equilibrium with its vapor. We keep the total number of molecules constant; i.e., $\delta n_v = -\delta n_s$ or an increase in the number of molecules in the vapor is accompanied by a corresponding decrease in the number of molecules in the solid. We also keep the total internal energy constant; i.e., $U_v + U_s$ is fixed. The question, of course, is how do the atoms distribute themselves between the two phases. Note that the idea of constant total energy rigorously demands that the two partition functions for the solid and the vapor be referred to the same zero of energy, for otherwise the movement of molecules between adjacent levels, say from ε_j to ε_j' in distributions A and B of Fig. 7.1 would represent a change in total energy, which, of course, it does not.

First, from the viewpoint of classical thermodynamics, we can write the Clausius-Clapeyron equation for the sublimation of solid as

$$\frac{d \ln p}{dT} = \frac{\Delta H(\text{sub})}{RT^2} = \frac{\Delta U + RT}{RT^2} \tag{7.2}$$

wherein we have neglected the volume of the solid relative to that of the vapor and we have taken the molar volume of the vapor as being adequately represented by RT/p. This is because we can rigorously evaluate the partition function for an ideal gas but not so rigorously for a real gas. The energy change upon sublimation at any temperature may be rigorously related to that at 0 K by

$$\Delta U(\text{sub}) = \Delta U_0(\text{sub}) + \int_0^T \Delta C_v \, dT$$

This may be inserted into Eq. (7.2) which, on integration, yields

$$\ln p = -\frac{\Delta U_0(\text{sub})}{RT} + \int \left[\frac{1}{RT^2}\int_0^T \Delta C_V \, dT\right] dT + \int \frac{dT}{T} + i$$

and, since the vapor is monatomic with only translational degrees of freedom,

$$\ln p = -\frac{\Delta U_0(\text{sub})}{RT} + \frac{5}{2}\ln T - \int \left[\frac{1}{RT^2}\int_0^T C_V(\text{sol}) \, dT\right] dT + i \tag{7.3}$$

An accurate value of $\Delta U_0(\text{sub})$ can be obtained from one measurement of the heat of sublimation together with heat capacity measurements on the solid phase down to very low temperatures. Some typical values of ΔU_0(sub) appear in Table 7.1. We can also calculate C_V(solid) as a function of temperature. With a monatomic solid, this calculation will require only one measurement of C_V(solid) to enable an evaluation of the single parameter in the theory (see Chap. 6). Measurements of vapor pressure as a function of temperature, or indeed at some one temperature, would allow an evaluation of the integration constant i, and then the vapor pressure of the solid at all temperatures would be calculable. How does this compare with the microscopic theory?

First, we recall from Chap. 2 that the Helmholtz free energy may be expressed in terms of the partition function as

$$A_v = -n_v kT\left[\ln\left(\frac{f_v}{n_v}\right) + 1\right]$$

for the vapor phase, and as

$$A_s = -n_s kT \ln f_s$$

for the solid phase. At fixed temperature and volume of the mixture of two phases, equilibrium will exist when the Helmholtz free energy is minimized under the constraints of fixed total energy and fixed total number of molecules. Then,

$$A(\text{total}) = A(\text{vap}) + A(\text{sol})$$

and

$$\frac{\partial A(\text{total})}{\partial n_v} = -kT\left[\ln\frac{f_v}{n_v} + 1\right] + kT - (-1)kT\ln f_s = 0 \tag{7.4}$$

where we recognize (1) that n_v cancels from the ratio f_v/n_v and (2) that $\partial n_s/\partial n_v = -1$. Note that equating $\partial A(\text{total})/\partial n_v$ to zero is equivalent to equating the chemical potentials in the solid and vapor phases. To deduce the distribution of molecules among the two alternate sets of energy levels, the two partition functions must be evaluated relative to the same arbitrary zero of energy. With the lowest state of the solid assigned zero energy, each term in the partition function of the vapor

TABLE 7.1 Heat of Sublimation at 0 K for Several Monatomic Species

Element	ΔU_0(sub), J/mol	Element	ΔU_0(sub), J/mol
Li	150.5	Xe	16.1
Mg	144.1	Pt	521.7
Fe	403.8	Hg	64.6
Ni	410.6	Pb	195.5
Cu	342.0	Ag	290.3

must be multiplied by $\exp(-s/kT)$, where s is the difference between the lowest energy levels of each set of levels (see Fig. 7.1). Then Eq. (7.4) may be immediately rearranged to

$$f(\text{sol}) = \frac{f(\text{vap})}{n_v} e^{-s/kT} \tag{7.5}$$

For ease of notation, consider a monatomic species and an Einstein solid. Then,

$$f(\text{sol}) = g_s(1 - e^{-h\nu_E/kT})^{-3}$$

and

$$\frac{f(\text{vap})}{n_v} = g_v\left(\frac{2\pi mkT}{h^2}\right)^3\left(\frac{kT}{p}\right)e^{-s/kT}$$

Here g_s and g_v are the electronic ground-state multiplicities of the solid and vapor species, respectively. These multiplicities are not always the same, as is evident when one considers a metal with its free-electron gas (see Chap. 6). The Einstein lattice frequency of the solid is ν_E, and s is just the energy of sublimation at 0 K per molecule. With these partition functions, the statement of equilibrium, Eq. (7.5), becomes

$$\ln p = -\frac{s}{kT} + \frac{5}{2}\ln T + 3\ln(1 - e^{-h\nu_E/kT}) + \ln\left[\frac{g_v}{g_s}\left(\frac{2\Pi m}{h^2}\right)^{3/2} k^{5/2}\right] \tag{7.6}$$

In general, of course, $U - U^0 = nkT^2(\partial \ln f/\partial T)_V$ and $C_V = (\partial U/\partial T)_V$, and we note with a little algebra that the third term in the above expression is identically equivalent to the third term in the wholly classical expression for the vapor pressure, Eq. (7.3). The difference in ground-state energies of the two sets of levels is exactly the sublimation energy at 0 K, that is, $s = \Delta U_0(\text{sub})/n_A$, and the last term of Eq. (7.6) then represents an evaluation of a constant of integration that appeared in our purely classical thermodynamic development by using the ideas of molecular engineering.

It is also interesting to note that the integration constant is, of course, independent of temperature, and the vapor pressure equation [Eq. (7.6)] is applicable at all temperatures for monatomic species under the assumptions of an Einstein solid and an ideal vapor.

When the temperature is sufficiently low that the intramolecular vibrators are frozen out, the vapor pressure constant for all diatomic and linear polyatomic species becomes

$$i = \ln\frac{g_v}{g_s}\left(\frac{2\pi m}{h_2}\right)^{3/2}\frac{8\pi^2 I k^{7/2}}{\sigma h^2} \tag{7.7}$$

The vapor pressure is then written

$$\ln p = \frac{s}{kT} + \frac{7}{2} \ln T + 5 \ln (1 - e^{-h\nu_E/kT}) + i \tag{7.8}$$

Here the term ⁷⁄₂ ln T appears rather than ⁵⁄₂ ln T for the monatomic species because of the temperature dependence of f(rot). This expression also uses an effective ν_E to represent both the three oscillations of the center of mass of the linear species about its equilibrium lattice site and its two rocking motions about the two axes around which it has a moment of inertia (see the discussion of the heat capacity of molecular solids in Chap. 6).

Exactly similarly for a rigid nonlinear polyatomic species, the vapor pressure constant is

$$i = \ln \frac{g_v}{g_s} \left(\frac{2\pi m}{h^2} \right)^{3/2} \left(\frac{\pi^{3/2} (8\pi^2)^{3/2} (I_A I_B I_C)^{1/2} k^4}{\sigma h^3} \right) \tag{7.9}$$

and the vapor pressure is written

$$\ln p = -\frac{s}{kT} + 4 \ln T + 6 \ln (1 - e^{-h\nu_E/kT}) + i \tag{7.10}$$

Table 7.2 presents several comparisons of the experimental vapor pressure constants with the purely theoretical values. These constants are for Eqs. (7.6), (7.8), or (7.10), but in terms of the logarithm to base 10 for the vapor pressure in units of atmospheres. The electronic ground state of Na, K, etc. is a ^{2}S; g_v is, of course, 2, and this is the value used in evaluating i(calc). However, in each instance, the electronic multiplicity of Na or K in the solid state has been taken to be one. In species such as sodium, when in the solid phase, the single

TABLE 7.2 Some Experimental and Theoretical Vapor Pressure Constants for log *p* in atmospheres

Species	g_v	$I \times 10^{47}$, kg · m^2	i(calc)	i(exp)
Hg	1		1.866	1.83 ± 0.03
Ne	1		0.37	0.39 ± 0.04
Na	2		0.756	0.78 ± 0.1
N_2	1	13.8	− 0.18	− 0.16 ± 0.03
O_2	3	99.15	0.53	0.55 ± 0.02
Cl_2	1	116	1.35	1.66 ± 0.08
NH_3	1	2.815	− 1.55	− 1.50 ± 0.04
CH_4	1	5.3	− 1.94	− 1.97 ± 0.05
CO	1	15.0	− 0.141	− 0.07 ± 0.05

SOURCE: Adapted from H. S. Taylor and S. Glasstone, *Physical Chemistry*, Vol. 1, Van Nostrand, New York, 1942.

unpaired electron is free to roam the volume of the metal, leaving behind a sodium ion, Na^+, having a rare gas configuration of $g_s = 1$, or a ^1S state, which would appear much as does neon. It is this resulting free-electron gas that gives rise to the electrical conductivity of a metal. We had also accounted for the heat capacity of this free electron gas in molecular theories of the heat capacity of metals (see Chap. 6).

It is interesting that the value of i(calc) in Table 7.2 for oxygen, which agrees well with i(exp), has been obtained using $g_s = 1$ rather than 3. The explanation of this is uncertain.

It is also interesting that the data on CO were obtained using $g_s = 2$. This may be explained by realizing that CO has a weak dipole moment and the molecule is spatially similar on both ends. This value is then the multiplicity of the arrangement of molecules in the lattice. Actually, the multiplicity within the lattice is always the product of the electronic multiplicity of the ground state and the multiplicity, if any, in the physical or geometric arrangement. Thus, when the crystal forms, the CO molecules may enter the crystal as CO or as OC; that is, there will be disorder in solid CO. If this disorder were completely random, it would produce an entropy of $-R \ln 0.5$ or 5.76 J/mol · K. Third-law studies yield a value of 193.38 J/mol · K, while the statistical value based on an ordered solid is 198.03 J/mol · K, or 4.65 J/mol · K greater. So the disorder in the crystal is not random (see Chap. 6); some order to the approximate extent of 1.11/5.76 or 19 percent exists. A similar situation occurs with NO. Monodeuteromethane, CH_3D, can assume four orientations in the solid which would suggest a residual entropy of $-R \ln 0.25$, or 11.53 J/mol · K. The experimental entropy is actually 11.72 J/mol · K less than the statistical value.

The molecular description of any physical equilibrium is developed analogously to that just shown for the vapor pressure of a solid. The above development was simple and it gave good predictions of the experimental facts. But we note that this is true because of our rather restricted view in which the partition functions were both simple and accurate. Generalization from the ideal gas to the real gas in our example would have quickly produced enormous difficulties.

High- and Low-Temperature Modifications of a Solid

The equilibrium between two modifications of a solid is of a totally different kind than that between a solid and its vapor because of the lack in the solid-solid case of a difference in the partition function. Let Fig. 7.1 represent a solid phase transition like, for example, that from rhombic to monoclinic sulfur. The fundamental condition of equilibrium $\delta \ln W_A W_B = 0$ states as before

$$f_A = f_B e^{-s/kT} \tag{7.11}$$

where f_A and f_B are the partition functions for the lower energy phase A and the higher energy phase B respectively. Each partition function is based on its own zero of energy, and the two zeros are exactly s J apart. The transition temperature from A to B occurs at a fixed temperature, given rigorously by

$$T = \frac{s}{k \ln (f_B/f_A)} \tag{7.12}$$

The solid phase of lowest ε_0 is obviously the modification that exists at the lower temperature. Note also that, since s is always a positive number, a real transition temperature can exist only if f_B is greater than f_A. So another insight into problems of equilibrium may be inferred. If equilibrium is to exist, the partition functions associated with each set of levels must be equal when referred to the same energy zero as is stated by Eq. (7.11). This can occur only if the high-temperature modification has a greater density of energy levels. That is, the greater density of levels compensates for the higher zero of energy, and if this were not the case, the transition from one set of levels to another would not occur at any temperature.

This state of affairs is characteristic of all equilibria. Consider the equilibrium phenomenon of solubility. If there are N sites in the solution where the solute molecule can have an energy more or less corresponding to a single level in the pure solute, the density of states is high in solution, and the solute will dissolve even though the energy difference between the ground states in the pure solute and solution situations may be even several times kT.

Adsorption-Desorption Equilibria

The adsorption of molecules from the gas phase onto the surface of a solid can be developed from a molecular point of view through the partition functions. If we can write the partition functions of the gaseous phase and the adsorbed phase, we can develop the chemical potential of species in each phase. These may then be equated to give a perspective on the extent of adsorption as a function of T and p. For ease of consideration, let us imagine adsorption from a pure ideal gas. The chemical potential here is just the molar Gibbs free energy,

$$\mu^0(T,p) = G^0(T,p) = -n_A kT \ln \left[g_0 \left(\frac{2\pi mkT}{h^2} \right)^{3/2} \left(\frac{kT}{p} \right) f_{\text{vib,rot}} \right] + n_A u \tag{7.13}$$

where we have taken the gas to be ideal and we have set the zero of energy at the lowest energy of the adsorbed species. This adsorption energy is u J/molecule.

The partition function for the adsorbed species is more complex. Let us imagine that adsorbed molecules are free to move over the surface of the solid. That is, we imagine a two-dimensional gas. The factor V/n in the translational partition function of the three-dimensional gas becomes A/n_s or σ_0, or the area per site available for adsorption, rather than the volume per molecule. The truthfulness of this substitution will be evident upon reflecting on the original development of the translational partition function in Chap. 3. We also imagine an Einstein vibrator perpendicular to the surface arising from the van der Waals bonding to the surface. This bond is much weaker than the intramolecular bonds, and its characteristic frequency is low. The resulting contribution to the partition function can then be approximated,

$$f(\mathrm{ad}) = (1 - e^{-h\nu/kT})^{-1} \cong \frac{kT}{hc\omega} \tag{7.14}$$

Finally, there is a degeneracy on the surface that must be included in the total partition function just as, for example, the ground-state electronic degeneracy g_0 appears as a multiplicative factor. There are n_s total sites on the surface, but at equilibrium we assume that only n_a are occupied. How many distinguishable ways may these n_a sites be filled? There are n_s choices for the first adsorbed species, $(n_s - 1)$ for the second, etc., or a total of $n_s!/(n_s - n_a)!$ ways to place n_a species on n_s sites. But all species are identical, so this total must be divided by the total number of possible permutations, or $n_a!$. The degeneracy of the adsorbed n_a species is then

$$\frac{(n_s)!}{(n_s - n_a)!(n_a)!}$$

With all of this as background, we can now write the molar Gibbs free energy of the adsorbed phase as

$$G_{\mathrm{ad}}(T, p) = -n_A kT \ln\left[\left(\frac{n_s!}{(n_s - n_a)!n_a!}\right) g_0 \left(\frac{2\pi mkT}{h^2}\right)\left(\frac{kT}{hc\omega}\right)(\sigma_0) f_{\mathrm{vib,rot}}\right] \tag{7.15}$$

With Stirling's approximation, this can be written

$$G_{\mathrm{ad}}(T, p) = -n_A kT \ln\left[\left(\frac{n_s - n_a}{n_a}\right) g_0 \left(\frac{2\pi mkT}{h^2}\right)\left(\frac{kT}{hc\omega}\right)(\sigma_0) f_{\mathrm{vib,rot}}\right]$$

Just as the chemical potential for a pure gas is the molar Gibbs free energy, so the chemical potential of a pure adsorbed species is the molar Gibbs free energy of that adsorbed species, i.e.,

$$\mu_{\mathrm{ad}}(T, p) = G_{\mathrm{ad}}(T, p)$$

Equating μ_g^0 to μ_{ad} and rearranging, one obtains

$$\frac{\theta}{1-\theta} = \left[\left(\frac{h^2}{2\pi mkT}\right)^{1/2} \frac{\sigma_0}{hc\omega} e^{U/RT}\right] p \tag{7.16}$$

where θ is the fraction of surface that is covered, or $\theta \equiv n_a/n_s$. This is the equation for the well-known Langmuir adsorption isotherm,

$$\frac{\theta}{1-\theta} = [C]p \tag{7.17}$$

where we have produced a theoretical evaluation or, better, theoretical interpretation of the constant. The bonding or adsorption energy U is typically 1 to 10 kcal/mol. The Langmuir model certainly oversimplifies the nature of bonding onto a surface. We have ignored lateral bonding, multilayer formation, the dependence of u on temperature, and more. Nevertheless, the model remains generally useful. It is also the starting place for more sophisticated treatments.

Chemical Equilibria

Chemical equilibria may also be understood from the point of view of molecular engineering. Consider first a single reaction occurring in an ideal gas phase. Both of these restrictions will be subsequently relaxed. The problem is one of predicting the extent of reaction at equilibrium. As an example, consider the dissociation of a perhaps complex molecule into two parts,

$$\nu_{AB}AB \leftrightarrows \nu_A A + \nu_B B \tag{7.18}$$

where the ν_i are stoichiometric coefficients.

From classical thermodynamics, we know that equilibrium will exist when the Gibbs free energy is minimized. From Chap. 2, the free energy of each pure species is

$$G = -nkT\left(\ln\frac{f}{n} + 1\right) + nkTV\left(\frac{\partial \ln f}{\partial V}\right)_T$$

If the number of molecules, n, in this equation is Avogadro's number, G will have units of J/mol. Then for our reaction, Eq. (7.18), in an ideal gas phase of however many moles,

$$G^0(\text{mix}, T, p) = \Sigma n_i G_i^0(T, p) + \Delta G^0(\text{mix})$$

or $$G^0(\text{mix}) = -kT\sum_i n_i \ln\left[\left(\frac{2\pi m_i kT}{h^2}\right)^{3/2}\left(\frac{kT}{p}\right) f_{i,\text{vib,rot}}\right] + kT\sum n_i \ln\left(\frac{n_i}{n}\right) \tag{7.19}$$

This $G^0(\text{mix})$ has its minimum value at equilibrium. Here $n \equiv \Sigma_i n_i$, and p is the total pressure of the mixture. The requirement of minimum free energy also implies that

$$\Sigma \nu_i \mu_i = 0 \tag{7.20}$$

at equilibrium, where μ_i is the chemical potential. The extent of reaction at equilibrium may be developed either by a direct minimization of Eq. (7.19) or from Eq. (7.20), which leads to what is called the law of mass action. The choice of techniqué is one of computational convenience. For single reactions or perhaps for two simultaneous reactions, Eq. (7.20) is frequently more convenient when the need is for only one or two calculations. For more thorough studies of conversion and certainly for more complex systems, the direct minimization will be preferable. For our simple dissociation reaction,

$$\mu_A^0 = \left(\frac{\partial G^0(\text{mix})}{\partial n_A}\right)_{T,p,n_j} = -kT\ln\left[\left(\frac{2\pi m_A kT}{h^2}\right)^{3/2}\left(\frac{kT}{p}\right) f_{A,\text{ vib, rot}}\left(\frac{1}{x_A}\right)\right] \tag{7.21}$$

and similarly for the other two species. Substituting these three expressions for the chemical potential into Eq. (7.20), we obtain

$$\frac{(x_A p)^{\nu A}(x_B p)^{\nu B}}{(x_{AB} p)^{\nu AB}} = \frac{\left[\left(\dfrac{2\pi mkT}{h^2}\right)^{3/2} kTf_{\text{vib,rot}}\right]_A^{\nu A}\left[\left(\dfrac{2\pi mkT}{h^2}\right)^{3/2} kTf_{\text{vib,rot}}\right]_B^{\nu B}}{\left[\left(\dfrac{2\pi mkT}{h^2}\right)^{3/2} kTf_{\text{vib,rot}}\right]_{AB}^{\nu AB}} \tag{7.22}$$

which we may compare with the result from purely classical thermodynamic arguments,

$$\Delta G^0 = -RT\ln K_p = -RT\ln\left[\frac{(x_A p)^{\nu A}(x_B p)^{\nu B}}{(x_{AB} p)^{\nu AB}}\right] \tag{7.23}$$

where $x_A p$ etc. within the square bracket are partial pressures. Clearly then,

$$K_p = \frac{\left[g_0\left(\dfrac{2\pi mkT}{h^2}\right)^{3/2} f_{\text{vib,rot}}\right]_A^{\nu A}\left[g_0\left(\dfrac{2\pi mkT}{h^2}\right)^{3/2} f_{\text{vib,rot}}\right]_B^{\nu B}(kT)^{\nu A+\nu B-\nu AB}}{\left[g_0\left(\dfrac{2\pi mkT}{h^2}\right)^{3/2} f_{\text{vib,rot}}\right]_{AB}^{\nu AB}} \tag{7.24}$$

where $f_{\text{vib,rot}}$ is the internal partition function for each species which arises from rotation, vibration, and the internal rotation of the molecule.

There is, however, an essential part of the partition function of each species that has been omitted, and that arises from the necessity that each of the separate partition functions of the species be referred to the same energy zero. (See earlier discussion about physical equilibrium). That is, just as before, the energy levels of each partition function must be reckoned relative to the same zero of energy if we are to determine that distribution among alternate sets of levels that can occur in the most number of ways. That distribution is called, in this case, chemical equilibrium. The choice of zero is, as always, arbitrary, and we might imagine the separated atoms at rest as having zero energy, as is schematically evident in Fig. 7.2. In this scheme, all molecules are exothermic, since all atoms, specifically here atoms of C, H, and O, N, and S descend in total energy on formation of molecules of CO, H_2O, etc. In Chap. 8 we will discuss a quantum mechanical scheme for the calculation of the total energy of a molecule relative to the separated atoms, but certainly the more generally accepted and useful scheme is to define the zero of energy, or, more properly, the zero of enthalpy of the elements, not as separated atoms, but for their normal state at 25°C. Then a schematic of the relative enthalpies would appear as in Fig. 7.3, where we see both endothermic and exothermic molecules relative to the elements in their standard states. Clearly the energies of the schemes of Figs. 7.2 and 7.3 are simply related through quantities such as the dissociation energy of H_2 and the heat of sublimation of graphite. And this is the reason why such quantities are of such significance in thermochemistry.

For our example of a simple dissociation in the gaseous phase, the partition function of species A is then

$$f(A) = g_0\left(\frac{2\pi mkT}{h^2}\right)^{3/2} V f_{\mathrm{vib,rot}} e^{-\Delta H_0^0/RT} \tag{7.25}$$

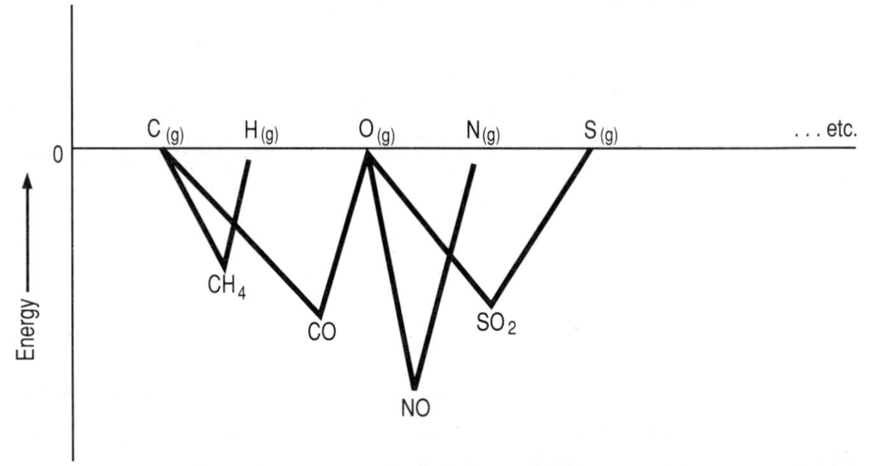

Figure 7.2 Schematic representation of relationships between ground state energies of gaseous atoms and molecules (not to scale).

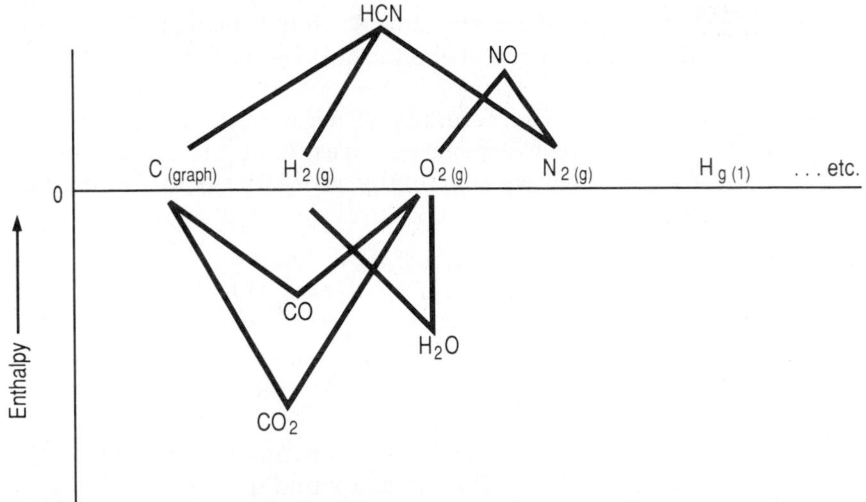

Figure 7.3 Schematic representation of relationships between the energies of the ground-state elements in their standard states and gaseous molecules (not to scale).

where ΔH_0^0 is the heat of formation of A from its atomic elements, all as ideal gases at 0 K. And this ΔH_0^0 will always have a negative sign; that is, all molecules are exothermic relative to the separated atoms. Also note that this ΔH_0^0 is exactly the ΔU_0^0 of formation at 0 K, for

$$\Delta H_0^0 \equiv \Delta U_0^0 + (pV)_0^0 = \Delta U_0^0 + RT = \Delta U_0^0$$

For the more usual definition of the energy zeros, the partition function of A is written exactly as before, but now ΔH_0^0 is the heat of formation at 0 K from the elements in their perhaps more usual standard states, for example,

$$\text{C (graphite)} + 2\text{H}_2\text{ (g)} \rightarrow \text{CH}_4\text{ (g)}$$

rather than from C and H atoms as ideal gases at 0 K. The heats of formation of species will be different depending on the arbitrary choice of the energy zeros of the elements, i.e., their standard states. The heats of formation of CH_4 (g) from graphite and from atomic C will obviously be different. Now for some reaction of interest, say our dissociation reaction, the partition function of each species will be multiplied by exp [− heat of formation at 0 K/RT], which ensures that each energy level of the species has its value relative to the common zero of energy. In formulating the equilibrium constant for some reaction, these exponential terms will obviously add as

$$[\Sigma \nu_i \Delta H_0^0(\text{products})_i - \Sigma \nu_i \Delta H_0^0(\text{reactants})_i]$$

which is merely the heat of reaction at 0 K. Note that regardless of the arbitrary standard state of the elements, this heat of reaction $\Delta H_0{}^0$ is invariant.

With this insight into the necessity of a common zero if multiple sets of energy levels are available as a result of the molecular arrangements of the atoms, we can finally write the equilibrium constant, Eq. (7.24), as

$$K_p = \frac{\left[g_0\left(\frac{2\pi mkT}{h^2}\right)^{3/2} f_{\text{vib,rot}}\right]_A^{\nu_A}\left[g_0\left(\frac{2\pi mkT}{h^2}\right)^{3/2} f_{\text{vib,rot}}\right]_B^{\nu_B}(kT)^{\nu_A+\nu_B-\nu_{AB}}}{\left[g_0\left(\frac{2\pi mkT}{h^2}\right)^{3/2} f_{\text{vib,rot}}\right]_{AB}^{\nu_{AB}}} e^{-\Delta H_0^0/RT} \tag{7.26}$$

where $\Delta H_0{}^0$ in this equation is the heat of reaction at 0 K, that is, the sum of the heats of formation of each compound at 0 K.

Such heats of formation, or heats of any reaction for that matter, are calculated from only one experimental measurement at some more reasonable temperature using

$$\left(\frac{\partial \Delta H}{\partial T}\right) = \Delta C_p$$

or $$\Delta H_0^0 = \Delta H^0(T)_{\text{exp}} - \int_0^T \Delta C_p^0\, dT = \Delta H^0(T)_{\text{exp}} - \sum_i \nu_i(H^0 - H_0^0)_i$$

The heat capacities or enthalpies are available from molecular thermodynamics. As we will see in Chap. 8, it is even possible to calculate such heats of reaction from quantum mechanics and frequently to within chemical accuracies. Thereby we escape the need of any experimental calorimetric data at all.

As a more specific example, we note that the iodine atom laser has been proposed as the basis of a laser weapons system. Let us calculate the equilibrium dissociation of I_2 as a function of temperature. Here $\omega = 215\ \text{cm}^{-1}$, $B = 1130$ MHz; the spectroscopic dissociation energy is 1.5417 eV; by inspection, $\sigma = 2$; $g_0(I_2) = 1$; and for the atoms in the ^2P ground state $J = 3/2$ and $1/2$; and therefore since $g = 2J + 1$, $g_0 = 4$ and $g_1 = 2$. The low-lying electronic state of the atom may be populated at the high temperatures required to significantly dissociate the molecule,

$$f_{\text{elect}} = g_0 + g_1 e^{-\varepsilon_1/kT} + \cdots = 4 + 2e^{-7603hc/kT} + \cdots$$

where the first excited energy level is at 7603 cm^{-1}. The temperature must be 2800 K before this higher term can contribute even 1 percent to the value of f_{elect}. The electronic partition function is then essentially equal to 4 at temperatures of 800 to 2000 K. Using this value and simplifying Eq. (7.26) to be specific to the iodine dissociation, we get

$$K_p = \frac{32B\,(\pi m_I kT)^{3/2}}{h^2}(1 - e^{-h\nu/kT})e^{-D/kT} \tag{7.27}$$

At 1000°C, we deduce $K_p = 1.78 \times 10^4$ Pa or 0.176 atm, which compares well with a value of 0.165 atm from measurements of equilibrium composition of the dissociating vapor at several temperatures.

Reaction equilibrium may be deduced by first calculating the ΔG^0(reaction) using the techniques of molecular engineering as discussed later in this chapter. However, it is possible to gain interesting insights into the relationship between molecular properties and reaction equilibrium by casting the problem as the equilibrium constant in terms of the several partition functions as exemplified by Eq. (7.26). The required input data are the same for both procedures. The approach is one of personal preference.

As an example of an equilibrium between more complex molecules, consider a linear molecule reacting with a nonlinear species to give a linear and a nonlinear product (for ease of notation, we take all stoichiometric coefficients to be 1):

$$A_i + B_j \leftrightarrows C_k + D_1$$

Then,

$$K_p = \frac{g_C g_D}{g_A g_B} \times \frac{\left(\frac{2\pi m_C kT}{h^2}\right)^{3/2}\left(\frac{2\pi m_D kT}{h^2}\right)^{3/2}}{\left(\frac{2\pi m_A kT}{h^2}\right)^{3/2}\left(\frac{2\pi m_B kT}{h^2}\right)^{3/2}}$$

$$\times \frac{\frac{8\pi^2 I_C kT}{\sigma_C h^2}\cdot\frac{\pi^{1/2}(8\pi^2 kT)^{3/2}(I_A I_B I_C)_D^{1/2}}{\sigma_D h^3}}{\frac{8\pi^2 I_A kT}{\sigma_A h^2}\cdot\frac{\pi^{1/2}(8\pi^2 kT)^{3/2}(I_A I_B I_C)_B^{1/2}}{\sigma_B h^3}}$$

$$\times \frac{\prod(1 - e^{-h\nu_i/kT})_C^{-1}(1 - e^{-h\nu_i/kT})_D^{-1}}{\prod(1 - e^{-h\nu_i/kT})_A^{-1}(1 - e^{-h\nu_i/kT})_B^{-1}}\,e^{-\Delta H_0^0/RT} \tag{7.28}$$

Simplifying, we get

$$K_p = \frac{g_C g_D}{g_A g_B} \times \left(\frac{m_C m_D}{m_A m_B}\right)^{3/2} \times \frac{\sigma_A \sigma_B}{\sigma_C \sigma_D} \times \frac{I_C (I_A I_B I_C)_D^{1/2}}{I_A (I_A I_B I_C)_B^{1/2}}$$

$$\times \prod(1 - e^{-h\nu_i/kT})_{A,B}(1 - e^{-h\nu_i/kT})_{C,D}^{-1}\,e^{-\Delta H_0^0/RT} \tag{7.29}$$

and we see the simple dependence of K_p on ratios of the several microscopic characteristics of the reactants and products. Equation (7.29) allows a number of interesting inferences. For instance, other things being equal, a conversion will be greater if the reactants are more symmetric than the products; that is, K_p is larger if there is a decrease in symmetry in the reaction. Similarly, K_p is greater if the products have a greater electronic multiplicity than do the reactants. The moment of inertia plays a greater role if the species involved are linear, but, whatever, K_p is greater if the products have larger moments, that is, if the products are larger and structurally more complex than are the reactants. The conversion will be greater, the smaller are the vibration frequencies of the product molecules relative to those of the reactant molecules. If a product molecule has a free rotation while the reactants do not, the conversion will be greater, all other factors notwithstanding. It is interesting to note that the equilibrium constant varies with temperature as $e^{-\Delta H_0^0/RT}$ and that the temperature variation of the conversion is different from this only to the extent that the internal partition function of each species varies with temperature.

It is impressive that the molecular formalism enables an accurate calculation of an equilibrium. In the dissociation of I_2, we have required only the rotational constant of the molecule, the vibration frequency of the molecule, the mass of the atom, and the dissociation energy of the molecule to allow a calculation of the equilibrium dissociation at all temperatures. In Chap. 8 we will see something of the successes of at least one quantum-mechanical scheme in calculating B, ω, and D from first principles. This, in a sense, represents one of the ultimate aims of theoretical chemistry, that is, the calculation of the equilibrium extent of reaction for any species under any conditions. And impressive successes have been achieved.

Example Problem 7.1 Calculate the equilibrium constant for the dissociation of cyanogen at 2100 K. The heats of formation at 0 K of C_2N_2 and CN are 73.428 and 103.2 kcal/mol respectively (JANAF).

$$C_2N_2 \leftrightarrows 2CN$$

and
$$K = \left[\frac{g^2_{\mathrm{CN,\,elect}}}{g_{\mathrm{C_2N_2,elect}}}\right]\left[\frac{(2\pi m_{\mathrm{CN}}kT/h^2)^3}{(2\pi m_{\mathrm{C_2N_2}}kT/h^2)^{3/2}}\right]\left[\frac{(8\pi^2 I_{\mathrm{CN}}kT/\sigma_{\mathrm{CN}}h^2)^2}{(8\pi^2 I_{\mathrm{C_2N_2}}kT/\sigma_{\mathrm{C_2N_2}}h^2}\right]$$

$$\times\left[\frac{\prod_{i=1}^{7}(1-e^{-h\nu_i/kT})_{\mathrm{C_2N_2}}}{(1-e^{-h\nu/kT})^2_{\mathrm{CN}}}\right](kT)^{2-1}e^{-\Delta H_0^0/RT}$$

The ground-state multiplicity of CN is 2, and there is a low-lying state at 9118 cm^{-1} with a multiplicity of 4. At 2100 K, the electronic partition function for CN is then 2.0077. The low-lying state is insignificant, contributing 0.4 percent at 2100 K. The ratio of electronic partition functions is 4.031, the ratio of translational partition functions becomes $[(26)^2 10^{-3}/52]^{3/2}[2\pi kT/n_A h^2]^{3/2}$ or 8.474×10^{32}, the ratio of symmetry numbers is $(2)/(1)^2$ or 2, the ratio of moments is 1.216×10^{-47}, and the remaining constant in the rotational partition function, $[8\pi^2 kT/h^2]$, is $5.214 \times 10^{48}\ \mathrm{J^{-1} \cdot s^{-2}}$. The single frequency of CN is 2069 $\mathrm{cm^{-1}}$, so the denominator in the above expression for K becomes 0.7579. There are three nondegenerate frequencies of C_2N_2: 2322 $\mathrm{cm^{-1}}$, 848 $\mathrm{cm^{-1}}$, and 2149 $\mathrm{cm^{-1}}$; and two doubly degenerate frequencies: 506 $\mathrm{cm^{-1}}$ and 226 $\mathrm{cm^{-1}}$. The term in the numerator of the above expression for K that arises from each of these frequencies is 0.7964, 0.4408, 0.7708, 0.2931, and 0.1435 respectively. The heat of reaction at 0 K is $\Delta H_0^0 = 2(103.2) - 73.428 = 132.97$ kcal/mol. With all quantities in SI units, the units of K will be $\mathrm{N/m^2}$ or Pa.

Inserting all of these values, we obtain

$$K = (4.031)(8.474 \times 10^{32})[(2)(1.216 \times 10^{-47})(5.214 \times 10^{48})]$$
$$\times \left[\frac{(0.7964)(0.4408)(0.7708)(0.2931)^2(0.1435)^2}{(0.7579)^2}\right](kT) \exp\left[\frac{-132{,}970}{1.987(2100)}\right]$$
$$= (4.031)(8.474 \times 10^{32})(12.68 \times 10^{1})(8.333 \times 10^{-4})$$
$$\times (2.9 \times 10^{-20})(1.447 \times 10^{-14})$$
$$= 1.515 \times 10^{-1}\ \mathrm{Pa}$$
$$= 1.515 \times 10^{-1}\ \mathrm{Pa} \times \frac{1\ \mathrm{atm}}{101.3 \times 10^3\ \mathrm{Pa}} = 1.495 \times 10^{-6}\ \mathrm{atm}$$

This is 2 orders of magnitude smaller than an earlier value of 1.54×10^{-4} atm, largely because of a now very different value of the heat of reaction. This value of ΔH_0^0 appears exponentially, of course, in the expression for K, so its value is critical (see Rutner[1]).

Ionic Equilibria

The equilibrium degree of ionic dissociation is important in some lasers, in magnetohydrodynamics (MHD), in electric propulsion in space, and the like. Ionic seeding of hot gases in MHD problems frequently employs cesium due to its very low ionization energy of $D = 3.88$ eV. For this ionization, $g(\mathrm{Cs}) = 2$, $g(e) = 2$, $g(\mathrm{Cs}^+) = 1$, and the equilibrium constant may be written

$$K_p = \left(\frac{2\pi mkT}{h^2}\right)^{3/2}(kT)e^{-D/kT} \tag{7.30}$$

where m is the mass of the electron. At the high temperature required for ionization, kT must be of such magnitude relative to D that the exponential has a significant value and we can be confident in our assumption that the gas phase is ideal. Temperatures sufficiently high to produce ionization are also and clearly sufficiently high to produce electronic excitation as well, and the electronic partition functions for the charged and the uncharged species may each be of larger magnitude than just the ground-state multiplicity, g_0. A summary of the results for hydrogen atoms at a series of pressures and temperatures is given in Table 7.3, wherein the maximum principle quantum number due to the large charge density has been estimated by using a semi-empirical formalism. It is clear that the enthalpy of the excited atoms is an insignificant fraction of the total enthalpy. From a molecular thermodynamic point of view, the atom would tend to be ionized rather than electronically excited, and indeed this ionization is essentially complete at temperatures low enough for I to be somewhat greater than kT, that is, at temperatures corresponding to a very small Boltzmann factor. This phenomenon is best considered as an entropy effect. For example, hydrogen at 1 atm and 15,000 K is over half ionized, yet I/kT is somewhat greater than 10. Although e^{-10} is a small number, the ΔS of the reaction is some + 23 cal/K, which will yield a ΔG of some − 3 K/cal. Such values are characteristic of promising reactions.

When it is recalled that the radius of the atomic orbitals will vary approximately as the square of the principal quantum number, it is clear that the excited atoms will be enormous in size although small in number. The effect of these species on the transport properties of the plasma requires further study, but some general comments may be made. In kinetic theory, the transport coefficients of viscosity,

TABLE 7.3 Population Density of Electronically Excited Hydrogen Atoms in cm^{-3}

n_T is total particle density, n_{max} is the maximum principal quantum number, n_1 is the particle density of species with principal quantum number of 1, etc. Occupation levels less than $10^{-6} \times n_T$ have been neglected.

	Pressure, atm		
Temperature	10	1	0.01
8,000 K	$n_T = 9.18 \times 10^{19}$	$n_T = 9.18 \times 10^{17}$	$n_T = 9.18 \times 10^{15}$
	$n_1 = 9.17 \times 10^{19}$	$n_1 = 9.13 \times 10^{17}$	$n_1 = 8.75 \times 10^{15}$
	$n_2 = 1.38 \times 10^{14}$	$n_2 = 1.37 \times 10^{12}$	$n_2 = 1.32 \times 10^{10}$
	$n_{max} = 8$	$n_{max} = 10$	$n_{max} = 14$
10,000 K	$n_T = 7.34 \times 10^{19}$	$n_T = 7.34 \times 10^{17}$	$n_T = 7.35 \times 10^{15}$
	$n_1 = 7.29 \times 10^{19}$	$n_1 = 6.98 \times 10^{17}$	$n_1 = 4.68 \times 10^{15}$
	$n_2 = 2.11 \times 10^{15}$	$n_2 = 2.02 \times 10^{13}$	$n_2 = 1.36 \times 10^{11}$
	$n_3 = 5.31 \times 10^{14}$	$n_3 = 5.09 \times 10^{12}$	$n_3 = 3.41 \times 10^{10}$
	$n_4 = 4.38 \times 10^{14}$	$n_4 = 4.20 \times 10^{12}$	$n_4 = 2.81 \times 10^{10}$
	$n_5 = 4.80 \times 10^{14}$	$n_5 = 4.60 \times 10^{12}$	$n_5 = 3.08 \times 10^{10}$
	$n_{max} = 5$	$n_6 = 5.46 \times 10^{12}$	$n_6 = 3.66 \times 10^{10}$
		$n_7 = 6.62 \times 10^{12}$	$n_7 = 4.43 \times 10^{10}$
		$n_8 = 8.02 \times 10^{12}$	$n_8 = 5.37 \times 10^{10}$
		$n_{max} = 8$	$n_9 = 6.45 \times 10^{10}$
			$n_{10} = 7.68 \times 10^{10}$
			$n_{max} = 11$

SOURCE: From H. A. McGee, Jr., and G. Heller, *Progress in Astronautics and Aeronautics*, Vol. 9, *Electric Propulsion Development*, Academic Press, New York, 1963, p. 443.

thermal conductivity, and diffusivity vary as the mean free path (see Chap. 9). And the mean free path varies inversely as the square of the collision diameter. Thus larger molecules will have smaller mean free paths and therefore smaller transport coefficients.

As we saw in Chap. 3, entropy effects are so large that ionization is much more likely than electronic excitation. Therefore the gross effect of the electronically excited atoms on the thermodynamic properties is small. It has been shown by this analysis that detailed use of the electronic energy levels in forming the partition function is unnecessary for many purposes, such as determination of the enthalpy as a function of temperature and pressure. In addition, in regions where the excited state occupation is approaching even 1 part per 1000, the charge density has so depressed the ionization energy that most of the electronic levels have disappeared.

Temperature measurements are often made from the relative intensities of emission lines from transitions involving these relatively sparsely populated levels. It is then clear that the spectroscopic analysis involves and depends on a very small fraction of the total particle number density. In order to draw conclusions from measurements associated with the excited levels, these anomalously large species must also fulfill the requirement of thermal equilibrium with the bulk of the gas, which is, of course, the basic assumption of all thermodynamic calculations.

In calculations concerning plasmas, we understand that the free electron gas that is always present strictly obeys Fermi-Dirac statistics, while we have used Bose-Einstein statistics. There will never be a problem with these very hot gases, however, since the two statistics are indistinguishable when the number of available energy levels is much greater than the number of species, i.e., electrons, that are available to occupy these energy levels. This situation is always comfortably satisfied.

Isotopic Equilibria

Isotopic exchange is important in separation processes in the nuclear industry and in following the course of reactions in fundamental research in chemistry. Experiments suggest that the dissociation energies of isotopically substituted bonds are essentially equivalent, so that the heat of reaction is reasonably approximated by the appropriate combination of the zero-point vibrational energies. Let us consider the hydrogen-deuterium exchange

$$H_2 + D_2 \leftrightharpoons 2\mathrm{HD} \tag{7.31}$$

for which we may write the equilibrium constant as

$$K_p = \frac{g_{HD}^2}{g_{H_2}g_{D_2}} \frac{m_{HD}^2}{m_{H_2}m_{D_2}} \frac{\sigma_{H_2}\sigma_{D_2}}{\sigma_{HD}^2} \frac{B_{HD}^2}{B_{H_2}B_{D_2}} \frac{(1 - e^{-h\upsilon/kT})_{H_2} (1 - e^{-h\upsilon/kT})_{D_2}}{(1 - e^{-h\upsilon/kT})_{HD}^2}$$

$$\times \exp\left(\frac{h\upsilon_{HD} - \frac{1}{2}h\upsilon_{H_2} - \frac{1}{2}h\upsilon_{D_2}}{kT}\right) \qquad (7.32)$$

The electronic ground states are all singlets, the dissociation energies are essentially the same, and the appropriate molecular data on all three species are as follows:

	H_2	HD	D_2
m	2.01556	3.02131	4.02706
B, cm^{-1}	59.309	44.655	29.913
υ, cm^{-1}	4395.2	3817.1	3118.5
σ	2	1	2

We see here why the molecular calculation of equilibrium is at its best when applied to isotopic exchange reactions. The largest source of error in the calculation of a reaction equilibrium is the heat of reaction, ΔH_0^0. Fortunately, for isotopic exchanges the ΔH_0^0 is, to a good approximation, just the algebraic sum of the zero-point vibrational energies which are accurately determined spectroscopically.

With all of these molecular data in hand, we can write the expression for K_p in the hydrogen/deuterium exchange as

$$K_p = \frac{(3.02131)^2}{(2.01556\,(4.02706)} \left[\frac{(2)\,(2)}{(1)^2}\right] \left[\frac{(44.655)^2}{(59.309)\,(29.913)}\right]$$

$$\times \frac{(1 - e^{-h\upsilon/kT})_{H_2} (1 - e^{-h\upsilon/kT})_{D_2}}{(1 - e^{-h\upsilon/kT})_{HD}} \exp\left(\frac{\frac{1}{2}h\upsilon_{H_2} + \frac{1}{2}h\upsilon_{D_2} - h\upsilon_{HD}}{kT}\right) \qquad (7.33)$$

As will be immediately evident, the frequencies of all three vibrators are of such magnitude that the vibrational partition functions will make little or no contribution at temperatures of up to 1000 K, so neglecting them, the equilibrium constant may be written

$$K_p = 5.06e^{-86.4/T} \qquad (7.34)$$

At 670 K, we calculate $K_p = 4.45$, which may be compared with an experimental value from actual equilibrium measurements of 3.78. The lack of agreement is most likely due to inaccuracies in the equilibrium measurements, for the molecular data may be determined with high precision.

As one more, perhaps somewhat more complex, example of the calculation of equilibrium constants from molecular data, let us consider the technically important exchange reaction in the production of heavy water:

$$H_2O + HD \leftrightharpoons HDO + H_2$$

The molecular data for each species are summarized below:

	H_2O	HD	HDO	H_2
ω, cm^{-1}	3825	3817	2818	4395
	1654		1450	
	3935		3883	
σ	2	1	1	2
I,	0.6117	0.6264	0.7248	0.4716
$kg \cdot m^2 \times 10^{-47}$	1.1531		1.8422	
	1.7649		2.5667	
m	18.01556	3.02131	19.02131	2.01556

This exchange has been widely studied, so let us calculate K_p at 100°C and compare the result with that from equilibrium composition measurements. At 100°C, the vibrational partition function at $\omega = 1450$ cm^{-1} is 1.0037, and of course it is much more nearly equal to one for each of the other vibrators, which all have higher frequencies. Neglecting all of the vibrational contributions to the partition function, we write

$$K_p = \frac{(19.02131)(2.01556)}{(18.01556)(3.02131)} \frac{(2)(1)}{(1)(2)} \frac{[(0.7248)(1.8422)(2.5667)]^{1/2}\,(0.4716)}{[(0.6117)(1.1531)(1.7649)]^{1/2}\,(0.6264)}$$

$$\times \exp\left(\frac{hc}{2\,kT}\right)(3825 + 1654 + 3935 + 3817 - 2818 - 1450 - 3883 - 4395)$$

At any temperature below the onset of significant vibrational contributions, then,

$$K_p = 0.88e^{493/T} \tag{7.35}$$

and specifically at 100°C, $K_p = 3.30$, which compares with $K_p = 2.6$ from equilibrium composition measurements. This low-temperature solution represents no restriction, since it is only necessary to include the vibrational contributions at temperatures sufficiently high for that vibrator to become active.

Although one could imagine circumstances that would argue otherwise, it is interesting to note that equilibrium will favor the movement of a heavy isotope in a more simple molecule onto a more complex molecule. Stated the other way round, a heavy isotope within a more complex molecule will increase the moments and lower the frequencies to push the equilibrium toward that species.

All isotopic exchange equilibria may be handled similarly to those discussed here.

Chemical Equilibrium from Free Energy Calculations

Instead of expressing K_p in terms of the several partition functions, it is equally proper to calculate ΔG^0 for the reaction and evaluate K_p from exp $(-\Delta G^0/RT)$. In this perhaps more usual scheme,

$$\frac{\Delta G^0}{T} = \sum_i \frac{G^0 - H_0^0}{T} + \frac{\Delta H_0^0}{T} \tag{7.36}$$

wherein the quantites $(G^0 - H_0^0)/T$ are readily calculable, as has been explained in detail in Chap. 3, and these have even been tabulated in handbooks and computer-readable data bases for many species. The JANAF tables are extensive such compilations.[5] Note that this computational technique demands exactly the same input information as was required by the techniques discussed earlier in this chapter. The difference is merely the point of view that one brings to the problem, and advantages of one over the other are simply matters of opinion and choice. Depending on the particular problem, perhaps there is more insight or clarity or relationship to reality in one or the other scheme. The relative influence of the molecular parameters is more evident in the former procedure. The former procedure will also be more helpful in developing a theory of reaction kinetics, as we shall see in Chap. 10.

As an example of this more conventional computational scheme, consider the formation of acetylene from methane,

$$2CH_4 \rightleftarrows C_2H_2 + 3H_2$$

The free energies $(G^0 - H_0^0)/T$ are readily calculated by the techniques of Chap. 3. An experimental heat of combustion of each species may be readily reduced to the heat of formation of each species at 0 K. For example, for CH_4 at 298.15 K,

$$\Delta H_{(\text{comb})} = \Delta H_f^0\,(CO_2) + 2\,\Delta H_f\,(H_2O) - \Delta H_f^0\,(CH_4)$$

and

$$\Delta H_f^0(298.15) = \Delta H_0^0 + \sum_i (H^0 - H_0^0)_{298.15}$$

These sorts of numbers are tabulated in the JANAF tables (and elsewhere), where we read at 900 K

$$\frac{\Delta G^0}{T} = \left(\frac{G^0 - H_0^0}{T}\right)_{\text{acet}} + 3\left(\frac{G^0 - H_0^0}{T}\right)_{\text{hyd}} - 2\left(\frac{G^0 - H_0^0}{T}\right)_{\text{met}} + \frac{\Delta H_0^0(\text{acet}) + 3\,\Delta H_0^0(\text{hyd}) - 2\,\Delta H_0^0(\text{met})}{T}$$

$$= -53.464 + 3(-34.250) - 2(-49.098) + \frac{54.325 + 3(0) - 2(-15.991)}{900}(1000)$$

$$= 37.879$$

or $$\Delta G^0 = 34{,}091 \text{ cal/mol}$$

and $$K_p = 5.2 \times 10^{-9}$$

Some additional examples of the accuracy that might be expected for several industrially important reactions are evident from Table 7.4. However, in fairness, Table 7.4 summarizes data on reactions in the ideal gas phase, and comparisons involving reactions in the dense gas and liquid regions would require approximations to handle the real gas problems which appear formally, of course, in how we elect to calculate the fugacity of each species. Molecular engineering can address these concerns, and we do so in Chap. 5, but the practical utility of the results is poor. Any difference in ΔG^0 of reaction as might be calculated using data from the JANAF table and from Table 7.4 must arise from differences in the spectroscopic and/or calorimetric data on which each calculation was based.

We will refer to the data in Table 7.4 again in Chap. 8. But an additional interesting point about these calculated data is that in each case all of the vibration frequencies and moments were computed by using a semiempirical quantum-mechanical scheme that is discussed in Chap. 8. That is, the molecular calculations of the ΔG^0 of reaction have been made using quantum-mechanically predicted data rather than experimental spectroscopic data. Fortunately, these quantum schemes have been developed to the point of practical and friendly utility in molecular engineering. The user need not be an expert in theoretical chemistry.

Chemical Equilibria from Direct Minimization of Free Energy

The technique for computing an equilibrium conversion has nothing to do with molecular engineering. It is, however, appropriate to note an alternative mechanical approach to calculating a chemical equilibrium conversion through the direct minimization of the free energy

TABLE 7.4 Free Energies of Reaction for Several Industrially Important Reactions as Calculated from Molecular Parameters but Using Experimental Heats of Formation

		Free energy of reaction, ΔG^0, kcal/mol					
		300 K		900 K		1500 K	
Reaction	ΔH^0_{298}	Observed	Calculated	Observed	Calculated	Observed	Calculated
$N_2 + 3H_2 \rightarrow 2NH_3$	-21.94	-7.74	-7.82	24.24	24.15	58.36	57.55
$2CH_4 \rightarrow CH \equiv CH + 3H_2$	88.80	68.01	68.04	31.65	31.91	-7.43	-7.01
$CH_2 = CH_2 + 0.5O_2 \rightarrow$ H_2C—CH_2 (bridged by O)	-25.12	-19.45	-19.72	-7.70	-8.07	3.88	5.00
$CH_4 + 2H_2O \rightarrow 4H_2 + CO_2$	39.45	27.07	27.03	-2.02	-1.80	-34.25	-33.72
$CH_4 + NH_3 + 1.5O_2 \rightarrow HCN + 3H_2O$	-112.33	-118.02	-119.02	-129.48	-133.81	-139.83	-143.43
$CH_4 + Cl_2 \rightarrow CH_3Cl + HCl$	-24.82	-25.64	-25.60	-27.48	27.52	-29.46	-29.78

rather than through the law of mass action. Of course, the formalism leading to the definition of the equilibrium constant K is itself also a particular way to find the composition corresponding to minimum free energy. When one requires extensive calculations for chemistries involving more than one or two reactions occurring simultaneously, it is usually computationally more convenient to merely write the expression for the free energy of the phase, and then by some numerical scheme and the computer, find that set of mole fractions of species within the phase that will minimize this free energy. The formal techniques of lagrangian undetermined multipliers is satisfactory. A random walk scheme to map the free energy surface is similarly satisfactory. Studies suggest that there is no best method, but rather the "best" procedure will depend on the specific characteristics of the problem at hand.[6] Thus we can minimize the free energy or solve simultaneously the R independent mass action equations as we think best. Usually the number of these equations, R, will be the difference between the number of species present, N, and the number of chemical elements present, n; that is, R is usually $N - n$. The composition of the equilibrium phase can also be fixed by specifying only C independent variables, where in general C is equal to $N - R$ and usually C is equal to n.[7] In both solution techniques, the equilibrium set of mole numbers will minimize the Gibbs free energy of the phase subject to the constraints of a constant or fixed mass of each chemical element that was present initially.

In the calculational scheme involving minimization of free energy, we must express the free energies of each pure species in terms of its partition function in the usual way.

Let us compare the results obtained from the minimization of free energy with those obtained from equilibrium constants. Table 7.5 is such a comparison of a complex equilibrium that exists when air is heated to a very high temperature as in, for example, the reentry of a spacecraft into the atmosphere. The differences in the mole fractions result from minor differences in the physical properties that were input to each scheme as well as lack of precision in the numerical computational procedures themselves. In both calculations of Table 7.5, whether by direct minimization (McGee and Heller) or by equilibrium constants (Gilmore), the entropy of each species was taken to be zero at 0 K, and the entropy of the hot gas mixture is the same by either scheme. However, sharply different assignments of the zeros of enthalpy were used, and consequently there is a sharp difference in the total enthalpy and free energy of this very hot phase. Is this of any consequence? Absolutely not, for it is the difference in properties between two states that is important in any real calculation. The two multidimensional surfaces of free energy versus all other independent

TABLE 7.5 Equilibrium Composition and Properties of Air at 12,000 K and 9.0615 atm

Component	Moles present in plasma	
	Minimization of free energy†	Equilibrium constants‡
N	1.4848	1.4835
N^+	0.06586	0.066
O	0.39627	0.4080
O^+	0.023048	0.01087
A	0.8960×10^{-2}	0.896×10^{-2}
A^+	0.3720×10^{-3}	0.372×10^{-3}
N_2	0.49975×10^{-2}	0.575×10^{-2}
N_2^+	0.20138×10^{-3}	0.245×10^{-3}
NO	0.31663×10^{-3}	0.374×10^{-3}
NO^+	0.26067×10^{-3}	0.315×10^{-3}
e	0.089746	0.0774
Enthalpy, kcal	173.97	370.45
Entropy, cal/K	109.59	109.21
Gibb's free energy, kcal	-1.1411×10^3	-940.07

†H. A. McGee, Jr., and G. H. Heller, *Progress in Astronautics and Aeronautics*, Vol. 9, *Electric Propulsion Development*, Academic Press, New York, 1963, p. 443.
‡F. R. Gilmore, "Equilibrium Composition and Thermodynamic Properties of Air to 24,000K," Rand report RM-1543, August 1955.

variables are merely displaced by this scale factor and each value of the free energy at the equilibrium composition is at the minimum on its respective surface.

Either calculational scheme can sometimes lead to nonunique solutions for the conversion. And this is an interesting and important subject in its own right, but it falls outside the scope of our discussion of molecular engineering.[8]

Finally, we should recall that if we seek an equilibrium conversion at a fixed temperature and volume, rather than pressure, we can most conveniently minimize the Helmholtz rather than the Gibbs free energies. This, of course, presents no complications whatsoever from a molecular thermodynamic perspective.

The Chemical Potential

Whether we are concerned with chemical or with physical equilibrium, the primary quantity of concern in classical thermodynamic treatments is always the chemical potential. The ideal solution model relates the chemical potential of a component in such a solution to that of the pure species. But this clearly is a macroscopic approximation that cancels out all of the effects of interactions between dissimilar species. The inability of the techniques of molecular engineering to satisfactorily account for these interactions is the major fundamental cause of our inability to evaluate the chemical potential in real systems. This inability is a collection of operational problems, for there are no difficulties in principle.

The chemical potential μ_i is equivalent to the partial molar Gibbs free energy which is just $(\partial G/\partial n_i)_{T,p,n_j}$. For a mixture of ideal gases, this is trivial to evaluate, and for a real mixture of whatever the phase, it is an intractable problem. For n moles of an ideal gas mixture,

$$G^0(\text{mix}, T, p) = \sum_i n_i G_i^0(T, p) + RT\sum_i n_i \ln\left(\frac{n_i}{n}\right)$$

The chemical potential is defined

$$\mu_i^0(T, p) = \left(\frac{\partial G^0(\text{mix})}{\partial n_i}\right)_{T,p,n_j} = G_i^0(T, p) + RT \ln x_i$$

where $G_i^0(T, p)$ is the molar Gibbs free energy of pure i at the T and p of the mixture. The expression for this Gibbs free energy from Table 2.2 allows the substitution

$$\mu_i^0(T, p) = -RT\left[\ln\frac{f_i}{n_i} + 1\right] + RTV\left(\frac{\partial \ln f_i}{\partial V}\right)_T + RT \ln x_i$$

Here the partition function for pure i, f_i, is, of course, evaluated at the T and p of the mixture, where $V_i = n_i kT/p$. Then

$$\mu_i^0 = -RT \ln\left[g_0\left(\frac{2\pi mkT}{h^2}\right)^{3/2}\left(\frac{kT}{p}\right)f_{\text{vib,rot}}\right]_i + RT \ln x_i \qquad (7.37)$$

where all of the symbols for molecular quantities have their usual significance. The reason for the inability to evaluate μ_i in real systems is, of course, the complexity of the forces between molecules and their effect on the partition function. That is, in a real gas mixture, there would be another quantity in the square bracket of Eq. (7.37) above. This is the partition function arising from the intermolecular potential energy. With this we had written in Chap. 5 a more descriptive complete partition function as,

$$f = g_0\left(\frac{2\pi mkT}{h^2}\right)^{3/2}(V - nb)f_{\text{vib,rot}}e^{\phi/2kT} \qquad (5.17)$$

Here ϕ is the potential energy of a molecule resulting from its nearness to all the other molecules. Although it cannot now be rigorously evaluated, it can be formally written as

$$\phi = \left(\frac{n}{V}\right)\int_0^\infty \phi(r)g(r)4\pi r^2\, dr \tag{5.18}$$

Here $\phi(r)$ may be any one of many intermolecular models such as Lennard-Jones, and $g(r)$ is called the radial distribution function. These problems were discussed in detail in Chap. 5. In mixtures, all these problems are much worse than for pure substances, for we must now also worry about forces between dissimilar molecules. For example, in a binary gas mixture, the second virial coefficient may be rigorously shown to be

$$B(T) = y_1^2 B_{11}(T) + 2y_1y_2B_{12}(T) + y_2^2 B_{22}(T) \tag{7.38}$$

where the three coefficients are to be evaluated using 1,1; 1,2; and 2,2 intermolecular interactions respectively. In other words, even in this very simple case, one requires three Lennard-Jones (or other) potential functions. Recall that arguments concerning second virial coefficients are devoted to gases of only moderate density, and the story enormously increases in complexity for dense gases and for liquids. Clearly, empirical approaches are necessary. These empirical approaches have always taken the form of the development of accurate equations of state. Using these and the methods of classical thermodynamics, it is readily possible to evaluate partial fugacity coefficients, from which the chemical reaction equilibrium constant may be written

$$K = e^{-\Delta G^0(T)/RT} = \prod_i \bar{f}_i^{\nu_i} = \prod_i \bar{\phi}_i^{\nu_i} \prod_i (y_i p)^{\nu_i}$$

or

$$K = \left(\prod_i \bar{\phi}_i^{\nu_i}\right) K_p = K_\phi K_p \tag{7.39}$$

where $\bar{\phi}_i = \bar{f}_i / y_i p$. For simplicity consider a two-component mixture reaction such as an isomerization. For a binary gas mixture described by the virial equation of state at densities such that the equation can be truncated after the second virial coefficient, the fugacity coefficients may be written

$$\ln \bar{\phi}_1 = \frac{p}{RT}[B_{11} + y_2^2(2B_{12} - B_{11} - B_{22})]$$

and

$$\ln \bar{\phi}_2 = \frac{p}{RT}[B_{22} + y_1^2(2B_{12} - B_{11} - B_{22})] \tag{7.40}$$

These expressions are readily generalized to mixtures of any number of components. The second virial coefficients for every possible binary interaction in the mixture will be involved. The values of $\bar{\phi}_i$ and $\bar{\phi}_2$ can be evaluated as soon as we select an intermolecular model and evaluate the parameters in that model. Similar evaluations to those of Eq. (7.40) above for other commonly used equations of state appear in textbooks on chemical engineering thermodynamics.[9]

In the ideal solution approximation,

$$\bar{f}_i(T,p) = y_i f_i(T,p) \qquad \text{or} \qquad \bar{\phi}_i(T,p) = \phi_i(T,p)$$

and

$$\ln \phi_1 = \frac{B_{11}p}{RT}$$

and

$$\ln \phi_2 = \frac{B_{22}p}{RT} \tag{7.41}$$

Here ϕ_1 and ϕ_2 have the same values regardless of the complexity of the mixture, for all interactions between dissimilar molecules are ignored at the ideal solution level of approximation. Again, corresponding expressions to those of Eq. (7.41) are readily developed for any equation of state, and many examples appear in textbooks on chemical engineering thermodynamics.[9]

Finally, at the ideal-gas level of approximation, $\bar{f}_i(T,p) = y_i p$, and the equilibrium constant becomes merely K_p.

The accurate prediction of reaction equilibria in real gas mixtures using molecular engineering alone seems a long way off. The problems are twofold: we are unable to rigorously describe intermolecular forces, and, even if we could, the mathematical complexity of deducing macroscopic behavior presents an intractable problem. At least we realize the molecular origins of the complexity, and molecular arguments provide the only illumination of our empirical path toward practically useful insights.

References

1. E. Rutner et al., *J. Chem. Phys., 24*, 173 (1956).
2. G. Ecker and W. Weizel, *Ann. Physik, 17*, 126–140 (1956).
3. B. F. Dodge, *Chemical Engineering Thermodynamics*, McGraw-Hill, New York, 1944.
4. P. J. Dickerman (ed.), *Optical Spectrometric Measurements of High Temperatures*, University of Chicago Press, Chicago, 1961.
5. D. R. Stull and H. Prophet (eds.), *JANAF Thermochemical Tables*, MBS37, June 1971.
6. M. A. Stadther and L. E. Scriven, *Chem. Eng. Sci., 29*, 1165 (1974).
7. S. R. Brinkley, *J. Chem. Phys., 14*, 563 (1946).
8. H. S. Caram and L. E. Scriven, *Chem. Eng. Sci. 31*, 163 (1976).
9. S. M. Walas, *Phase Equilibria in Chemical Engineering*, Butterworth, Stoneham, Mass., 1985.

Further Reading

1. R. W. Gurney, *Introduction to Statistical Mechanics*, McGraw-Hill, New York, 1949.
2. J. M. Prausnitz, R. N. Lichtenthaler, and E. Gomes de Azevedo, *Molecular Thermodynamics of Fluid Phase Equilibria*, 2d ed. Prentice-Hall, Englewood Cliffs, N.J. 1986.
3. E. A. Moelwyn-Hughes, *Physical Chemistry*, Pergamon, London, 1957.

Chapter

8

Quantum-Mechanical Calculation of Molecular Parameters

In all of our previous arguments, we have taken the molecular parameters of moments of inertia and vibration frequencies as experimentally determined quantities with which the formalism of statistical mechanics then allowed us to deduce all of the thermodynamic properties. We have marveled on several occasions at how purely optical measurements could be used to rigorously deduce calorimetric or thermal quantities. Statistical mechanics provides the bridge between these molecular properties on the one hand and the macroscopic behavior of matter on the other hand. And the essential component of this bridge is the partition function. A legitimate question now is, "Is it possible to calculate from first principles—from quantum mechanics—the molecular structure and vibration frequencies?" This is, of course, the ultimate goal of theoretical chemistry, to simply calculate all of the macroscopic properties of matter from first principles without having to experimentally measure anything. To do this, we must calculate the energy of the collection of atoms that is the molecule as a function of their geometrical configuration, for the arrangement of minimum energy will then describe the actual molecule. We require that this minimum energy be calculable to within chemically significant accuracy and that the calculation be cost-effective. The calculation must make economic sense. Clearly, if the calculation gives an absolutely precise energy, but at a greater cost than would be incurred

---

The discussion in this chapter was abstracted in part from a thesis by K. N. Shah, presented in partial fulfillment of requirements for the M.S. degree in chemical engineering, Virginia Polytechnic Institute and State University, August 1983.

by experimentally measuring, say, the heat of formation, we have accomplished little. Quantum mechanics, in principle, permits just such calculations, but obtaining accurate solutions for molecules of practical interest is an exceedingly difficult issue.

As we have seen, if we can describe the forces between molecules, the partition function can, in principle, be written in a way that will allow us to describe the thermodynamic properties of real gases, liquids, and solids, that is, substances in which intermolecular interactions play significant roles. But if we neglect for now those much more difficult real-gas problems, approximate quantum-mechanical insights of sufficient power and accuracy do now exist to enable calculations of molecular structure and vibration frequencies, and hence we are able to immediately calculate all of the thermodynamic properties of the ideal gas at all temperatures and pressures.

These are the techniques of so-called semiempirical quantum mechanics. The techniques have been highly developed, and it is possible to calculate observable properties to within chemically useful accuracies. Heat capacities good to within 1 J, absolute entropies to within about 1 J/mol · K, and heats of formation to within a few kilojoules or so are all not uncommon, as we shall see. The required input information is only an approximate structure of the molecule, the computer codes are available,† and they can be used more or less as "black boxes," for it is not at all necessary to understand the intricacies of the quantum-mechanical arguments in order to use the techniques. This is analogous perhaps to one's use of an infrared spectrophotometer for analysis without really understanding the chemical physics of molecular vibration and rotation.

An examination of the ideas and approximations of these semiempirical quantum-mechanical arguments would take us too far into theoretical chemistry. Suffice to merely note that, in exactly the same way that electrons about an atom are envisioned as occupying atomic orbitals, so we assume that the valence electrons of molecules occupy molecular orbitals (MOs). These MOs are described mathematically as a linear combination of the atomic orbitals (AOs) of each atom, and these may be reasonably approximated. We take these valence electrons to move in a fixed field formed by the nuclei and the inner core of electrons. These assumptions together form the so-called linear-combination-of-atomic-orbitals, molecular-orbital, self-consistent-field (or LCAO-MO-SCF) approach to molecular quantum mechanics. In quantum-mechanical arguments, one must deal with many complex integrals that may be difficult or impossible to accurately evaluate, so

---

†Request the MINDO, MNDO, or similar semiempirical code from the Quantum Chemistry Program Exchange (QCPE), University of Indiana, Bloomington, IN 47401.

one assigns values to integrals or parameterizes the formalism to force a fit to the known heats of atomization and geometries of a set of standard molecules which have been chosen to represent as wide a variety of types of structures and bonding as possible. To illustrate the idea in a greatly oversimplified setting, if we parameterize on the experimental facts of CH_4 and C_2H_6 to characterize the C-H and C-C bonds respectively, we can then use the now-calibrated theory to calculate the properties of all straight chain alkanes.

There have been a number of common criticisms of semiempirical methods. By their nature, such methods are curve-fitting, which should not be relied on outside the range of compounds used to determine the parameters. Also by their nature, such methods can be reparameterized indefinitely, and several versions of each technique have appeared. Semiempirical methods are like equations of state, of which hundreds of versions have appeared. Some early versions of both quantum methods and equations of state are then out of date. It appears that current parameterization is about the best that can be accomplished relative to the basic quantum-mechanical assumptions that are involved in each method. We expect little further improvement due to reparameterization until better theory itself is evolved. Certainly, it is necessary to use discretion in interpreting results, one must reason analogously, one must use one's chemical intuition, and one must be skeptical.

Normal Mode Analysis

It would take us too far afield to develop the details of the wholly theoretical determination of vibration frequencies. However, it is worthwhile to understand the gist of the arguments and to see that the analysis, although complex in mathematical detail, is nonetheless straightforward in principle.

Consider, for example, the bent asymmetric molecule HOCl,

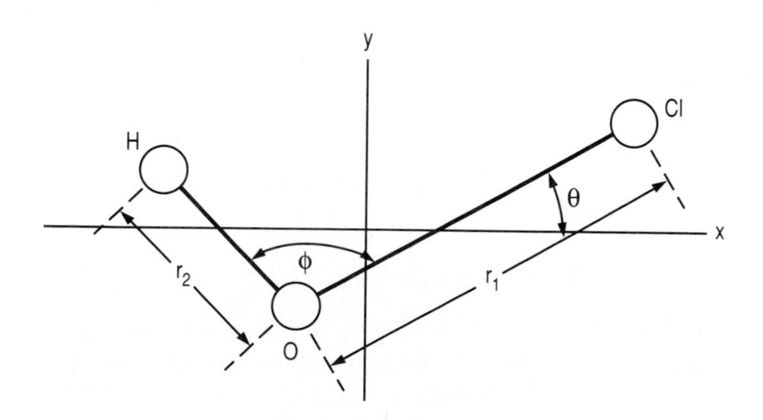

with the origin of the cartesian system at the center of mass. Clearly, the (x, y) coordinates of each atom may be expressed in terms of the two angles θ and ϕ and the two interatomic distances r_1 and r_2. Similarly, the moments of inertia about each of the two axes can be developed in terms of r_1, r_2, θ, and ϕ, since, for example, the moment about the x axis is just

$$I_x = \Sigma m_i y_i^2$$

where i is over the H, O, and Cl masses and their positions off the x axis. The MNDO or other semiempirical quantum-mechanical codes output the molecular structure, i.e., the set of r_i's and θ_i's (just two distances and one angle in the bent triatomic case) that minimizes the total energy of the molecule. With this as input, the calculation of the three moments is clear.

The kinetic energy of the three atoms in the xy plane as shown, i.e., $z = 0$, is also clear:

$$T = \tfrac{1}{2}m_{\mathrm{H}}[(\dot{x}_{\mathrm{H}})^2 + (\dot{y}_{\mathrm{H}})^2] + \tfrac{1}{2}m_{\mathrm{O}}[(\dot{x}_{\mathrm{O}})^2 + (\dot{y}_{\mathrm{O}})^2] + \tfrac{1}{2}m_{\mathrm{Cl}}[(\dot{x}_{\mathrm{Cl}})^2 + (\dot{y}_{\mathrm{Cl}})^2] \tag{8.1}$$

After some algebra, it is possible to also express this kinetic energy in terms of r_1, r_2, and θ. Instead of working with the velocity of each atom, it is more convenient to consider the total energy T as made up of the translation of the center of mass of the three-atom system, the rotation of the rigid structure, and the motion of the parts of the molecule relative to each other. Thus, instead of cartesian momenta, it is more convenient to consider three such moments for translation of the center of mass and three for rotation, leaving three for the motion of the atoms relative to each other. This rearrangement of Eq. (8.1) for the kinetic energy is again just a matter of some tedious algebra.

The equations of motion in any mechanical system are

$$\frac{d}{dt}\left(\frac{\partial T}{\partial q_i{}^0}\right) - \frac{\partial T}{\partial q_i} = -\left(\frac{\partial V}{\partial q_i}\right) \tag{8.2}$$

where V is the potential energy of the system and q is a positional coordinate. We take the potential energy to follow Hooke's law or to be quadratic in the displacement of q_i from its equilibrium position. In still other words, we limit attention to internal motions that are harmonic. Then

$$V = V_0 + \tfrac{1}{2}\Sigma\, k_i\, (q_i - q_i^0)^2$$

and the potential energy of the molecule is just a constant, V_0, when each q_i is in its equilibrium (i.e., minimum-energy) position, q_i^0. With $V = V_0$ and constant, the restoring force is zero, but the restoring force varies as the displacement $(q_i - q_i^0)$, just as for any Hooke's law vibration. The semiempirical quantum-mechanical schemes determine the internal energy of the molecule for any geometric configuration. Specifically, the energy of the molecular system at some small displacement from equilibrium is readily calculable using MNDO or other scheme, so the force constants, the k_i's, can be evaluated. A solution to the equation of motion, Eq. (8.2), is the familiar

$$q_i = A_i e^{-2\pi i \nu_i t}$$

and

$$\nu_i = \frac{1}{2\pi}\sqrt{\frac{k_i}{\mu_i}} \tag{8.3}$$

where μ_i is the reduced mass of the specific vibration. The frequencies are then at hand. The three so-called normal vibrations for HOCl are represented:

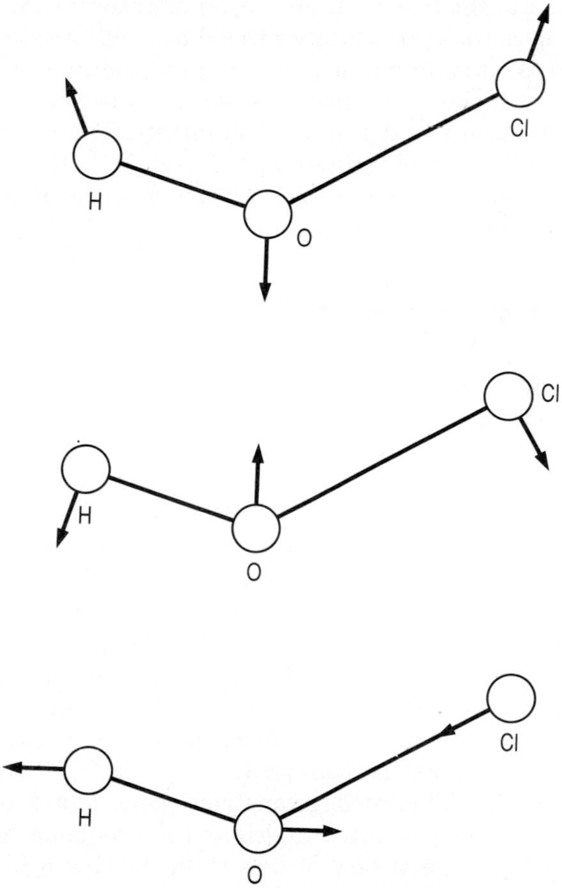

The above argument is only a cursory sketch of the so-called normal mode analysis. For complex molecules, the mathematical detail would be tedious in the extreme. It is work for a specialist. However, the user-oriented computer codes from QCPE are friendly, and with modest study any research engineer can use them to estimate the molecular data needed in thermodynamic, chemical, kinetic, and transport arguments.

In all thermochemical calculations, we must define a zero of energy someplace, and this standard state is almost always that normal state of aggregation of the pure element at 25°C and 101.325 kPa, that is, H_2 (ideal gas), C (graphite), Br_2 (liquid), etc. However, the quantum-mechanical arguments give us the energy of the molecule relative to its separated, or gaseous, atoms. Thus, to write the usual thermochemical heat of formation, we need also to know the heats of formation of atoms as ideal gases at 25°C and 101.325 kPa from the elements in their standard states. This involves quantities like the bond energy of H_2 and the heat of sublimation of graphite. And all of these sorts of numbers that are usually needed are well-known. If the energies of the separated atoms relative to the elements in their standard state are known, the difference between the energy of the molecule and that of its atoms is the heat of atomization, which can be easily converted into its heat of formation, ΔH_f^0, which is the same quantity that one would obtain from, say, a calorimetric measurement of the heat of combustion of the compound.

Some comparisons with experiment

The usefulness of any theory is gauged by its ability to reproduce and predict experimental observations. Figure 8.1 shows a plot of calculated versus experimental ΔH_f^0 (at 25°C) for 193 molecules. The agreement is quite good, with most of the compounds lying within ± 5 kcal/mol (± 20 kJ/mol) of the line of unit slope. The types of compounds included in Fig. 8.1 include ethane, ethylene, acetylene, propene, isomeric butanes, butadiene, cyclopropanes, cyclopropenes, cyclobutanes, cyclopentanes, cyclobutenes, cyclopentenes, cyclohexanes, substituted benzene derivatives, amines, aldehydes and ketones, nitrates, phenols, fluorinated hydrocarbons, silanes, phosphanes, sulfur-containing compounds, etc., and free radicals and carbonium ions. In short, the heats of formation of a wide variety of molecules have been well-calculated. Comparisons of experiment and theory as well as predictions for unknown molecules appear in Table 8.1.

In addition, the calculated geometries were found to be in good agreement with experiment. Experimental problems are such that discrepancies from the theory of less than 0.001 nm in bond length

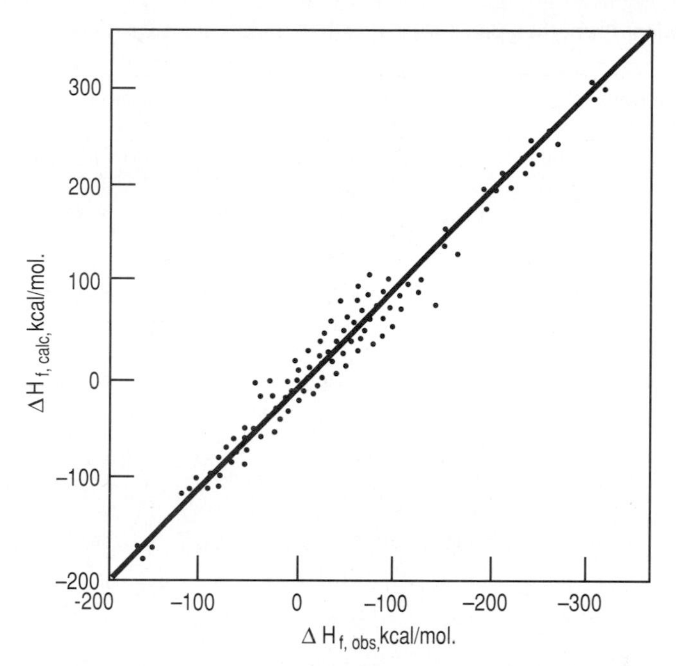

Figure 8.1 Calculated and observed heats of formation of 193 compounds derived from H, N, O, F, Si, P, S, and Cl. [*Reprinted with permission from R. C. Bingham, M. J. S. Dewar, and D. H. Lo, J. Am. Chem. Soc., 97, 1285 (1975). Copyright 1975 American Chemical Society.*]

and 1° in bond angle are not too meaningful. Mean errors are 0.0014 nm in bond length and 2.8° in bond angle in calculations using MNDO. Examples of agreement for several species appear in Table 8.2. Moments of inertia are, of course, immediately calculated from the geometry, and Fig. 8.2 and, in more detail, Table 8.3 show calculated versus experimental moments of inertia for many of the same molecules that were depicted in Fig. 8.1. In view of the variety of molecules, the results are remarkably good.

The computational scheme is to first input a trial geometry of the molecule of interest. Which atom is connected to which and its approximate distance and angle from the preceding atom is all that is required. Cartesian coordinates of each atom are not necessary. MNDO will optimize the geometry for minimum total energy, calculate the ΔH_f^0, and calculate the coordinates of each atom. A second code, such as *GEOMO/RV* (also available from QCPE), takes the MNDO geometry as input and calculates vibrational frequencies and moments of inertia. A third code, a statistical thermodynamics package, takes this output and calculates the thermodynamic properties.

Figure 8.3 shows a plot of nearly 500 calculated versus experimen-

TABLE 8.1 Some Physical Properties Calculated by MNDO versus the Experimental Values

Molecule	Heat of formation, kcal/mol		Ionization potential, eV		Dipole moment, debye	
	Calculated	Experimental	Calculated	Experimental	Calculated	Experimental
C_2H_6	- 19.7	- 20.2	12.7	12.1	0.0	0.0
$CH_3CH_2CH_3$	- 24.9	- 24.8	12.34	11.5	0.0	N/A
	32.0	31.99	9.0	8.57	0.18	0.40
	21.2	19.8	9.39	9.25	0.26	N/A
	68.2	66.2	9.89	9.86	0.48	0.44
	38.1	36.1	8.57	8.15	0.31	N/A
N N	72.5	79.0	7.98	N/A	1.55	1.56
B_2O_3	- 198.5	- 199.2	—	N/A	—	N/A
BO_2H	- 133.1	- 134.1	—	N/A	—	N/A
FBeOBeF	- 286.7	- 287.9	12.76	N/A	—	N/A
BeF_2	- 192.3	- 190.3	14.45	N/A	—	N/A

SOURCE: Taken from M. J. S. Dewar and W. Thiel, *J. Am. Chem. Soc.*, *99*, 4907 (1977).

tal vibrational frequencies for 34 molecules with an average error of about ± 10 percent. These include water, hydrogen sulfide, ammonia, carbon dioxide, carbon disulfide, HCN, formaldehyde, methanol, ethylene, methylamine, dimethylether, acetone, maleic anhydride, furan, pyrrole, thiophene, benzene, and a variety of heterocyclic compounds.

A wide variety of additional properties of molecules have been investigated, including dipole moments, first ionization potentials, polarizabilities, hyperpolarizabilities, nuclear quadrupole coupling constants, ESCA (electron spectroscopy for chemical analysis) chemical shifts, and the electronic band structure of polymers. In addition, the mechanisms predicted for several hundred reactions have also been consistent with available experimental data.

As a natural extension and the one most important for our purposes here, the heat capacities and absolute entropies were also calculated using the above predictions of moments and frequencies. A comparison of results is presented in Table 8.4 and Figs. 8.4 and 8.5, where the

TABLE 8.2 Optimized Structures for Several Molecules as Calculated by MNDO (Experimental Values Appear in Brackets)

Molecule	Bond length, Å	Bond angle, °
$C(CH_3)_4$	CC 1.554(1.539) CH 1.109(1.120)	HCC 111.7(110.0)
	CC 1.549(1.548) CH 1.105(1.133)	HCH 107.6(108.1) $C^1C^2C^4C^3$ 180.
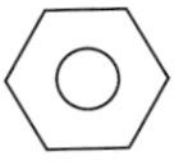	CC 1.407(1.397) CH 1.090(1.084)	
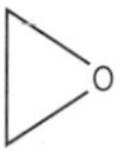 O	CO 1.417(1.435) CC 1.513(1.470)	HCH 111.7(116.3) $C\text{-}CH_2$ 159.2(158.1)
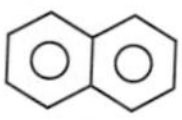	C^1C^2 1.382(1.364) C^2C^3 1.429(1.415) C^1C^9 1.439(1.421) C^9C^{10} 1.435(1.418)	

SOURCE: Taken from M. J. S. Dewar and W. Thiel, *J. Am. Chem. Soc.*, *99*, 4907 (1977).

agreement is seen to be excellent, being within 1 cal/mol · K (4.2 J/mol · K) at 298 K in almost every case. The plot of heat capacity, Fig. 8.5, does not extend below 8 cal/mol · K because there C_p^0 is not a function of the ω's, i.e.,

$$C_p^0 = \frac{3}{2}R + \frac{3}{2}R + \sum_i C_V^0(\text{vib}) + R$$

or

$$C_p^0 = \frac{8}{2}R + R\sum_i \frac{C_V^0(\text{vib})}{R}$$

Reaction Equilibria

Chemical reaction equilibria are immediately calculable from the molecular data predicted by either semiempirical or *ab initio* quantum-mechanical techniques. Some results for several industrially important reactions at three temperatures are summarized in Table 8.5. For comparison purposes, Table 8.5 also lists free energies of reaction calculated by using well-known experimental frequencies and moments. The tabulated values of ΔG^0 are for ideal gas reaction, of course, at 1 atm. Small errors in the theoretically predicted heats of formation can accumulate to produce errors in ΔH^0 of reaction that might strongly influence the expected ΔG^0 of reaction and hence the equilibrium conversion.

As an example, consider from Table 8.5,

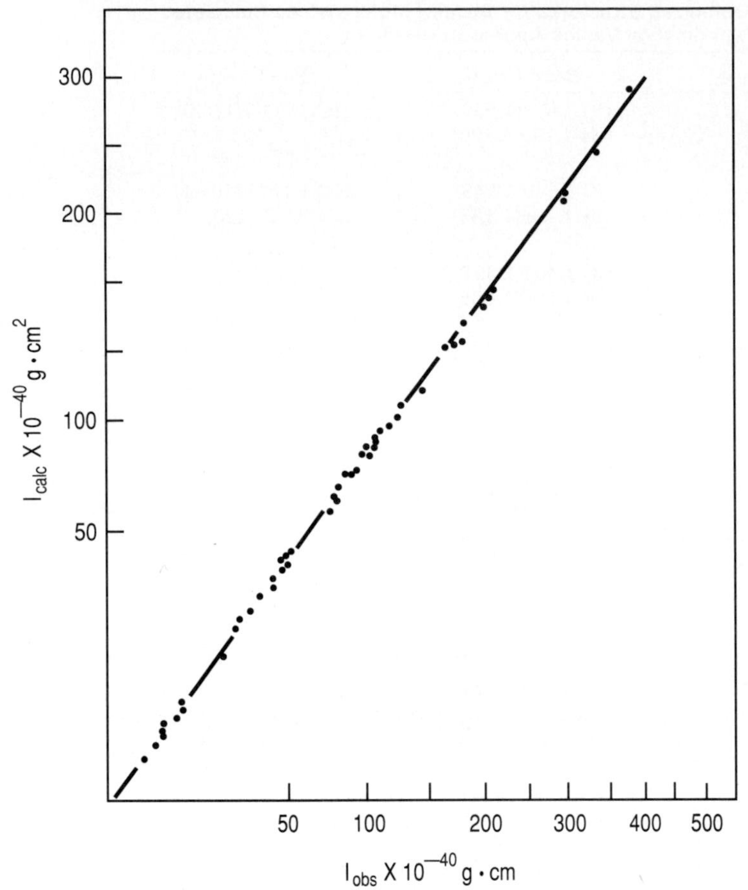

Figure 8.2 Calculated and observed moments of inertia for 27 compounds. [*Reprinted with permission from M. J. S. Dewar and G. P. Ford, J. Am. Chem. Soc., 99, 7822 (1977). Copyright 1977 American Chemical Society.*]

$$2CH_4 \leftrightharpoons C_2H_2 + 3H_2$$

at 1500 K. The ΔG^0 of reaction is

$$\frac{\Delta G^0}{T} = \sum_i \left(\frac{G_0 - H_0^0}{T} \right)_i + \sum_i \frac{\Delta H_0^0 f}{T}$$

and we will first calculate ΔG^0 using predicted molecular data and an experimental ΔH_f^0 at 25°C, and then repeat the calculation using theoretical values of ΔH_f^0 of each of the three species. The MINDO/3 scheme yields:

TABLE 8.3 Moments of Inertia of Several Molecules Calculated from MNDO Geometries

Molecule	$I \times 10^{-40}$ g/cm²					
	I_A		I_B		I_C	
	Observed	Calculated	Observed	Calculated	Observed	Calculated
H_2O	1.004	1.016	1.928	1.808	3.015	2.884
HCN	—	—	18.93	18.97	—	—
CH_4	—	—	5.34	5.418	—	—
C_2H_2	—	—	23.78	23.58	—	—
CH_2CO	2.974	2.851	81.50	82.41	84.61	85.27
C_2H_4	5.751	5.478	27.95	28.20	33.78	33.67
	32.92	30.73	37.92	38.34	59.50	58.29
	66.58	67.52	41.76	41.77	—	—
	259.1	299.9	147.6	150.0	—	—
	88.81	89.09	90.73	90.30	179.6	179.4
NH_3	4.645	4.431	2.814	2.96	—	—

SOURCE: Taken from M. J. S. Dewar and G. P. Ford, *J. Amer. Chem. Soc.*, *100*, 7822 (1977).

	ω (exp),[2] cm^{-1}	ω (calc),[2] cm^{-1}
CH_4		
$I = 5.418 \times 10^{-40}$ g · cm² (Ref. 1)	2917	(1) 3505
Experimental: $\Delta H_f^0(298) = -17.9$	1534	(2) 1326
Theoretical: $\Delta H_f^0(298)$[3] $= -11.9$	3019	(3) 3551
	1306	(3) 1270
C_2H_2		
$I = 23.56 \times 10^{-40}$	3374	(1) 3827
Experimental: $\Delta H_f^0(298) = 54.3$	1974	(1) 2237
Theoretical: $\Delta H_f^0(298)$[3] $= 57.3$	3289	(1) 3770
	612	(2) 488
	730	(2) 885
H_2		
$I = 0.473 \times 10^{-40}$	4395	(1) 5116
Experimental: $\Delta H_f^0(298) = 0$		
Theoretical: $\Delta H_f^0(298)$[3] $= 0.7$		

The column of integers in parentheses is the multiplicity of that vibrational mode.

We calculate the several contributions to the free energy in the

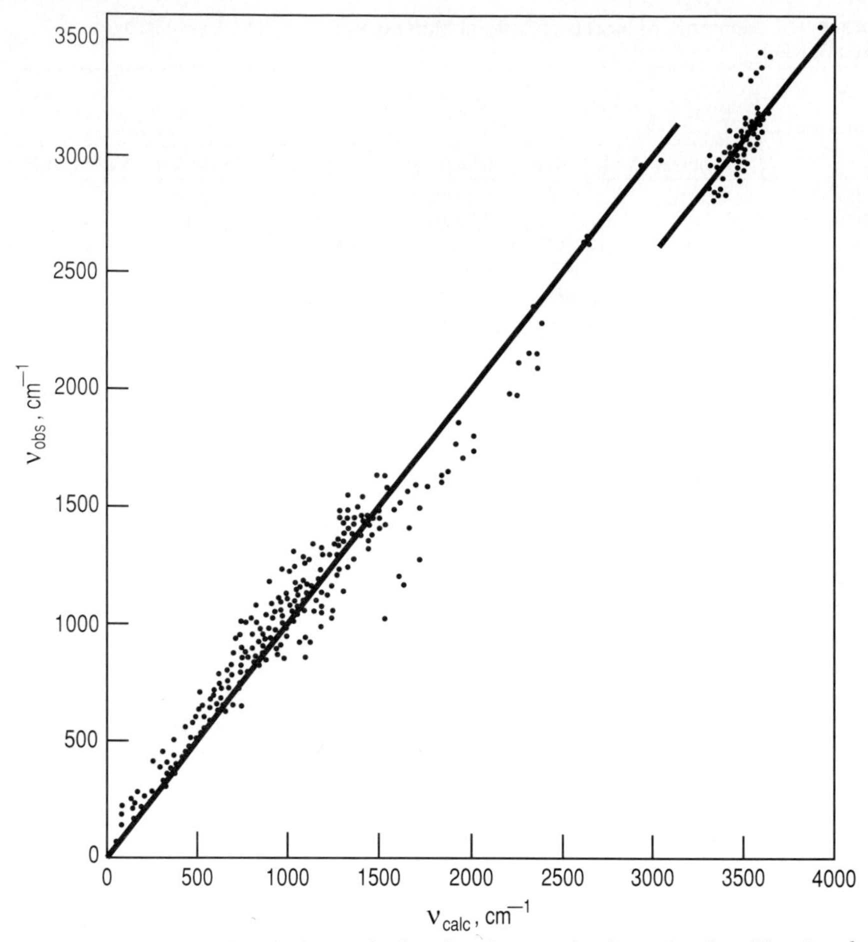

Figure 8.3 Calculated and observed vibration frequencies for molecules. [*Reprinted with permission from M. J. S. Dewar and G. P. Ford, J. Am. Chem. Soc., 99, 1685 (1977). Copyright 1977 American Chemical Society.*]

usual way. First concentrating on methane, we find the electronic contribution is

$$\left(\frac{G^0 - H_0^0}{RT}\right)_{\text{elect}} = -\ln g_0 = -\ln 1 = 0$$

The translational contribution is

$$\left(\frac{G^0 - H_0^0}{RT}\right)_{\text{trans}} = \frac{5}{2} - \frac{3}{2}\ln M - \frac{5}{2}\ln T + \ln p - 3.4533$$

TABLE 8.4 Calculated and Observed Entropies and Heat Capacities at 298 K

Compound	S^0, cal · mole^{-1} · K^{-1}			C_p^0, cal · mole^{-1} · K^{-1}		
	Observed	Calculated	Error	Observed	Calculated	Error
H_2O	45.1	45.0	0.1	8.0	8.0	0.0
H_2S	49.1	49.2	− 0.1	8.2	8.5	− 0.3
NH_3	46.0	46.0	0.0	8.5	8.4	0.1
CO_2	51.1	51.4	− 0.3	8.9	9.2	− 0.3
CS_2	56.8	57.4	− 0.6	10.9	11.3	− 0.4
HCN	47.9	48.1	− 0.2	7.8	8.4	− 0.6
H_2CO	55.3	55.0	0.3	10.1	10.5	− 0.4
CH_2N_2	58.0	58.7	− 0.7	12.5	13.0	− 0.5
CH_3Cl	55.9	56.1	− 0.2	9.7	10.4	− 0.7
CH_4	44.5	44.6	− 0.1	8.5	8.7	− 0.2
CH≡CH	48.0	48.3	− 0.3	10.5	10.6	− 0.1
CH_2=C=O	57.8	58.3	− 0.5	12.4	12.9	− 0.5
CH_3CN	58.1	58.1	0.0	12.5	12.7	− 0.2
CH_2=CHCl	63.1	63.9	− 0.8	12.8	13.8	− 1.0
CH_2=CH_2	52.4	52.6	− 0.2	10.2	11.0	− 0.8
(three-membered ring, O)	58.0	58.0	0.0	11.4	11.6	− 0.2
(three-membered ring, S)	61.9	61.3	0.6	13.3	13.7	− 0.4
(three-membered ring, NH)	59.8	60.0	− 0.2	12.2	12.9	− 0.7
H_2C=C=CH_2	58.2	58.6	− 0.4	14.1	14.7	− 0.6

$$= \frac{5}{2} - \frac{3}{2}\ln 16 - \frac{5}{2}\ln 1500 + \ln 101.325 - 3.4533$$

$$= -18.777$$

The rotational contribution will require the rotational constant in megahertz. The calculated moment of inertia from the calculated structure is converted into B:

$$B = \left(\frac{h}{8\pi^2 I}\right)10^{-6} = \left[\frac{6.626 \times 10^{-34}}{8\pi^2(5.418 \times 10^{-47})}\right]10^{-6}$$

$$= 1.549 \times 10^5 \text{ MHz}$$

The value of I from the JANAF table is 5.313×10^{-47} kg · m^2, which corresponds to a rotational constant of 1.580×10^5 MHz. The contribution to the free energy from rotation is then

$$\left(\frac{G^0 - H_0^0}{RT}\right)_{\text{rot}} = \frac{3}{2} - \frac{3}{2}\ln T + \ln \sigma + \frac{1}{2}\ln B_x B_y B_z - 16.9890$$

TABLE 8.4 Calculated and Observed Entropies and Heat Capacities at 298 K (*Contd.*)

Compound	S^0, cal · mole^{-1} · K^{-1}			C_p^0, cal · mole^{-1} · K^{-1}		
	Observed	Calculated	Error	Observed	Calculated	Error
$CH_3C{\equiv}CH$	59.3	59.3	0.0	14.5	14.5	0.0
	56.6	56.9	- 0.3	13.1	13.7	- 0.6
O O O	72.6	74.5	- 1.9	20.7	21.4	- 0.7
O	63.8	64.8	- 1.0	15.6	17.0	- 1.4
S	67.9	68.3	- 0.8	17.4	19.3	- 1.9
N H	65.8	66.4	- 0.6	16.8	19.7	- 1.9
H O	64.8	65.0	0.0	14.5	14.8	- 0.3
	64.3	65.2	- 0.9	19.6	20.9	- 1.3

SOURCE: From M. J. S. Dewar and G. P. Ford, *J. Am. Chem. Soc.*, *99*, 7822 (1977).

The moment of inertia of CH_4 about each axis is the same; so then is the rotational constant. And then

$$\left(\frac{G^0 - H_0^0}{RT}\right)_{\text{rot}} = \frac{3}{2} - \frac{3}{2} \ln 1500 + \ln 12 + \frac{3}{2} \ln (1.549 \times 10^5) - 16.9890$$

$$= -\ 6.048$$

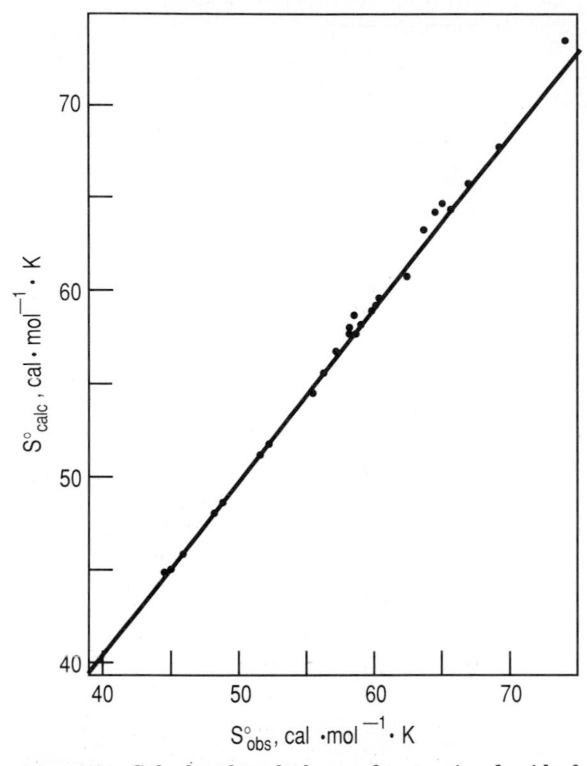

Figure 8.4 Calculated and observed entropies for ideal gases at 298.15 K. S^0(calc) are each wholly theoretical predictions based on no input data. [*Reprinted with permission from M. J. S. Dewar and G. P. Ford, J. Am. Chem. Soc., 99, 7822 (1977). Copyright 1977 American Chemical Society.*]

This rotational contribution, according to the experimental rotational constant, is

$$\left(\frac{G^0 - H_0^0}{RT}\right)_{\text{rot}} = -6.018$$

The vibrational contribution is calculated from each of the theoretical and the experimental frequencies:

ω (theo)	$(\omega/T)_{\text{theo}}$	ω (exp)	$(\omega/T)_{\text{exp}}$	g	Theoretical $-(G^0 - H_0^0/RT)_{\text{vib}}$	Experimental $-(G^0 - H_0^0/RT)_{\text{vib}}$
3505	2.34	2917	1.94	1	0.0352	0.0635
1326	0.88	1534	1.02	2	0.3321	0.2626
3551	2.37	3019	2.01	3	0.0337	0.0571
1270	0.85	1306	0.87	3	0.3502	0.3381
				Totals	1.8509	1.7744

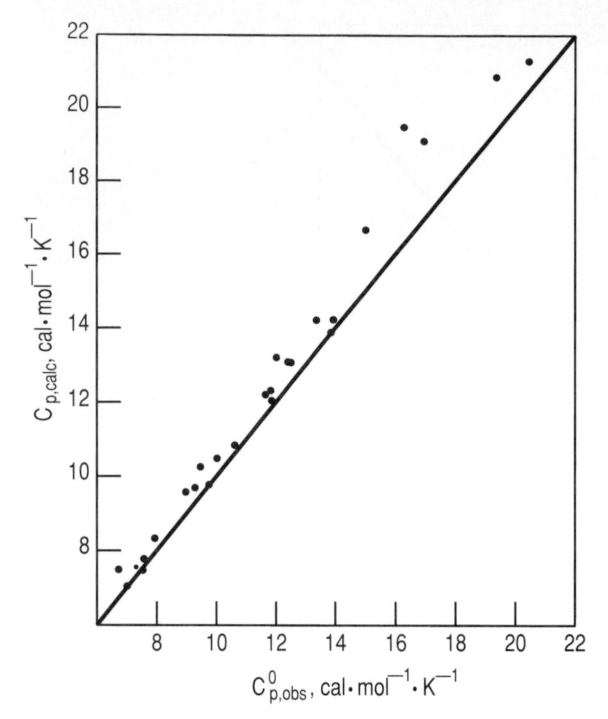

Figure 8.5 Calculated and observed heat capacities for ideal gases at 298.15 K. C_p^0(calc) are each wholly theoretical predictions based on no input data. [*Reprinted with permission from M. J. S. Dewar and G. P. Ford, J. Am. Chem. Soc., 99, 7822 (1977). Copyright 1977 American Chemical Society.*]

The total free energy function is then

$$\frac{G^0 - H_0^0}{RT} = 0 + (-18.777) + (-6.048) + (-1.851)$$

$$= -26.676$$

for the calculated molecular data. This may be compared with

$$\frac{G^0 - H_0^0}{RT} = 0 + (-18.777) + (-6.018) + (-1.774)$$

$$= -26.569$$

for the experimental molecular data. The wholly theoretical molecular data have led to a free energy that is only 0.4 percent different from that obtained by using experimental data. Both of these results may be further compared with the value of the free energy function as read

TABLE 8.5 Free Energies of Reaction for Several Industrially Important Reactions Calculated from MINDO/3 Molecular Orbital Theory and Experimental Heats of Formation at 298 K

Reaction	ΔH^0_{298}	Free energy of reaction ΔG^0, kcal/mol					
		300 K		900 K		1500 K	
		Observed	Calculated	Observed	Calculated	Observed	Calculated
$N_2 + 3H_2 \rightarrow 2NH_3$	- 21.94	- 7.74	- 7.82	24.24	24.15	58.36	57.55
$2CH_4 \rightarrow CH{\equiv}CH + 3H_2$	88.80	68.01	68.04	31.65	31.91	- 7.43	- 7.01
$CH_2{=}CH_2 + 0.5O_2 \rightarrow \overset{O}{H_2C-CH_2}$	- 25.12	- 19.45	- 19.72	- 7.70	- 8.07	3.88	5.00
$CH_4 + 2H_2O \rightarrow 4H_2 + CO_2$	39.45	27.07	27.03	- 2.02	- 1.80	- 34.25	- 33.72
$CH_4 + NH_3 + 1.5O_2 \rightarrow HCN + 3H_2O$	- 112.33	- 118.02	- 119.02	- 129.48	-133.81	- 139.83	- 143.43
$CH_4 + Cl_2 \rightarrow CH_3Cl + HCl$	- 24.82	- 25.64	- 25.60	- 27.48	27.52	- 29.46	- 29.78

SOURCE: From M. J. S. Dewar and G. P. Ford, *J. Am. Chem. Soc.*, *99*, 7822 (1977).

from the JANAF table, where the reference enthalpy is at 298.15 K rather than 0 K. Then, for methane,

$$\frac{G^0 - H_0^0}{RT} = \frac{G^0 - H_{298}^0}{T} + \frac{H_{298}^0 - H_0^0}{T} = \frac{G^0 - H_{298}^0}{T} + \frac{2.396 \times 10^3}{T}$$

and at 1500 K, and in units of cal/mol · K,

$$\frac{G^0 - H_0^0}{T} = -54.387 + \frac{2.396 \times 10^3}{1500} = -52.790$$

where − 54.387 and 2.396 were read from the JANAF table. Then

$$\frac{G^0 - H_0^0}{RT} = -26.568$$

This tabulated value from JANAF must, of course, agree with the value calculated above from experimental molecular data, and it does.

Using the corresponding molecular data from the MINDO/3 calculations, one can deduce the following result for acetylene:

$$\left(\frac{G^0 - H_0^0}{RT}\right) = -28.850$$

where the rotational contribution must be evaluated for linear acetylene. This value may be compared with the result for experimental data from the JANAF table:

$$\left(\frac{G^0 - H_0^0}{RT}\right) = -28.879$$

The values are different by 0.1 percent. For hydrogen,

$$\left(\frac{G^0 - H_0^0}{RT}\right) = -17.812$$

which may be compared with the result for experimental data from the JANAF table:

$$\left(\frac{G^0 - H_0^0}{RT}\right) = -17.910$$

The difference is 0.5 percent, but the free energy was calculated here by utilizing the experimental rotational constant and a theoretical frequency.

The values of ΔH_f^0 from the MINDO/3 calculation have been reported[3] at 298 K, whereas we need the heat of reaction at 0 K to combine with the above free energy functions to obtain the desired

ΔG^0 of reaction. The correction is made by using the following identity:

$$\Delta H_0^0(\text{rx}) = \Delta H_{298}^0(\text{rx}) - RT\sum_i\left(\frac{H^0 - H_0^0}{RT}\right)_{298}$$

The enthalpy at 298 K relative to 0 K is evaluated for CH_4 by using the MINDO/3 frequencies. The vibrational contribution at 298 K is evaluated as:

ω (theo)	ω/T	g	$(H^0 - H_0^0)/RT$
3505	11.76	1	—
1326	4.45	2	0.0107
3551	11.92	3	—
1270	4.26	3	0.0134
		Total	0.0616

The total enthalpy function for CH_4 is then

$$\frac{H^0 - H_0^0}{RT} = 4 + \sum_i\left(\frac{H^0 - H_0^0}{RT}\right)_{\text{vib}} = 4 + 0.0616 = 4.062$$

which may be compared with the value from the JANAF table, from experimental frequencies, of

$$\frac{H^0 - H_0^0}{RT} = \frac{2.396 \times 10^3}{1.987(298.15)} = 4.044$$

Using the corresponding theoretically calculated molecular data and procedure, one calculates for linear acetylene

$$\frac{H^0 - H_0^0}{RT} = 3.5 + \sum_i\left(\frac{H^0 - H_0^0}{RT}\right)_{\text{vib}} = 3.5 + 0.614 = 4.114$$

which may be compared with

$$\frac{H^0 - H_0^0}{RT} = \frac{2.393 \times 10^3}{1.987(298.15)} = 4.039$$

from the JANAF table. For H_2, the vibrator is not excited at room temperature, and

$$\frac{H^0 - H_0^0}{RT} = 3.5$$

which may be compared with

$$\frac{H^0 - H_0^0}{RT} = \frac{2.024 \times 10^3}{1.987(298.15)} = 3.416$$

from the JANAF table. Combining all of these results, one calculates the heat of reaction at 0 K from the identity

$$\Delta H_0^0(\text{rx}) = \Delta H_{298}^0(\text{rx}) - RT\sum_i \left(\frac{H^0 - H_0^0}{RT}\right)_{298}$$

When using all-theoretical values one obtains

$$\begin{aligned}\Delta H_0^0\,(\text{rx}) &= [57.3 + 3(0.7) - 2(-11.9)] - 1.987(298)10^{-3}[4.114 \\ &\qquad + 3(3.500) - 2(4.062)] \\ &= 83.2 - 3.8 \\ &= 79.4 \text{ kcal/mol}\end{aligned}$$

The corresponding number for all-experimental data rather than the theoretical data is

$$\begin{aligned}\Delta H_0^0(\text{rx}) &= [54.3 + 3(0) - 2(-17.9)] + \frac{1.987(298.15)}{1000}[4.039 \\ &\qquad + 3(3.416) - 2(4.044)] \\ &= 93.8 \text{ kcal/mol}\end{aligned}$$

The discrepancy is large. The error in the MINDO/3 scheme for ΔH_f^0 of CH_4 and C_2H_2 accounts for 15 kcal/mol of the difference. In this case the errors add. In other examples, the errors might be opposite in sign and cancel. In any event, the major uncertainty in predicting a conversion from first principles, i.e., using no input data, results from the uncertainty in ΔH_f^0 from the MINDO/3 scheme. Conclusions will be similar for MNDO or some other scheme as well.

The $\Delta G^0(\text{rx})$ of the reaction forming acetylene from methane at 1500 K and 1 atm for all-theoretical data is

$$\begin{aligned}\Delta G^0(\text{rx}) &= RT\sum_i \left(\frac{G^0 - H_0^0}{RT}\right) + \Delta H_0^0(\text{rx}) \\ &= 1.987(1500)[-28.850 + 3(-17.812) \\ &\qquad - 2(-26.676)]10^{-3} + 79.4 \\ &= -86.2 + 79.4 = -6.8 \text{ kcal/mol}\end{aligned}$$

Using the experimental heat of formation with the wholly theoretical free energies, this becomes

$$\Delta G^0(\text{rx}) = -86.2 + 93.8 = 7.6 \text{ kcal/mol}$$

The major source of error is the theoretically predicted heats of formation. The best possible calculation of $\Delta G^0(\text{rx})$ is obtained by using experimental molecular data to evaluate the $(G0 - H_0^0)/RT$ functions combined with calorimetric heat-of-formation data. In that case,

$$\Delta G^0(\text{rx}) = 1.987(1500)[-28.879 + 3(-17.910) - 2(-26.569)]10^{-3} + 93.8$$

$$= -87.8 + 93.8 = 6.0 \text{ kcal/mol}$$

Not surprisingly, this result is very near to that just obtained by using free energies from wholly theoretical molecular data and calorimetric heat-of-formation data. This same calculation reported earlier (see Table 8.5) lists the ΔG^0 of reaction to be -7.01 kcal/mol. Presumably the difference is attributable to the use of somewhat different standard values for the heats of formation of CH_4 and C_2H_2, though these values are well-known.

Whatever the source of discrepency with the earlier calculations, we must conclude that uncertainty in the quantum-mechanically calculated heats of formation, whether by *ab initio* or semiempirical techniques, prevents their credible use in calculating conversions in chemical reactions. A small error in ΔG^0 results in a thoroughly erroneous calculation of conversion because of the exponential relationship. Either experimental or better estimates of heats of formation do allow reasonably credible calculations of conversion, since the values $(H^0 - H_0^0)$ and $(G^0 - H_0^0)/RT$ from theoretical molecular data are reasonably accurate. In this example, the differences with the tabulated JANAF values were usually only a few tenths of a percent.

Propellant Chemistry†

Many N-O-F compounds can be imagined, a few are known, and most of them could, in the absence of experimental data, be reasonable candidates as oxidizers in propellant systems. Synthesis procedures are complex, and it would be desirable to have some sort of guide as to the expected endothermicity and stability of candidate compounds that are presently unknown. Endothermic fuels and oxidizers are desirable because the efficiency of a propellant combination depends on the specific enthalpy of the flame gas. Unfortunately, endothermicity frequently implies lack of stability, so storage is troublesome and the candidate propellant compound may then be impractical. In any event, it would be desirable to have an estimate of the heat of formation of can-

†Adapted from Ganguli and McGee.[4]

didate compounds which can serve as a sort of screen before embarking on a long experimental synthesis program.

The MINDO scheme was calibrated to a number of relevant compounds and then used to calculate the heat of formation of all known N-O-F compounds. The results appear in Table 8.6, where the agreement is reasonably good. No data on any compound in Table 8.6 were used in parameterizing the MINDO code.

The heats of formation of several postulated, but unknown, N-O-F compounds were then calculated using the same scheme as was employed with the compounds of Table 8.6. The results appear in Table 8.7. This table also shows the sum of the heats of formation of the products for the particular decomposition pathway that minimizes this total energy of the products. The heats of formation of these individual species in the products from the expected decomposition are all reasonably well-known, and they are summarized in Table 8.8.

Arguments that cannot be supported as a matter of principle can nonetheless sometimes serve as a useful guide. Such an argument is the Thomsen-Berthelot principle,[5] where we imagine a species to be thermodynamically stable if its heat of formation is lower than the sum of that for all the decomposition products. Of all imaginable decompositions, that which appears in Table 8.7 has the lowest $\Sigma \Delta H_f^0$ (products). A comparison of the quantities in columns 2 and 4 of Table 8.7 suggests those species labeled m would more likely be stable than those labeled u. Interestingly, after this analysis had been completed, FONO was reportedly synthesized, which is, of course, in keeping with expectations from this approximate analysis.

Although it does not denigrate the above analysis, actually none of these N-O-F compounds will likely be interesting as propellant oxidizers because of their expected low endothermicity.

Better quantum-mechanical codes and parameterizations have occurred since this analysis first appeared. It is unlikely that use of any

TABLE 8.6 Theoretically Predicted Heats of Formation of Known N-O-F Compounds by MINDO

Compound	ΔH_f^0, kcal/mol, theoretical	ΔH_f^0, kcal/mol, experimental
FNO_2 (planar)	− 31.5	− 25.8 to − 33.8
$FONO_2$ (planar)	1.5	2.5
ONF_3	− 33.3	− 34.1
$ONNF_2$ (planar)	18.2	18.9
O_2NNF_2 (nonplanar)	− 14.7	0
FNNF (chair)	19.5	19.4
F_2NNF_2 (nonplanar)	− 2.8	− 5.0

SOURCE: Adapted from Ganguli and H. A. McGee.[4]

TABLE 8.7 Theoretically Predicted Heats of Formation of Some Unknown N-O-F Compounds and a Postulated Decomposition

Compound	ΔH_f^0, kcal/mol	Expected decomposition	$\Sigma\Delta H_f^0$, kcal/mol, products	Stability
FONO	7.6	$\rightarrow F + NO_2$	23.8	m
FO_2NO	- 31.1	$\rightarrow FO_2 + NO$	27.7	m
$F_2N\text{-}O\text{-}NF_2$	21.7	$\rightarrow NF_2 + ONF_2$	2.6	u
$F_2N\text{-}OO\text{-}NF_2$	37.1	$\rightarrow 2\ ONF_2$	- 11.8	u
ONFN-NFNO	78.8	$\rightarrow 2NO + N_2 + F_2$	43.2	u
$F_2N(OF)$	- 20.5	$\rightarrow F + ONF_2$	10.0	m
$FN(OF)_2$	- 9.9	$\rightarrow F_2 + FNO_2$	- 25.8	u
$N(OF)_3$	6.6	$\rightarrow F_2 + FNO_3$	2.5	u
$F_2N(O_2F)$	7.9	$\rightarrow NF_2 + O_2F$	14.6	m
$FN(O_2F)_2$	74.0	$\rightarrow FNO + OF + O_2F$	22.3	u
$N(O_2F)_3$	140.0	$\rightarrow NO_3 + 3\ OF$	113.0	u

SOURCE: Adapted from Ganguli and McGee.[4]

TABLE 8.8 Heats of Formation of N-O-F Decomposition Products at 298 K

Species	ΔH_f^0, kcal/mol	Species	ΔH_f^0, kcal/mol	Species	ΔH_f^0, kcal/mol
F	15.9	FNO	- 15.8	O_2F	6.1
NO	21.6	FNO_2	- 31.5	NF_2	8.5
NO_2	7.9	FNO_3	2.5	ONF_2	- 5.9
NO_3	17.0	OF	32.0		

SOURCE: Adapted from Ganguli and McGee.[4]

of these more recent formulations would result in any substantive change in the overall conclusions.

Metal-Organic Compounds

Semiempirical codes can be used with metal-organic compounds. The MNDO code, for example, has been parameterized for beryllium, and has allowed a number of interesting calculations and comparisons with experiment.[6]

Beryllium bridges to the face of a cyclopentadienyl ring to form compounds reminiscent of ferrocene, for example:

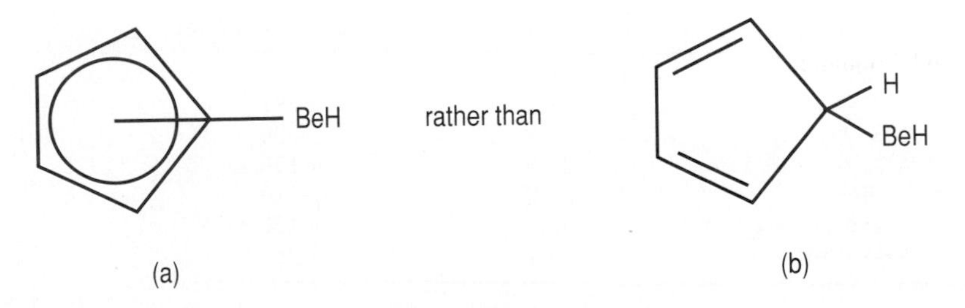

The MNDO calculated structure of lowest energy is *a*, which is also the experimental structure as revealed by microwave spectroscopy. The structure *a* has a ΔH_f^0 of 12.8 kcal/mol, while *b* has a ΔH_f^0 of 21.9 kcal/mol. The calculated ionization energy of *a* is 9.88 eV (compare 9.64 eV from experiment), and the calculated and experimental dipole moments are 2.02 and 2.08 debyes, respectively.

BeO is isoelectronic with C_2, and it is interesting to speculate on the stability of certain substituted compounds reminiscent of the corresponding hydrocarbon. HBeOH is predicted to be linear like acetylene and to have ΔH_f^0, I, and μ of − 73.9 kcal/mol, 12.7 eV, and 1.96 D respectively. FBeOH is also linear but both HBeOF and FBeOF are predicted to be bent. The napthalene analog

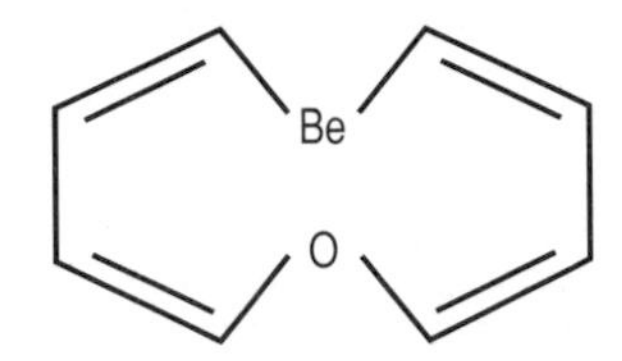

is predicted to be planar with ΔH_f^0, I, and μ of − 21.3 kcal/mol, 8.87 eV, and 3.75 D respectively. This compound has been synthesized. The benzene analog $(HBeOH)_3$ is predicted to be planar with a ΔH_f^0 = − 325 kcal/mole.

The heats of formation of a number of molecules for which comparison with experiment is possible are summarized in Table 8.9.

Various sorts of experimental data have led to at least five proposed structures of BeB_2H_8. The double-hydrogen-bridged beryllium borohydride has been predicted by MNDO to be optimum and to have a predicted structure of

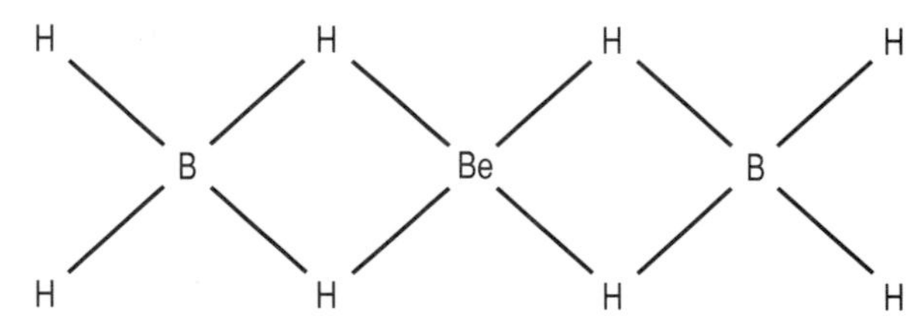

TABLE 8.9 Heats of Formation of Some Compounds of Beryllium

Compound	ΔH_f^0(calc), kcal/mol	ΔH_f^0(exp), kcal/mol
BeH_2	5.6	− 8
BeO	38.2	31
Be_2O_2	− 49.3	− 98
$Be(OH)_2$	− 136.6	− 156.4
BeF	− 52.9	− 48
BeF_2	− 192.3	− 190.3
FBeOBeF	− 286.7	− 287.9

and a calculated $\Delta H_f^0 = -56.0$ kcal/mol. Although the list is incomplete, where calculation is possible, several observed frequencies may be compared with those predicted by the MNDO technique. These comparisons appear in Table 8.10.

Calculations to serve as a guide can also be developed for other isoelectronic systems. There are many BN isoelectronic analogs of C_2 compounds; that is, the chemistry of BN species should be analogous to organic chemistry. An important practical result from such isoelectronic reasoning is behind the discovery of a cubic form of boron nitride analogous to cubic carbon, i.e., diamond. Cubic boron nitride is unknown in nature, but since amorphous carbon could be converted into diamond by very high pressure techniques, it seemed reasonable to try to similarly convert amorphous boron nitride into a cubic form. The experiments were successful. Cubic boron nitride is better than diamond for high-speed drilling and cutting applications because it does not oxidize at high temperatures.

Interestingly, borazene, the BN isoelectronic analog of benzene, even smells like benzene.

Aminoborane, the BN isoelectronic analog of ethylene,

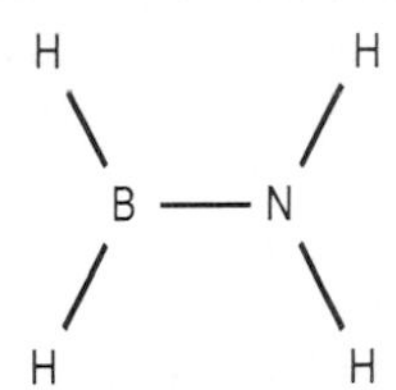

has been produced by a radio-frequency discharge in borazine in a fast-flowing system at low pressures followed by a rapid quench to 77 K.[7] The ionization potential of aminoborane was determined to be 11.0 ± 0.1 eV. The species slowly polymerizes to form a white insoluble solid that is presumably an inorganic polyethylene.

Dipole Moments

The dipole moment, so important in concerns for real-gas phenomena, transport properties, and chemical reaction rate, can also be obtained

TABLE 8.10 Experimental and Theoretical Vibration Frequencies, in cm^{-1}, for Beryllium Borohydride as Predicted by the MNDO Scheme

ω (obs)	ω (calc)	ω (obs)	ω (calc)
2500	2922	1548	1564
1615	1532	1000	893
588	494	2550	2977
2071	2283	2000	2189
		1650	1595

from quantum-mechanical calculations. Some insight into the agreement with experiment that may be expected is evident in Table 8.11.

Barriers to Internal Rotation

Most molecules are not rigid, and restricted or hindered internal rotations about single bonds occur. As was discussed in Chap. 4, such hindered internal rotations significantly affect the thermodynamic properties of such species. Hindered rotation may be taken into account by the methods described in detail in Chap. 4, wherein the required input data are the barrier to free rotation and the moment of inertia of the rotating fragment of the molecule. The assumptions of the method were two: the barrier was shaped as a cosine [see Eq. (4.4)] and the rotating fragment was a symmetric top. A symmetric top is an internal rotator that has no effect on the moments of inertia of the molecule as a whole regardless of the position of the rotator between 0 and 2π radians. Thus the $-CH_3$ in toluene is a symmetric top, but $-CH_2Cl$ is not. We saw in Chap. 4 that the methods of calculating the thermodynamic properties of molecules with symmetric tops could often be applied to molecules with asymmetric tops with reasonable success.

The quantum-mechanical schemes allow estimates of this barrier to rotation, and some examples are collected in Table 8.12. Here the computed values have been obtained using the semiempirical code called MINDO/3. As we have already seen, the optimum structure of a molecule, i.e., the collection of bond angles and bond distances that min-

TABLE 8.11 Calculated and Experimental Values of the Dipole Moment

Compound	μ, D	
	Theoretical	Experimental
NH_3	1.76	1.47
C_6H_5—NH_2	1.48	1.53
O_3	1.18	0.53
H_2O	1.78	1.85
C_2H_5OH	1.40	1.69
C_6H_5—OH	1.67	1.45
CH_3CHO	2.38	2.69
CH_3COOH	1.68	1.74
N_2O	0.76	0.17

SOURCE: Adapted from Dewar and Thiel.[3]

TABLE 8.12 Theoretical and Experimental Barriers to Internal Rotation

Molecule	Barrier, kJ/mol, theoretical	Barrier, kJ/mol, experimental
CH_3-NH_2	4.6	8.4
CH_3-OH	2.9	4.6
HO-OH	0	4.6
CH_3-CH_3	4.2	12.1

SOURCE: Adapted from Dewar and Thiel.[3]

imize the total energy of the molecule, can also be predicted by using the MINDO/3 code. Other codes, whether *ab initio* or semiempirical, of course also predict structure. In any event, the internal moment of inertia I_r is readily calculated from the optimum structure, for $I_r = \Sigma m_i r_i^2$ where each r_i is just the perpendicular distance from the axis of rotation to the mass m_i.

Even though the percentage errors in the examples from Table 8.12 are large, this may or may not make a large difference in the calculation of some desired thermodynamic property. For example, if we take the internal rotation constant of $-CH_3$ to be a nominal value of 5.5 cm^{-1}, which corresponds to a moment of 5×10^{-47} $kg \cdot m^2$, we calculate an internal partition function of $f_{vib,rot} = 3.6$ at 300 K. Table 4.5 gives the contribution to $(H^0 - H_0^0)$ for ethane arising from internal rotation, with $1/f_{vib,rot} = 1/3.6 = 0.28$ and $V/RT = 4200/R(300) = 1.7$, as about $1.4(300) = 1800$ J/mol. For comparison, using the experimental barrier one obtains a V/RT of $12{,}100/R(300) = 4.9$, which leads to $H^0 - H_0^0 = 1.3(300) = 1600$ J/mol.

The theory is impressive, but caution in its use is essential. Its use as a guide in some engineering task is best. Relative comparisons are also sensible. Specific numbers should be used only with realization of the possible error as well as an engineering judgment as to the seriousness of that possible error.

Ionization Potentials

The quantum-mechanical computational schemes allow the calculation of the total energy of the neutral molecule as well as its energy when missing one of its electrons. These two numbers allow the calculation of the heat of formation of both the neutral and the charged species. The difference in the two is the ionization potential, and values of $(H^0 - H_0^0)$ for neutral, ion, and electron, permit the calculation of the ionization energy at any temperature. Some examples of ionization energies appear in Table 8.13.

TABLE 8.13 Calculated and Experimental Ionization Energies for Several Species

Compound	IP, eV		Compound	IP, eV	
	Calculated	Experimental		Calculated	Experimental
	8.57	8.15	O_3	12.71	12.75
	9.39	9.25	CH_3CHO	10.88	10.21
CH_4	13.87	14.0	HCOOH	11.74	11.51
NH_3	11.19	10.85	N_2	14.88	15.60

SOURCE: Adapted from Dewar and Thiel.[3]

These numbers are needed in calculations of the enthalpy of plasmas, in calculations of the extent of ionization in some high-temperature process, in arguments concerning the upper atmosphere, in spacecraft reentry problems, and in general in all high-temperature processes.

The agreement is good, but note that the unit of energy is large, i.e., $0.1 \text{ eV} \cong 2500 \text{ cal/mol} \cong 10 \text{ kJ/mol}$.

Summary

Quantum mechanics and statistical mechanics have come a long way toward providing answers to practical problems. The values calculated for properties like ΔH_f^0, S^0, C_p^0, and ΔG^0 are obviously very important. But even more to the point, these results have shown how molecular concepts can be taken out of the "in principle" category and placed into the "in practice" category.

References

1. M. J. S. Dewar and A. Kormornicki, *J. Am. Chem. Soc., 99*, 7822 (1977).
2. M. J. S. Dewar and G. P. Ford, *J. Am. Chem. Soc., 99*, 1685 (1977).
3. M. J. S. Dewar and W. Thiel, *J. Am. Chem. Soc., 99*, 4907 (1977).
4. P. S. Ganguli and H. A. McGee, Jr., *Inorg. Chem., 11*, 3071 (1972).
5. H. S. Taylor, *A Treatise on Physical Chemistry*, Van Nostrand, New York, 1925.
6. M. J. S. Dewar and H. S. Rzepa, *J. Am. Chem. Soc., 100*, 777 (1978).
7. H. A. McGee, Jr., and C. T. Kwon, *Inorg. Chem., 9*, 2458 (1970).

Further Reading

1. J. P. Lowe, *Quantum Chemistry*, Academic Press, New York, 1978.

Chapter

9

Kinetic Theory and the Transport Properties

Molecular events must be described quantum-mechanically. Translational energies are, however, sufficiently close together to constitute a continuum, and classical descriptions are satisfactory. Molecular translation is the central phenomenon in mass, momentum, and energy transport, and a purely classical mechanical insight gives useful results.

The Boltzmann distribution law from Chap. 2 may be cast in classical form. Recall that a collection of n molecules of fixed total energy will automatically assume a distribution, or a sharing, of the total energy that can occur in the most number of ways. The driving force is simple chance. In Chap. 2, the conventional lagrangian scheme for maximizing the number of configurations w subject to fixed total molecules n and sharing a fixed total energy U was employed to yield Eq. (2.20),

$$\frac{n_i}{n} = Ke^{-\varepsilon_i/kT} \tag{9.1}$$

where n_i/n is the fraction of molecules having energy ε_i and K is a temperature- and volume-dependent proportionality constant. The reciprocal of K was evaluated; it was named the partition function, and much of this book has been concerned with its evaluation. The ratio n_i/n is the fraction of molecules that have energy ε_i. It is also the probability or the chance that any one molecule selected at random will have energy ε_i. For initial simplicity, consider a mole of a monatomic ideal gas. Then Eq. (9.1) becomes

$$n_i/n = Ke^{-m(u_x^2 + u_y^2 + u_z^2)/2kT} \tag{9.2}$$

Obviously, there is an infinity of combinations of u_x, u_y, and u_z that, when squared and summed, will yield the same total energy ε_i. From Fig. 9.1, these combinations are all radius vectors c that terminate in the spherical shell of volume $4\pi c^2\, dc$ where $c^2 = u_x^2 + u_y^2 + u_z^2$. The much smaller fraction of molecules with *specific* velocity components is just the number that terminate in the elemental volume of Fig. 9.1, $c^2 \sin\theta\, d\theta\, d\phi\, dc$. Equation (9.2) for this fraction of molecules with specific velocity components u_x, u_y, u_z may then be rewritten

$$d\left(\frac{n_{u_x,u_y,u_z}}{n}\right) = Ke^{-mc^2/2kT}c^2 \sin\theta\, d\theta\, d\phi\, dc \tag{9.3}$$

In most applications, the concern is for total molecular speed rather than specific components of velocity. This distribution of speeds is readily obtained by just integrating over all θ and ϕ,

$$d\left(\frac{n_c}{n}\right) = Ke^{-mc^2/2kT}c^2\, dc \int_0^{\pi} \sin\theta\, d\theta \int_0^{2\pi} d\phi$$

or

$$d\left(\frac{n_c}{n}\right) = 4\pi Ke^{-mc^2/2kT}c^2\, dc \tag{9.4}$$

The final integration over all c must yield the total fraction of all molecules, or one. This boundary condition allows evaluation of K,

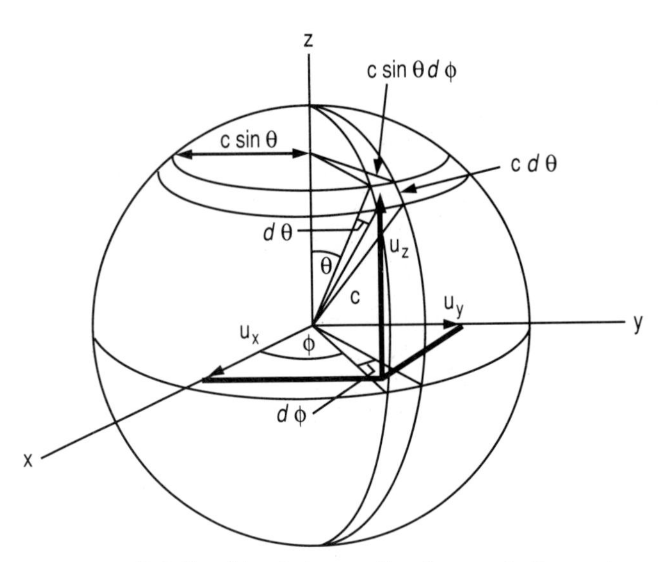

Figure 9.1 Relationships between the three velocity vectors and speed.

$$\int_0^1 d\left(\frac{n_c}{n}\right) = 1 = 4\pi K \int_0^\infty e^{-mc^2/2kT} c^2\, dc = 4\pi K\left[\frac{\sqrt{\pi}}{4}\left(\frac{2kT}{m}\right)^{3/2}\right]$$

and

$$K = \left(\frac{m}{2\pi kT}\right)^{3/2}$$

With this value of K, the original distribution of speeds, Eq. (9.4), becomes

$$d\left(\frac{n_c}{n}\right) = 4\pi\left(\frac{m}{2\pi kT}\right)^{3/2} e^{-mc^2/2kT} c^2\, dc \tag{9.5}$$

This is the Maxwell-Boltzmann distribution of molecular speeds. Clearly the fraction that has a speed of zero is zero and the fraction with infinite speed is also zero. In between, the most probable speed, c_{mp}, is found by differentiating Eq. (9.5) and setting this derivative equal to zero:

$$\frac{d}{dc}\left(\frac{dn_c/n}{dc}\right) = 0$$

or

$$c_{mp} = \left(\frac{2kT}{m}\right)^{1/2} \tag{9.6}$$

The distribution for ethylene, C_2H_4, is plotted in Fig. 9.2. The curve is characteristically narrower and higher at lower temperature. The most probable velocity of C_2H_4 at 25°C is 421 $m \cdot s^{-1}$, and it increases with temperature to 780 $m \cdot s^{-1}$ at 1025°C. Although any high speed is possible, a speed of 5 times c_{mp} at 25°C of 2100 $m \cdot s^{-1}$ has a probability of 1.7×10^{-12} $s \cdot m^{-1}$ or about 10^9 less than the probability of c_{mp} of 2×10^{-3} $s \cdot m^{-1}$. High speeds and thus high energies, although unlikely, are important in producing high-energy collisions so vital in chemical kinetics.

The average translational speed of a molecule is not the most probable speed. Rather it is just the fraction of molecules with speed c, multiplied by the total number of molecules, multiplied by c, summed over all c, and divided by the total number of molecules. That is,

$$\bar{c} = 4\pi\left(\frac{m}{2\pi kT}\right)^{3/2} \int_0^\infty \exp\left(\frac{-mc^2}{2kT}\right) c^3\, dc$$

On integration, one finally obtains

$$\bar{c} = \left(\frac{8kT}{\pi m}\right)^{1/2} \tag{9.7}$$

and the average speed is greater than the most probable speed by (8/

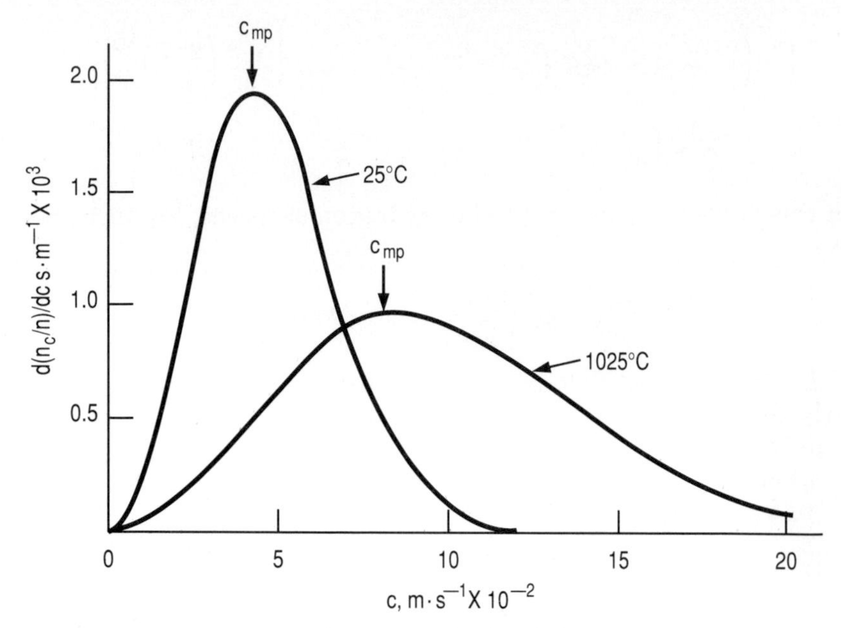

Figure 9.2 Distribution of molecular speeds for ethylene at two temperatures. Note the characteristic broadening and flattening as the temperature increases.

$\pi)^{1/2}/(2)^{1/2}$ or a factor of 1.13. The average speed of a molecule depends on its mass. The energy of a molecule with the most probable speed is $\frac{1}{2}\, m(2kT/m)$ or kT, which is 0.025 eV at 15°C, while at the same temperature, the molecule of average speed has an energy of 0.032 eV, and the molecule of root-mean-square speed $(3kT/m)^{1/2}$ has an energy of $\frac{3}{2}kT$ or 0.037 eV.

Example Problem 9.1 Calculate the average speed of O_2 gas at 300 K and at 1000 K.

$$\bar{c}_{300} = \left[\frac{8(1.38 \times 10^{-23})(300)(6.02 \times 10^{23})}{(3.14)(32)(10^{-3})}\right]^{1/2}$$

$$= 4.46 \times 10^2 \text{ m/s}$$

$$\bar{c}_{1000} = 4.46 \times 10^2 \left(\frac{1000}{300}\right)^{1/2}$$

$$= 8.15 \times 10^2 \text{ m/s}$$

Instead of the most probable speed, let us calculate the most probable velocity in, say the x direction, u_x. Then Eq. (9.3) in cartesian coordinates becomes

$$d\left(\frac{n_{u_x,u_y,u_z}}{n}\right) = Ke^{-m(u_x^2 + u_y^2 + u_z^2)/2kT}\, du_x\, du_y\, du_z$$

As before, the fraction of molecules with velocity components u_x, u_y, u_z, when summed over all velocity components, must equal 1, and

$$K = \frac{1}{\displaystyle\iiint_{-\infty}^{\infty} e^{-m(u_x^2 + u_y^2 + u_z^2)/2kT}\, du_x\, du_y\, du_z}$$

Equation (9.3) in cartesian coordinates becomes

$$d\left(\frac{n_{u_x,u_y,u_z}}{n}\right) = \frac{e^{-m(u_x + u_y + u_z)/2kT}\, du_x\, du_y\, du_z}{\displaystyle\iiint_{-\infty}^{\infty} e^{-m(u_x^2 + u_y^2 + u_z^2)/2kT}\, du_x\, du_y\, du_z} \tag{9.8}$$

We are interested only in u_x regardless of the corresponding components u_y and u_z. Performing the indicated three integrations in the denominator and two similar integrations over u_y and u_z in the numerator, one obtains the distribution of velocities in the x direction regardless of the components u_y and u_z,

$$d\left(\frac{n_{u_x}}{n}\right) = \left(\frac{m}{2\pi kT}\right)^{1/2} e^{-mu_x^2/2kT}\, du_x \tag{9.9}$$

The integrations over each velocity component are identical, with each equal to $(2\pi kT/m)^{1/2}$. The most probable velocity, u_x, is found, as before, by merely differentiating with respect to u_x and setting this equal to zero,

$$\frac{d}{du_x}\left[\frac{dn_{u_x}/n}{du_x}\right] = 0$$

or the most probable velocity in the x direction is zero. Similarly the average velocity in the x direction is

$$\bar{u}_x = \left(\frac{m}{2\pi kT}\right)^{1/2} \int_{-\infty}^{\infty} e^{-mu_x^2/2kT} u_x\, du_x$$

$$\bar{u}_x = 0$$

Note that Eq. (9.9) is already normalized. The result of $\bar{u}_x = 0$ is in keeping with intuitive expectation. That is, the molecules are moving with equal probability in both the $+x$ and $-x$ directions, so the average is zero.

A velocity space such as depicted in Fig. 9.1 is an example of multidimensional spaces useful in molecular theory that are called *phase-spaces*. The complete motion of one monatomic species is given by a single point in the phase-space of Fig. 9.1. If position is important, we imagine a six-dimensional phase-space of x, y, z, u_x, u_y, and u_z, and then both the position and motion of a monatomic species is indicated by a single point in this phase-space. The motion of a polyatomic species can be similarly represented in a phase-space of increasing dimension. A diatomic molecule would require a 12-dimensional space, and a molecule of n atoms would require a $6n$-dimensional phase-space.

The Maxwell-Boltzmann distribution of speeds, Eq. (9.5), is immediately converted into a translational energy distribution,

$$d\left[\frac{n(\varepsilon)}{n}\right] = \exp\left(\frac{-\varepsilon}{kT}\right)\frac{2\varepsilon^{1/2}\,d\varepsilon}{\sqrt{\pi}(kT)^{3/2}} \tag{9.10}$$

Here $d[n(\varepsilon)/n]$ is the fraction of molecules with energy ε. The right-hand side of Eq. (9.10) is dimensionless as expected. When Eq. (9.10) is integrated over all energy, the result is, of course, one.

The average energy of a monatomic species is determined just as was the average speed,

$$\bar{\varepsilon} = \frac{1}{n}\int_0^\infty \varepsilon\left\{\exp(-\varepsilon/kT)\frac{2\varepsilon^{1/2}\,d\varepsilon}{\sqrt{\pi}(kT)^{3/2}}\right\}dn \tag{9.11}$$

which is just the number of molecules of energy ε multiplied by the energy ε, summed over all ε, and divided by the total number of molecules. The integral of Eq. (9.11) may be written

$$I = \int_0^\infty e^{-a\varepsilon}\varepsilon^b\,d\varepsilon$$

where $a = 1/kT$ and $b = 3/2$. This is very close to the gamma function defined as

$$\Gamma(b + 1) = \int_0^\infty e^{-t}t^b\,dt$$

The gamma function is a generalized factorial function, and two characteristics may be readily shown by successive integration by parts

$$\Gamma(b + 1) = b\,\Gamma(b)$$

and $$\Gamma(b + 1) = b! \qquad \text{for } b = 1, 2, 3, \ldots$$

for all b greater than 0. The gamma function is tabulated in handbooks, but $\Gamma(1/2)$ occurs frequently in applications, and its value is readily shown to be $\Gamma(1/2) = \sqrt{\pi}$. Returning to Eq. (9.11), let $\varepsilon/kT = t$; then $dt = d\varepsilon/kT$ and with $b = 3/2$,

$$I = \int_0^\infty e^{-\varepsilon/kT}\varepsilon^{3/2}\,d\varepsilon = (kT)^{5/2}\int_0^\infty e^{-t}t^{3/2}\,dt$$

$$= (kT)^{5/2}\,\Gamma(5/2) = (kT)^{5/2}\,(3/2)\Gamma(3/2)$$

$$= (kT)^{5/2}\left(\frac{3}{2}\right)\left(\frac{1}{2}\right)\Gamma\left(\frac{1}{2}\right) = \frac{3\sqrt{\pi}}{4}(kT)^{5/2}$$

The average energy is then

$$\bar{\varepsilon} = \frac{2}{\sqrt{\pi}(kT)^{3/2}}\,\frac{3\sqrt{\pi}}{4}\,(kT)^{5/2} = \frac{3}{2}\,(kT) \tag{9.12}$$

Note that the average energy is not $\frac{1}{2}m\bar{c}^2$. Unlike the average speed, which depended on the mass of the molecule, the average energy depends on the temperature only. Recalling that $R = n_a k$, we find the thermodynamic internal energy per mole of a monatomic gas is

$$U^0 - U_0^0 = \tfrac{3}{2}RT$$

This is the same result, as expected, that was earlier obtained by using the quantum-mechanical translational energy levels.

Equipartition Law

The previous section revealed the average energy of monatomic species moving in three dimensions to be $\frac{3}{2}kT$. The same analysis in two dimensions, for an ideal gas adsorbed on a surface, but free to roam over the surface, would yield an average energy of $\frac{2}{2}kT$. There is a contribution $\frac{1}{2}kT$ to the average energy per degree of freedom of translational motion. Can this result be generalized?

Imagine a complex molecule, but nonetheless one in which each contribution to the energy arises from the square of some variable. For example, and as already shown, the translational energy in the x direction varies as $\frac{1}{2}mu_x^2$. Similarly, the energy of a one-dimensional rigid rotator in classical mechanics is $\frac{1}{2}I\omega^2$, where I is the moment of inertia of the rotator and ω is its angular velocity. Finally, the displacement of an atom in the molecule from its equilibrium position can be considered to result in a restoring force proportional to that displacement. This is, of course, the harmonic oscillator approximation, or Hooke's law. The potential energy or restoring energy due to such a displacement is $\frac{1}{2}kq^2$, where q is the displacement and k is a so-called force constant. The force constant is defined in terms of the frequency of oscillation ν as

$$\nu = \frac{1}{2\pi}\left(\frac{k}{\mu}\right)^{1/2} \tag{9.13}$$

The motion of the oscillator also produces a kinetic energy of $\frac{1}{2}\mu\dot{q}^2$ where μ is the so-called reduced mass (see later). In any event, the total kinetic energy of the molecule can be seen to arise from $3n$ squared terms. The total classical-mechanical kinetic *and* potential energy of a molecule in the rigid-rotator, harmonic-oscillator approximation is

$$\varepsilon = \sum_{i=1}^{3} \frac{1}{2} m_i u_i^2 + \sum_{i=1}^{2(\text{or } 3)} \frac{1}{2} I_i \omega_i^2 + \sum_{i=1}^{3n-5(\text{or } 6)} \left[\frac{1}{2} k q_i^2 + \frac{1}{2} \mu \dot{q}_i^2\right] \tag{9.14}$$

Of course, the total kinetic energy of the molecule could also be expressed by stating three components of velocity of each constituent atom, or a total of $3n$ velocities, where n is the number of atoms that together compose the molecule. But it is more physically sensible to state this kinetic energy as three components of translation of the center of mass of the molecule; two (or three) kinetic energies of rotation, depending on whether there is a moment of inertia about two or three axes (i.e., is the molecule linear or not); $3n - 5$ (or 6) harmonic vibrations with each contributing a term to the kinetic energy; and $3n - 5$ (or 6) potential energy terms, that is, one for each vibrator. The sum of both kinetic and potential terms is, of course, the total energy of the molecule. Note that each term in Eq. (9.14) involves a squared quantity: velocity, rotation frequency, displacement, or rate of change of displacement. The details are irrelevant, but it is obvious that new variables could be defined such that the total energy is

$$\varepsilon = \sum_{i=1}^{s} \rho_i^2 \equiv R^2 \tag{9.15}$$

with the sum extending over a total of s squared terms.

Imagine an s-dimensional phase-space of these variables, analogous to three-dimensional space. A single point in this space fixes all s variables, which, when squared and summed, fix the total energy of the molecule. Avogadro's number of molecules is represented by a swarm of 10^{23} points in this phase-space. In this hyperspace, the energy of a single molecule is just the radius vector squared, R^2, as is clear from Eq. (9.15). In the earlier monatomic case and in velocity space rather than energy space, the volume element was, after integration over all angles, represented as the spherical shell $4\pi c^2\, dc$ where $c^2 = u_x^2 + u_y^2 + u_z^2$ and where positive or negative velocity components were irrelevant since they were squared. We can similarly write a spherical shell in s dimensions as

$$dV = A\, dR = \left[\frac{2\pi^{s/2} R^{s-1}}{\Gamma(s/2)}\right] dR$$

This expression for area A in $A\, dR$ can be considered as a generalization of the known areas of "spherical" shapes in one and two dimensions.

Dimension of sphere (s)	2	3	s
Surface area	$2\pi R$ (circumference of a circle)	$4\pi R^2$ (area of a sphere)	$\frac{2\pi^{s/2}R^{s-1}}{\Gamma(s/2)}$

This generalized surface area follows from the defined length in a multidimensional Euclidean space as

$$R \equiv \left[\sum_{i=1}^{s} \rho_i^2\right]^{1/2}$$

in, for example, an s-dimensional space.

In three dimensions, of course, $\varepsilon = \frac{1}{2}mc^2 \equiv R^2$. In multidimensional space, with the radius vector similarly defined, R is still $\varepsilon^{1/2}$ and

$$dR = \tfrac{1}{2}\varepsilon^{-1/2}\, d\varepsilon$$

With the area of a "sphere" in this space as

$$A = \frac{2\pi^{s/2}R^{s-1}}{\Gamma(s/2)} = \frac{2\pi^{s/2}\varepsilon^{(s-1)/2}}{\Gamma(s/2)}$$

the differential volume of a spherical shell will be

$$dV = A\, dR = \left[\frac{2\pi^{s/2}\varepsilon^{(s-1)/2}}{\Gamma(s/2)}\right]\frac{1}{2}\varepsilon^{-1/2}\, d\varepsilon$$

As always, the Maxwell-Boltzmann distribution is exp $(-\varepsilon/kT)$ multiplied by the number of states with energy ε, divided by this same factor integrated over all energies,

$$d\left[\frac{n(\varepsilon)}{n}\right] = \frac{e^{-\varepsilon/kT}\, dV}{\int_0^\infty e^{-\varepsilon/kT}\, dV} = \frac{e^{-\varepsilon/kT}\, 4\pi R^2\, dR}{\int_0^\infty e^{-\varepsilon/kT}\, 4\pi R^2\, dR}$$

In three dimensions, this relationship becomes

$$d\left[\frac{n(\varepsilon)}{n}\right] = \frac{e^{-\varepsilon/kT}\varepsilon^{1/2}\, d\varepsilon}{\int_0^\infty e^{-\varepsilon/kT}\varepsilon^{1/2}\, d\varepsilon} = \frac{e^{-\varepsilon/kT}\varepsilon^{1/2}\, d\varepsilon}{(kT)^{3/2}\,(\frac{1}{2})\,(\sqrt{\pi})} \tag{9.16}$$

where we have recognized the denominator to be $(kT)^{3/2}\Gamma(\frac{3}{2})$. This result is, of course, the same as Eq. (9.10). With R considered as the radius vector, not in a three-dimensional space but in an s-dimensional space,

$$d\left[\frac{n(\varepsilon)}{n}\right] = \frac{e^{-\varepsilon/kT}\,dV}{\int_0^\infty e^{-\varepsilon/kT}\,dV}$$

$$= \frac{e^{-\varepsilon/kT}\,(2\pi^{s/2}\varepsilon^{(s-1)/2}/\Gamma(s/2))\,(\tfrac{1}{2}\varepsilon^{-1/2})\,d\varepsilon}{\int_0^\infty e^{-\varepsilon/kT}\,(2\pi^{s/2}\varepsilon^{(s-1)/2}/\Gamma(s/2))\,(\tfrac{1}{2}\varepsilon^{-1/2})\,d\varepsilon}$$

which may be simplified to

$$d\left[\frac{n(\varepsilon)}{n}\right] = \frac{e^{-\varepsilon/kT}\varepsilon^{(s-2)/2}\,d\varepsilon}{\int_0^\infty e^{-\varepsilon/kT}\varepsilon^{(s-2)/2}\,d\varepsilon} \tag{9.17}$$

The denominator is again proportional to a gamma function, and the final expression for the Maxwell-Boltzmann distribution in s dimensions becomes

$$d\left[\frac{n(\varepsilon)}{n}\right] = \frac{e^{-\varepsilon/kT}\varepsilon^{s/2-1}\,d\varepsilon}{(kT)^{s/2}\Gamma(s/2)} \tag{9.18}$$

Equation (9.18) is the fraction of molecules of energy ε regardless of the complexity of the molecule. Equation (9.18) is the multiatom equivalent of Eq. (9.16) for a monatomic species. We should also remember that Eq. (9.18) is a classical result; we have treated all of the motions, including vibration, as classical rather than quantized motions.

Just as was done for a monatomic gas, we can now calculate the average energy of this complex molecule by multiplying the number that have energy ε by ε, sum over all ε, and divide by n,

$$\bar{\varepsilon} = \frac{1}{(kT)^{s/2}\Gamma(s/2)}\int_0^\infty \varepsilon^{s/2}e^{-\varepsilon/kT}\,d\varepsilon$$

This integral is proportional to the gamma function, and it is evaluated as before to yield

$$\bar{\varepsilon} = \frac{s}{2}kT \tag{9.19}$$

as the average energy of a molecule of whatever complexity.

This is the general statement of the *law of equipartitioning of energy, which states that each degree of freedom, regardless of whether it relates to translation, rotation, or vibration, contributes* $\tfrac{1}{2}kT$ *to the energy*. On a macroscopic scale,

$$U^0 - U^0_0 = \frac{s}{2}RT$$

or

$$C^0_V = \frac{s}{2}R$$

and

$$C^0_p = \left(\frac{s}{2} + 1\right)R$$

and both heat capacities are independent of temperature. One might prepare a table as follows:

Molecular type	Number of squared terms in the energy expression	C^0_p
Monatomic	3	$\frac{5}{2}R$
Diatomic	7 (6 kinetic, 1 potential)	$\frac{9}{2}R$
Triatomic (linear)	13 (9 kinetic, 4 potential)	$\frac{15}{2}R$

This result agrees with experiment for monatomic species. But for everything else, the experimental heat capacity is neither independent of temperature nor a multiple of $\frac{1}{2}R$. Comparisons with earlier quantum-mechanical arguments (see Chap. 3) make clear that the error in the above arguments is treating the vibration as a classical-mechanical motion rather than as a quantum-mechanical motion (compare Einstein theory). It is interesting that the high-temperature limiting value of the Einstein vibrator is exactly the above classical value of R. That is, the maximum heat capacity of a diatomic gas is $\frac{9}{2}R$, it is $\frac{15}{2}R$ for all linear triatomic species, etc.

Fraction of Molecules with High Energy

Let us now calculate the fraction of molecules that have energy greater than a certain threshold value ε_a. When energy is shared among s degrees of freedom, this fraction of high-energy molecules is just Eq. (9.18), the distribution function, integrated from ε_a to ∞,

$$\frac{n(\varepsilon > \varepsilon_a)}{n} = \frac{1}{(kT)^{s/2}\Gamma(s/2)} \int_{\varepsilon_a}^{\infty} e^{-\varepsilon/kT} \varepsilon^{s/2-1}\, d\varepsilon$$

By using successive integration by parts, beginning with $u = \varepsilon^{s/2-1}$ and $dv = e^{-\varepsilon/kT}\, d\varepsilon$ and continuing similarly, this integral can be evaluated in closed form for even values of s. The result is

$$\frac{n(\varepsilon > \varepsilon_a)}{n} = e^{-\varepsilon_a/kT} \left[\frac{(\varepsilon_a/kT)^{s/2-1}}{\Gamma(s/2)} + \frac{(\varepsilon_a/kT)^{s/2-2}}{\Gamma\left(\frac{s}{2} - 1\right)} + \cdots + 1 \right] \tag{9.20}$$

where there are $s/2$ total terms inside the square bracket, and s can, of course, be no smaller than 2. Equation (9.20) is well-behaved in the limit of $\varepsilon_a = 0$ where $n(\varepsilon > \varepsilon_a)/n$ is correctly predicted to be 1 for any s. The last integration in this successive integration by parts does not vanish for odd values of s. Rather, one obtains a final term of

$$(\pi kT)^{-1/2} \int_{\varepsilon_a}^{\infty} e^{-\varepsilon/kT} \varepsilon^{-1/2}\, d\varepsilon$$

which, with $\varepsilon/kT = x^2$, becomes

$$2\pi^{-1/2} \int_{x_a}^{\infty} e^{-x^2}\, dx = 2\pi^{-1/2} \left[\int_0^{\infty} e^{-x^2}\, dx - \int_0^{x_a} e^{-x^2}\, dx \right]$$

$$= [1 - \operatorname{erf}(x_a)] = \operatorname{erfc}(x_a)$$

For example, if $s = 3$ Eq. (9.20) becomes,

$$\frac{n(\varepsilon > \varepsilon_a)}{n} = e^{-\varepsilon_a/kT} \left(\frac{2}{\sqrt{\pi}} \right) \left(\frac{\varepsilon_a}{kT} \right)^{1/2} + 1 - \operatorname{erf}\left(\frac{\varepsilon_a}{kT} \right)$$

This result is well-behaved in the limit of $\varepsilon_a = 0$, for then $n(\varepsilon > \varepsilon_a)/n$ is correctly predicted to be 1.

In many practical problems, the high-energy threshold, ε_a, of interest is greater than thermal, i.e., about kT, then $\operatorname{erf}(x_a) \cong 1$, and the last integration essentially vanishes as it did exactly for even values of s. For example, at $x = 1.6$, that is, for a threshold energy only 60 percent greater than thermal, $\operatorname{erf}(1.6) \cong 0.98$, and the last two terms above effectively cancel each other. One then obtains the same expression for the fraction of high-energy molecules that is obtained when retaining only the leading term in the series of Eq. (9.20). For all values of s, the ratio of the lead term in the square bracket to the second term is

$$\frac{(\varepsilon_a/kT)^{s/2-1}\Gamma(s/2 - 1)}{\Gamma(s/2)(\varepsilon_a/kT)^{s/2-2}} = \frac{\varepsilon_a}{kT} \left(\frac{2}{s - 2} \right)$$

In all practical problems, ε_a is much greater than $(s/2 - 1)kT$; the lead term alone in the series of Eq. (9.20) will suffice, and we obtain

$$\frac{n(\varepsilon > \varepsilon_a)}{n} = \left[\frac{(\varepsilon_a/kT)^{s/2-1}}{\Gamma(s/2)} \right] e^{-\varepsilon_a/kT} \tag{9.21}$$

We can see here the makings of a theory of chemical kinetics. This expression has also been used to estimate the rate of evaporation of a liquid or a solid and then the vapor pressure of such a condensed phase.

Frequency of Bombardment on a Plane Surface

The rate of chemical reaction, just like the rate of heat conduction, depends on intermolecular collision frequency. As a preliminary to understanding this intermolecular collision frequency, it is convenient to first understand the rate of molecular collision on a plane surface immersed in an ideal gas. This would also be the rate at which molecules escape into vacuum from a container at p and T per unit area of the hole. This is also called *Knudsen flow* or *effusive flow*, but it is also essential that a quantity called the mean free path (see later) be large relative to the dimensions of the hole. Consider such a plane perpendicular to the x axis as shown in Fig. 9.3. The frequency will be

$$z = n\bar{u}_x A \tag{9.22}$$

that is, the product of the number density of molecules, n, in m^{-3}, the average molecular speed in the $+x$ direction, and the area of the plane. The speeds in the y and z directions are, of course, irrevelant, so

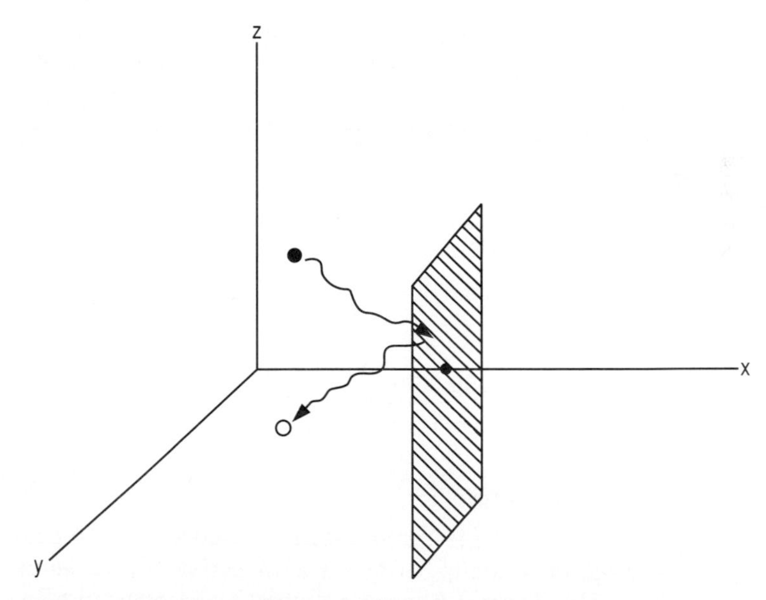

Figure 9.3 Molecular bombardment on a plane surface.

$$\bar{u}_x = \left(\frac{m}{2\pi kT}\right)^{3/2} \int_{-\infty}^{\infty} \exp\left(\frac{-mu_y^2}{2kT}\right) du_y \int_{-\infty}^{\infty} \exp\left(\frac{-mu_z^2}{2kT}\right) du_z \times \int_0^{\infty} \exp\left(\frac{-mu_x^2}{2kT}\right) u_x\, du_x$$

becomes, after integration over all u_y and u_z,

$$\bar{u}_x = \left(\frac{m}{2\pi kT}\right)^{1/2} \int_0^{\infty} \exp\left(\frac{-mu_x^2}{2kT}\right) u_x\, du_x \tag{9.23}$$

or

$$\bar{u}_x = \left(\frac{kT}{2\pi m}\right)^{1/2} \tag{9.24}$$

Clearly the average speed in any one direction is just one-fourth the overall average speed $\bar{c}$. The sought-for collision frequency in s^{-1} is then

$$z = n\left(\frac{kT}{2\pi m}\right)^{1/2} A = \frac{p}{kT}\left(\frac{kT}{2\pi m}\right)^{1/2} A \tag{9.25}$$

where p is in Pa, m in kg, and all the other symbols have their usual meaning. The number density n at standard temperature and pressure is called the Loschmidt number, and it is 2.69×10^{25} m^{-3}. The number density at 1 torr and 0°C is 3.54×10^{22} m^{-3}. In more conventional units one might rewrite Eq. (9.25) as

$$z = \frac{2.67 \times 10^{25} p}{(MT)^{1/2}} \quad \text{s}^{-1} \cdot \text{cm}^{-2} \tag{9.26}$$

where p is in atm, M is the molecular weight, and T, the temperature, is in K.

Example Problem 9.2 Determine the collision frequency on one face of a 1-cm^2 disk immersed in O_2 at 300 K and 0.1 bar. From Eq. (9.25),

$$z = \left[\frac{0.1 \times 10^5}{1.38 \times 10^{-23} \times 300}\right] \times \left[1.38 \times 10^{-23} \times 300 \times \frac{6.023 \times 10^{23}}{2 \times 3.14 \times 32 \times 10^{-3}}\right]^{1/2} \left[1 \times \left(\frac{1}{10^2}\right)^2\right]$$

$$= 2.7 \times 10^{22}\ \text{s}^{-1}$$

The collision frequency is very large.

To determine the frequency of hard collisions, we need only to find the average component of velocity in the + x direction above some threshold velocity u_a. This is readily accomplished by integrating Eq. (9.23) not from 0 to ∞, but from u_a to ∞. The result is

$$\bar{u}_x(u_x > u_a) = \left(\frac{kT}{2\pi m}\right)^{1/2} e^{-mu_a^2/2kT} \tag{9.27}$$

and, contrary to intuition, the average velocity decreases with increasing u_a. This result is, however, sensible when one recalls that the denominator in this calculation of an average, like that in the original development of Eq. (9.9), remains the total number of molecules. As before, the hard-collision frequency is now

$$z = n\bar{u}_x(u_x > u_a)A = \frac{p}{kT}\left(\frac{kT}{2\pi m}\right)^{1/2} e^{-mu_a^2/2kT}A \tag{9.28}$$

Here decreasing z with increasing u_a is intuitively reasonable.

Example Problem 9.3 Determine the hard-collision frequency on one face of a 1-cm^2 disk immersed in O_2 at 300 K and 0.1 bar when the threshold energy is 10 times thermal.

Thermal translation at 300 K in the x direction is $\frac{1}{2}kT$ as demanded by the principle of equipartitioning of energy. Then, from Eq. (9.28),

$$z = \left(\frac{p}{kT}\right)\left(\frac{kT}{2\pi m}\right)^{1/2} e^{-10(1/2)(kT)/(kT)}A$$

$$= 1.8 \times 10^{20}\ \text{s}^{-1}$$

The hard-collision frequency at this level of "hardness" is about 0.7 percent of the total collision frequency.

One can see here the beginnings of a theory of chemical reaction rates. Note that the hard-collision frequency, Eq. (9.28), is already in the macroscopically observed Arrhenius form.

Center-of-Mass Coordinate System

Consideration of the intermolecular collision frequency is greatly simplified by the introduction of what is called the center-of-mass coordinate system. For initial simplicity, consider the collision of two hard spheres of mass m_1 and m_2 which have total energy ε in the Lab system,

$$\varepsilon = \tfrac{1}{2}m_1(u_{x1}^2 + u_{y1}^2 + u_{z1}^2) + \tfrac{1}{2}m_2(u_{x2}^2 + u_{y2}^2 + u_{z2}^2)$$

Let us consider collisions in either the laboratory (Lab) coordinate system or the center-of-mass (CM) coordinate system.[1] The CM system moves with respect to the Lab system such that its origin is always at the center of mass of the colliding molecules. Obviously physical measurements are made in the Lab system, which is the frame of reference of the observer.

Assume a molecule m_1 moving toward a target molecule m_2 which is at rest in the Lab system. This condition holds, at least approximately, in most experimental studies of collision events. The situation both long before and long after the "collision" or point of nearest approach, b, the impact parameter, is schematically evident in Fig. 9.4. The scattering angles and changes in speed are our major concerns. The velocity of the center of mass, V_{CM}, is constant in magnitude and direction before, during, and after the collision, for we take the collision to be elastic.

For elastic collisions, momentum is conserved in magnitude, and then

$$(m_1 + m_2)V_{CM} = m_1 v_0$$

or

$$V_{CM} = v_0\left(\frac{\mu}{m_2}\right)$$

where μ is called the reduced mass and is clearly

$$\mu \equiv \frac{m_1 m_2}{m_1 + m_2}$$

Momentum is also conserved in direction, and then

$$m_1 v_0 = m_1 v \cos \phi + m_2 V \cos \theta$$

and

$$0 = m_1 v \sin \phi - m_2 V \sin \theta$$

where ϕ and θ are the scattering angles of m_1 and m_2, respectively. These expressions are based on defining velocities to the right and up in Fig. 9.4 to be positive. Kinetic energy is also conserved:

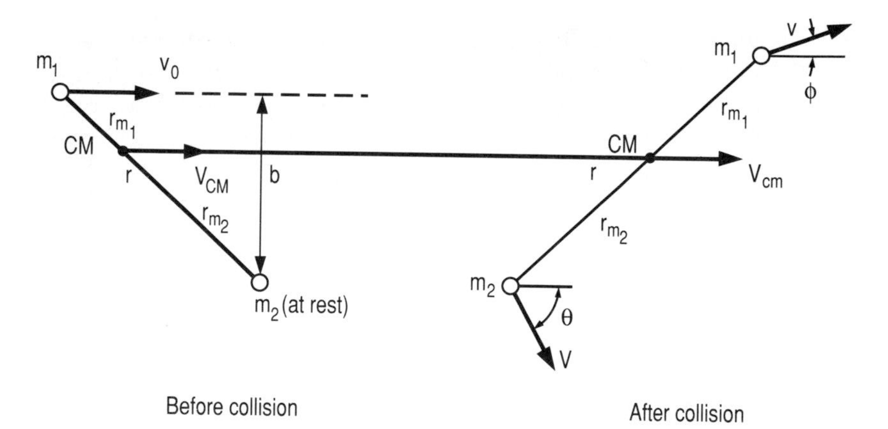

Figure 9.4 Asymptotic views of an elastic collision in the Lab system. The colliding particles of masses m_1 and m_2 are at distances r_1 and r_2, respectively, from the center of mass CM.

$$\tfrac{1}{2}m_1v_0^2 = \tfrac{1}{2}m_1v^2 + \tfrac{1}{2}m_2V^2$$

These relationships allow one to write the final velocity of each mass as

$$v = v_0\left[1 - \frac{4\mu^2\cos^2\theta}{m_1m_2}\right]^{1/2}$$

and
$$V = 2v_0\mu\,\frac{\cos\theta}{m_2}$$

Now consider the same collision, but occurring in the CM system as shown schematically in Fig. 9.5. Here the origin of the coordinate system is at the center of mass which is stationary. The velocity of approach is the same in the CM system as it was in the Lab system, v_0, but now the colliding molecules are moving toward each other. As before, we can equate the magnitudes of momenta of each molecule,

$$m_1(v_0 - V_{\mathrm{CM}}) = m_2V_{\mathrm{CM}}$$

The relative velocity of approach, v_0, is the same in both systems, but m_2 is now not stationary, but rather it must be moving toward m_1 with a velocity of V_{CM}. From the above equation, V_{CM} is $v_0(\mu/m_2)$, or the momentum of m_2 is just $v_0\mu$,

$$m_2V_{\mathrm{CM}} = v_0\mu$$

The momentum of m_1 is

$$m_1(v_0 - V_{\mathrm{CM}}) = m_1\left(v_0 - \frac{v_0\mu}{m_2}\right) = v_0\mu$$

Both molecules in the CM system have momentum $v_0\mu$. The linear

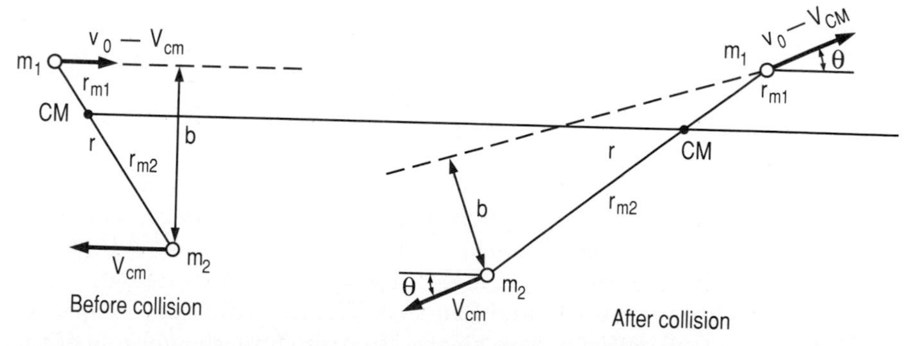

Figure 9.5 Asymptotic views of an elastic collision in the center-of-mass coordinate system.

momentum must be equal and opposite before and after collision, so the molecules are traveling in opposite directions before and after collision. There is no distinction between target and projectile. The necessarily same scattering angle is θ. Kinetic energy is conserved and so the speed of each molecule must be the same before and after collision. The total kinetic energy in the CM system is

$$T = \tfrac{1}{2}m_1(v_0 - V_{CM})^2 + \tfrac{1}{2}m_2V_{CM}^2$$

which may be rearranged to

$$T = \tfrac{1}{2}m_1v_0^2 - \tfrac{1}{2}(m_1 + m_2)V_{CM}^2$$

or to

$$T = \tfrac{1}{2}\mu v_0^2 \tag{9.29}$$

This useful relationship reveals that the total kinetic energy of a colliding pair of molecules in the CM system is of familiar form but involving the reduced mass μ and the relative velocity of approach, v_0. The motion of two molecules relative to each other can be replaced by the equivalent problem of the motion of a single mass μ in three-dimensional space. As before, the Maxwell-Boltzmann distribution law applies, and Eq. (9.5) that described a single molecule becomes

$$d\left[\frac{n_{v_0}}{n}\right] = \left(\frac{2}{\pi}\right)^{1/2}\left(\frac{\mu}{kT}\right)^{3/2} e^{-\mu v_0^2/2kT}\, v_0^2\, dv_0 \tag{9.30}$$

Equation (9.30) is the chance that two molecules taken at random will have a relative speed of approach of v_0. Or equally, with one molecule at rest, Eq. (9.30) is the chance that another molecule will be moving relative to it with speed v_0. Or equally, it is the fraction of all pairs of molecules that are moving toward each other with speed v_0.

Molecular Collision Frequency

With all this by way of background, let us consider the essential problem of transport processes and of chemical reaction rate, that is, the intermolecular collision frequency.

Let us approach the problem of intermolecular collision frequency in stepwise manner. First imagine all the molecules to be alike, each having radius σ, and let one molecule move through a swarm of all the others with each such target molecule held fixed in space. The wanderer, moving at a constant speed c, will trace out a zigzag cylinder as it collides with all of the target molecules whose centers lie within the traced cylinder of area $\pi\sigma^2$ and length c. The constant speed c means, of course, that all collisions are elastic. In unit time, the volume of the traced cylinder will be $\pi\sigma^2 c$, and volume corrections due to the kinks on each collision will be nil provided the distance of travel between collisions is large compared to the molecular diameter. The collisions suffered by the one wandering molecule, in s^{-1}, from this simplest perspective will then be

$$z = \pi\sigma^2 cn \tag{9.31}$$

where n is the target molecular density in m^{-3}. Alternatively, we could also imagine the wanderer as fixed and showered with a beam of molecules of number density n and all moving with speed c. The collision frequency is the same.

All of the molecules are, however, moving, and in our stepwise approach, we might next imagine all of the molecules to be moving with the same speed c. The collision frequency is now better determined as

$$z = \pi\sigma^2 n\, \bar{c}_r \tag{9.32}$$

where $\bar{c}_r$ is the average relative speed of a pair of colliding molecules. We can find this average relative speed with the help of Fig. 9.6. The relative speed c before and after the collision is the same, since collisions are elastic. With collision occurring at the center of the sphere of Fig. 9.6, the velocity vectors of the two receding molecules after collision will terminate on a sphere of radius c and separated by an angle θ. The desired relative velocity is the vector c_r that closes the isosceles triangle. All angles θ are equally probable, and if we can identify an average angle θ that we might call an average scattering angle, we could immediately determine the needed average relative

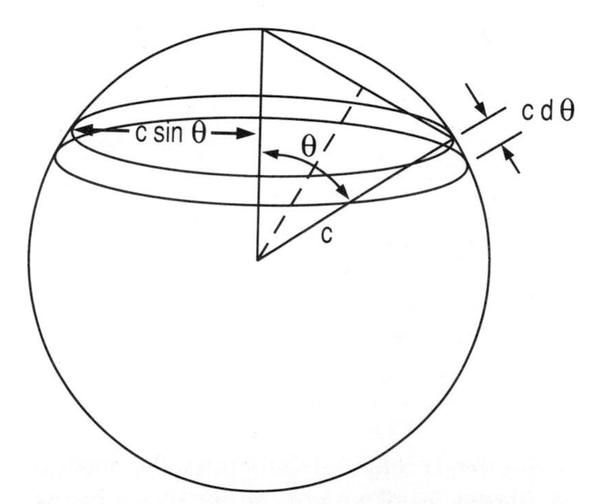

Figure 9.6 Velocity diagram to aid in determining the relative speed of two colliding molecules.

velocity. The chance of an angle between θ and $\theta + d\theta$, call it $p(\theta)\,d\theta$, is just the area of the thin circular rim divided by the total area of the sphere, or

$$p(\theta)\,d\theta = \frac{(2\pi c \sin\theta)\,(c\,d\theta)}{4\pi c^2} = \frac{\sin\theta\,d\theta}{2}$$

This probability function is normalized for its integral over all θ, i.e., from 0 to π, is obviously 1. The dashed line from the midpoint of c_r forms a right triangle and bisects θ such that the sine of one-half the average scattering angle is

$$\sin\frac{\overline{\theta}_r}{2} = \int_0^\pi \left(\sin\frac{\theta}{2}\right)\frac{\sin\theta\,d\theta}{2}$$

That is, the sine of half the average angle is just the sine of half of any angle, multiplied by the normalized probability of that angle, and integrated over all angles, or

$$\sin\frac{\overline{\theta}_r}{2} = \frac{2}{3}$$

But it is clear from Fig. 9.6 that $\sin \overline{\theta}_r/2$ is also $\frac{1}{2}\overline{c}_r/c$, then $\overline{c}_r = \frac{4}{3}c$.

The improved calculation of the collision frequency experienced by a single molecule is

$$z = \pi\sigma^2 n\overline{c}_r = \tfrac{4}{3}\pi\sigma^2 nc \tag{9.33}$$

which is 33 percent greater than was obtained from the simplest approach, Eq. (9.31). The value of c in both Eqs. (9.31) and (9.33) is the common fixed speed to which we can assign any value.

Finally, in our stepwise approach we recognize that the molecules are not all moving at relative speed $\overline{c}_r$ as assumed above, but rather there is a maxwellian distribution of speeds. The kinetic energy of a pair of colliding molecules in the Lab system is

$$T = \tfrac{1}{2}m_1c_1^2 + \tfrac{1}{2}m_2c_2^2$$

This was converted into the kinetic energy of the center of mass of the pair and their relative kinetic energy in the CM system [see Eq. (9.29)]:

$$T = \tfrac{1}{2}(m_1 + m_2)\,c_{\mathrm{CM}}^2 + \tfrac{1}{2}\mu c_r^2$$

where μ is the reduced mass, $m_1m_2/(m_1 + m_2)$. Clearly only the second term is relevant in collision problems, and we can think of a binary collision in terms of the motion of a single hypothetical particle of mass μ. Then the average relative speed of this hypothetical molecule is evaluated in the usual way from Eq. (9.30):

$$\bar{c}_r = \left(\frac{2}{\pi}\right)^{1/2}\left(\frac{\mu}{kT}\right)^{3/2}\int_0^\infty e^{-\mu v_0^2/2kT} v_0^3 \, dv_0$$

or

$$\bar{c}_r = \left(\frac{8kT}{\pi\mu}\right)^{1/2} \tag{9.34}$$

just as we had earlier obtained for the average speed of a real particle [see Eq. (9.7)]. When the colliding molecules are the same, μ is $m/2$ and $\bar{c}_r = \sqrt{2}\,\bar{c}$. In this final approximation, the collision frequency upon one molecule of species 1 that is immersed in pure species 1 of number density n is now

$$z = \pi\sigma^2 n\bar{c}_r = \sqrt{2}\pi\sigma^2 n\bar{c} \tag{9.35}$$

The difference in the numerical factors in these last two calculations of 4⁄3 or 1.33 and $\sqrt{2}$ or 1.41 is sometimes insignificant among other approximations when relating z to some macroscopic property.

In Eq. (9.35), z is the number of collisions per second suffered by one molecule in a "sea" of similar molecules of density n m^{-3}. The total collision frequency per unit volume is just

$$z_{11} = (\sqrt{2}\pi\sigma^2 n\bar{c})\,(n)\,(1/2) = n^2\sigma^2\left(\frac{4\pi kT}{m}\right)^{1/2} \tag{9.36}$$

where the division by 2 is to avoid the double count. That is, there are $\frac{1}{2}n^2$ pair of molecules of type 1. Using the expression just derived for c_r, in a binary mixture, the dissimilar collision frequency is

$$z_{12} = [\pi\sigma_{12}^2 n_1\bar{c}_r(12)](n_2) = n_1 n_2 \sigma_{12}^2\left(\frac{8\pi kT}{\mu}\right)^{1/2} \tag{9.37}$$

where there is now no division by two for the double count. That is, there are $n_1 n_2$ pairs of dissimilar molecules.

In a binary mixture, there would be 1,1 collisions at a frequency given by Eq. (9.36); there would be 2,2 collisions at a frequency given by a similar equation; and there would be 1,2 collisions at a frequency given by Eq. (9.37). The total collision frequency per unit volume is

$$z_{\text{tot}} = z_{11} + z_{22} + z_{12}$$

or

$$z_{\text{tot}} = n_1^2\sigma_1^2\left(\frac{4\pi kT}{m_1}\right)^{1/2} + n_2^2\sigma_2^2\left(\frac{4\pi kT}{m_2}\right)^{1/2} + n_1 n_2 \sigma_{12}{}^2\left(\frac{8\pi kT}{\mu}\right)^{1/2} \tag{9.38}$$

The total collision frequency on one molecule of type 1 is just this last relationship divided by the number density of 1, that is, n_1. Thus, for example, in pure chlorine gas at 1 bar and 300 K, one molecule of Cl_2 with σ = 0.4115 nm will be struck by other chlorine molecules at a rate of 5.4 GHz. In benzene gas at 1 bar and 300 K, with σ = 0.5270 nm, each C_6H_6 molecule will be struck by other benzene molecules at a rate of 8.5 GHz. In an equimolar mixture, each C_6H_6 molecule will be struck by Cl_2 molecules at a rate of 4.9 GHz, where we have used $\sigma_{12} = (\sigma_1 + \sigma_2)/2$ and recalled that $\mu \equiv m_1m_2/(m_1 + m_2)$. The total dissimilar collisions in $s^{-1} \cdot cm^{-3}$ is just the number of Cl_2 collisions on each C_6H_6, i.e., 4.9 GHz, multiplied by the number of C_6H_6 molecules per cubic centimeter, i.e., 0.5 (p/kT). Then

$$z_{1,2} = 4.9 \times 10^9 \times 1.2 \times 10^{19} = 5.9 \times 10^{28}\ s^{-1} \cdot cm^{-3}$$

This enormous number is only the dissimilar collision frequency in 1 cm^3 of equimolar mixture at 1 bar and 300 K.

Mean Free Path

Like the collision frequency, the mean free path is another useful concept in thinking about transport or kinetic processes. The mean free path is just the average distance traveled by a molecule between successive collisions. This picture makes sense only for intermolecular interactions that have sharp boundaries as in, for example, the rigid sphere model. It is not possible to define a single instant at which collision occurs between molecules with greatly varying interaction potentials which is, of course, exactly the case in collisions between real molecules. Nevertheless, the idea is useful.

Consider a pure species in the rigid sphere approximation. The mean free path is

$$\lambda \equiv \frac{\text{average distance traveled by one molecule in 1 s}}{\text{total number of collisions it experiences in 1 s}}$$

$$= \frac{\left(\dfrac{8kT}{\pi m}\right)^{1/2}}{\sqrt{2}c\pi\sigma^2 n}$$

or

$$\lambda = \frac{1}{\sqrt{2}n\pi\sigma^2} \tag{9.39}$$

The less rigorous derivations of collision frequency, Eqs. (9.31) and (9.33), will obviously yield somewhat different values of the mean free path. Minor technical differences in the definition of the free path by several workers in gas kinetic theory will also lead to differences in the constant in Eq. (9.39). But, in view of the approximate nature of the whole concept of a free path, these small differences are of no consequence.

Recall that the number density n is just p/kT, and then

$$\lambda = \frac{kT}{\sqrt{2}p\pi\sigma^2} \tag{9.40}$$

and λ increases directly with temperature and decreases inversely with pressure. In Eq. (9.40), with p in Pa and σ in m, λ will be in m.

It is instructive to have some feel for the size of the average speed, the mean free path, and the collision frequency. Such numbers for simple gases appear in Table 9.1. Note again that all of these numbers depend on an equivalent spherical molecular diameter σ, in nm, which has been obtained from experimental data on second virial coefficients, viscosity, etc. Numbers such as those from Table 9.1 are never absolute; they are rather indicative, but they are useful. For comparison, the muzzle velocity of a rifle is about 10^3 m · s^{-1} and the escape velocity from the earth is about 10^4 m · s^{-1}.

About a third of molecules have a free path in excess of the mean free path λ, while about 2 percent have free paths in excess of 4λ.

Duration of a Collision

The molecular dynamics underlying all macroscopic phenomena occur on collision. It is reasonable to ask about the approximate time that may be available for whatever interaction may occur on these collisions. The following chapter is wholly concerned with collisions wherein atomic rearrangement, that is, chemical reaction, occurs. But

TABLE 9.1 Mean Speed, Free Path, and Collision Frequency on One Molecule in Several Gases

Gas	MW	σ, nm	$\bar{c}$, $m \cdot s^{-1}$, 15°C	λ, nm, 15°C, 760 torr	z, GHz, 15°C, 760 torr
H_2	2.016	0.274	1740	117.7	14.8
CH_4	16.03	0.414	618	51.6	12.0
H_2O	18.02	0.460	582	41.8	13.9
C_2H_4	28.03	0.495	467	36.1	12.9
O_2	32.00	0.361	437	67.9	6.4
CO_2	44.00	0.459	372	41.9	8.8
A	39.94	0.364	391	66.6	5.9
Kr	82.9	0.416	271	51.2	5.3
Electron	5.49×10^{-4}		10.5×10^4		

SOURCE: Adapted from S. W. Benson, *Foundations of Chemical Kinetics*, McGraw-Hill, New York, 1960.

in a simple way, let us imagine that σ_{rx} is an effective radius from a hard-sphere target molecule within which the reaction event or process can occur. The distance traveled by a colliding molecule within this zone of reaction about the target is $2(\sigma_{rx} - \sigma_{\text{hs}})$, where σ_{hs} is the hard-sphere radius. The time for possible reaction is $2(\sigma_{rx} - \sigma_{\text{hs}})/\bar{c}_r$. If $\sigma_{rx} - \sigma_{\text{hs}}$ is about 0.1 nm and if $\bar{c}_r$ is about 400 m/s, then the reaction time is about 5×10^{-13} s, or about the same as the time of one cycle of a low-frequency Einstein vibration at $\omega \cong 100\ \text{cm}^{-1}$.

Distribution of Path Lengths

A prerequisite to understanding the transport processes from a molecular point of view is the understanding of the distribution of free paths between intermolecular collisions. Some are longer, some are shorter, and their mean is λ. Let $F(s)$ be the chance of a path of length s between successive collisions, that is, $F(s)$ is the chance of no collision during a path of length s. The chance that a collision occurs in the small length ds is proportional to the length of ds, that is, $\alpha\, ds$. The chance that a collision does *not* occur over the length ds is clearly $(1 - \alpha\, ds)$. And the chance that a collision does not occur in the total path length of $s + ds$ is just the product of the two above probabilities, for the events "no collision in length s" and "no collision in length ds" are independent of each other. Then,

$$\text{Chance of no collision in } s + ds = F(s + ds) = (F(s))(1 - \alpha\, ds)$$

The left-hand side may also be written as the first two terms in a Taylor series. Then the above expression may be rewritten

$$F(s + ds) = F(s) + \left[\frac{\partial F(s)}{\partial s}\right] ds = F(s) - \alpha F(s)\, ds$$

or

$$\frac{\partial F(s)}{\partial s} = -\alpha F(s)$$

which, on integration, yields

$$F(s) = e^{-\alpha s} \tag{9.41}$$

where the boundary condition of $F(s) = 1$ at $s = 0$ has been used. The chance of a path length between collisions of s is just $e^{-\alpha s}$. But what is the proportionality constant α? If, as before, $F(s)$ is the chance of no collision in a path of length s and $F(s + ds)$ is the chance of no collision in a path of length $s + ds$, then the chance of a path length ending in the length $s + ds$ is just the difference,

$$\text{Chance of path length } s \equiv f(s) = F(s) - F(s + ds)$$

or
$$f(s) = -\frac{\partial F(s)}{\partial s}\,ds = \alpha e^{-\alpha s}\,ds$$

Now the chance of a path of length s multiplied by that path length and multiplied by the total collision frequency and summed over all path lengths must be just the total distance traveled. Or, stated mathematically,

$$\int_0^\infty (\alpha e^{-\alpha s}\,ds)\,(s)\,(z) = \bar{c}$$

or
$$(\sqrt{2}\pi\sigma^2 n\bar{c})\,\alpha \int_0^\infty e^{-\alpha s}\,s\,ds = \bar{c}$$

where the term in parentheses is z, the collision frequency, Eq. (9.35). The integral is equivalent to the familiar gamma function, its value is $1/\alpha^2$, and then $\alpha = 1/\lambda$ with $\lambda = 1/\sqrt{2}n\pi\sigma^2$, Eq. (9.39). The distribution of path lengths between collisions is then

$$F(s) = e^{-s/\lambda} \tag{9.42}$$

A graph of this distribution function appears in Fig. 9.7. Numerically we see that 50 percent of the molecules have path lengths greater than 0.70λ, 36.5 percent greater than λ, 2 percent greater than 4λ, and 45 in 10^6 greater than 10λ. All of this will be a useful insight in developing a simplified view of the transport coefficients.

Molecular Transport

Consider a volume element $d\tau$ within a gas of uniform density n and arbitrarily located in the hemisphere above an area element of a con-

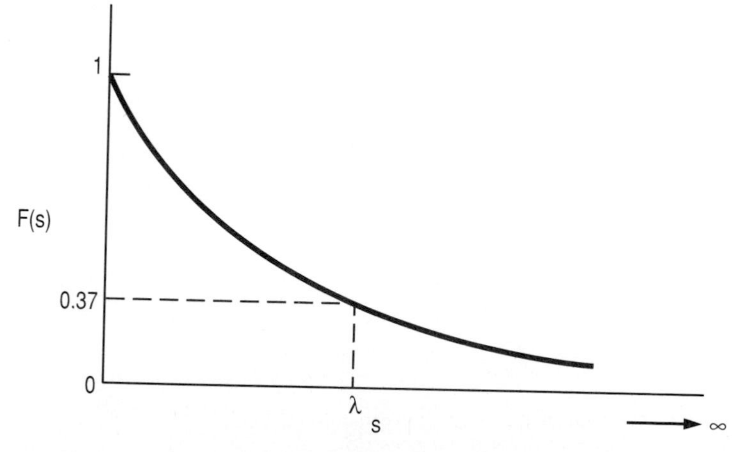

Figure 9.7 Chance of a free path of length s as a function of s.

taining wall, dS m^2. The geometry of such a configuration appears in Fig. 9.8, and we are concerned with the travel of molecules down upon dS resulting from motion in the $-z$ direction. The product $\bar{c}n\ d\tau$ is the total distance traveled by all molecules within volume $d\tau$ per unit time. If we divide this by the mean free path λ, we obtain the total collisions suffered by all molecules within $d\tau$ per unit time, $cn\ d\tau/\lambda$. Recall that this product was earlier divided by 2 to determine the number of collision events. The 2 is omitted here, so we have the total number of collisions suffered by all molecules.

After collision, molecules are emitted from $d\tau$ equally well in all directions. Specifically, those headed downward toward dS is just the fraction $\cos\theta\ dS/4\pi r^2$, i.e., the area subtended by dS located r away from $d\tau$ and divided by the total area about $d\tau$. The angle θ of course varies from 0 (looking straight down at dS) to $\pi/2$ (looking exactly sideways at dS), and $\cos\theta$ then varies from 1 to 0.

Finally, not all of the molecules leaving $d\tau$ and headed toward dS will collide with dS. Only those whose path length is at least r will make it to dS, and the fraction having path length of at least r is just $e^{-r/\lambda}$, as we have already seen.

The total number of molecules colliding with dS per unit area of dS, N, is then obtained by integrating the product of the total collisions

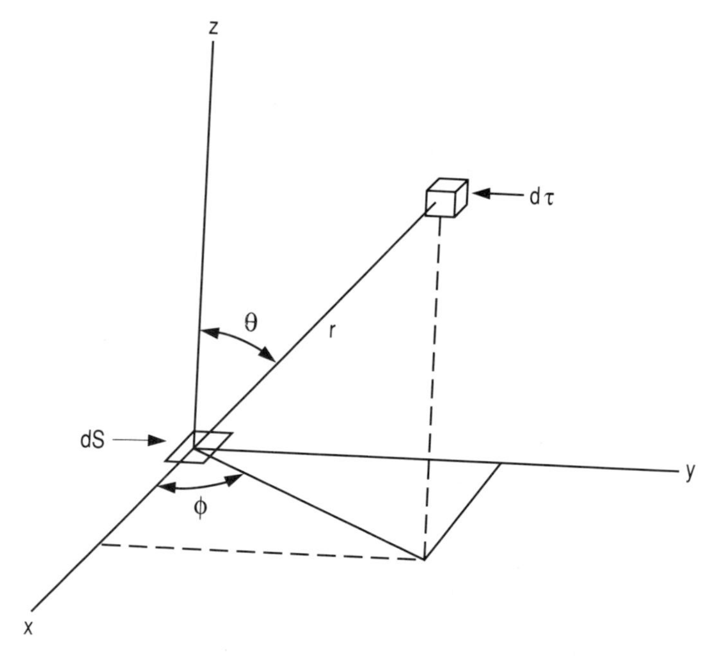

Figure 9.8 Schematic to aid in visualizing the collisionless transport of molecules from volume $d\tau$ through an area dS.

per second suffered by all molecules in the volume $d\tau$, the fraction of these collisions resulting in a particle headed toward dS, and the fraction of these that have a path length of at least r:

$$N = \int_\tau \left[\frac{\bar{c}n}{\lambda} d\tau\right] \left[\frac{\cos\theta}{4\pi r^2}\right] [e^{-r/\lambda}]$$

The volume element $d\tau$ in spherical coordinates is $d\tau = r^2 \sin\theta \, d\theta \, d\phi \, dr$, and then the collision frequency per unit area of dS due to molecules originating in the upper hemisphere is just

$$N(\text{from above}) = \frac{\bar{c}n}{4\pi\lambda} \int_0^{2\pi} d\phi \int_0^{\pi/2} \sin\theta \cos\theta \, d\theta \int_0^{\infty} e^{-r/\lambda} \, dr \qquad (9.43)$$

which, on integration, becomes

$$N\,(\text{m}^{-2} \cdot \text{s}^{-1}) = \frac{n\bar{c}}{4}$$

This is a result that we have already obtained in a much more simple way [see Eq. (9.24)]. We had earlier calculated the average speed in any direction to be one-fourth the total average speed. This speed multiplied by the number density is clearly the frequency of bombardment per unit area upon any plane surface. The units of N are $(\text{m} \cdot \text{s}^{-1})(\text{m}^{-3})$ or $\text{m}^{-2} \cdot \text{s}^{-1}$, just as they should be.

Why go through a conceptually more difficult derivation of an otherwise simple result? The only reason for doing so is the development of Eq. (9.43), which lends itself to calculations when the number density of molecules is not constant. The earlier simple calculation allowed no such insight.

Mass transfer in, say, the z direction, will occur when n is not constant but varies in the z direction. Suppose this variation is in the z direction only and is also small so that a Taylor series expansion of the number density above the xy plane at $z = 0$ may be truncated after the first derivative,

$$n_z = n_{z=0} + \left(\frac{\partial n}{\partial z}\right)_{z=0} z$$

The collision frequency of molecules per unit area that was just derived was conveniently done in polar coordinates. The above gradient in number density is expressed in cartesian coordinates, so we convert the Taylor expansion into polar coordinates (see Fig. 9.8),

$$n_{r\cos\theta} = n_{z=0} + \left(\frac{\partial n}{\partial z}\right)_{z=0} r\cos\theta$$

Returning to the statement of molecular flux from above, N(from above), we write a similar equation to describe the flux from the lower hemisphere, N(from below), and the difference is the net flux of molecules in the z direction because of the gradient in n is taken to be in the z direction. The flux will be in the direction opposite to the gradient. With n now inside the integral over all volume $d\tau$, Eq. (9.43) becomes

$$N(\text{from above}) = \frac{\bar{c}}{4\pi\lambda}\int_0^{2\pi} d\phi \int_0^{\pi/2} \sin\theta\cos\theta\, d\theta \int_0^{\infty} n e^{-r/\lambda}\, dr$$

$$\text{and}\quad N(\text{from below}) = \frac{\bar{c}}{4\pi\lambda}\int_0^{2\pi} d\phi \int_{\pi}^{\pi/2} \sin\theta\cos\theta\, d\theta \int_0^{\infty} n e^{-r/\lambda}\, dr$$

or the difference in these last two expressions is

$$N(\text{net}) = \frac{\bar{c}}{4\pi\lambda}\int_0^{2\pi} d\phi \int_0^{\pi} \sin\theta\cos\theta\, d\theta \int_0^{\infty} n e^{-r/\lambda}\, dr \qquad (9.44)$$

In Eq. (9.44), we have taken λ to be constant; that is, we take the gradient in n to be so small that λ is unaffected. Equation (9.44) also recognizes that the difference between the two integrals over θ is just the integral over all θ from 0 to π. We have now only to insert the Taylor expansion into Eq. (9.44) and perform the indicated integrations to determine the net flux,

$$N(\text{net}) = \frac{\bar{c}}{4\pi\lambda}\int_0^{2\pi} d\phi \int_0^{\pi} \sin\theta\cos\theta\, d\theta \int_0^{\infty} n_{z=0} e^{-r/\lambda}\, dr + \frac{\bar{c}}{4\pi\lambda}\int_0^{2\pi} d\phi \int_0^{\pi} \sin\theta\cos\theta\, d\theta \int_0^{\infty} \left(\frac{\partial n}{\partial z}\right)_{z=0} (r\cos\theta) e^{-r/\lambda}\, dr$$

The first term is zero, which states the obvious fact that if $n_{z=0}$ is constant over all space, there can be no net molecular flux. The second term becomes

$$N(\text{net}) = \frac{\bar{c}}{4\pi\lambda}(2\pi)\left(\frac{2}{3}\right)\left(\frac{\partial n}{\partial z}\right)_{z=0}(\lambda)^2$$

where $(\partial n/\partial z)$ is the constant gradient at $z = 0$. Then,

$$N(\text{net}) = -\frac{\bar{c}\lambda}{3}\left(\frac{\partial n}{\partial z}\right) \qquad \text{molecules/m}^2\cdot\text{s} \qquad (9.45)$$

Thus if n increases with z in some arbitrary manner, there will be a net flow of molecules in the opposite direction at a rate given by Eq.

(9.45). A minus sign has been added to reflect that the molecular flux N(net), is in a direction opposite to the gradient.

Equation (9.45) can be applied, for example, to the case of molecules diffusing through an array of molecules that are relatively fixed in space. A practical application would be the diffusion of slow neutrons through matter as in the design of a moderator or shield for a nuclear reactor.

More generally, the flux of *any* property, that is, any observable characteristic, can only be carried "on the backs" of the molecules. We have just derived that the flux of molecules depends on the gradient in molecular density. Now let us suppose that the flux of interest is not the molecules themselves but rather some property carried by the molecules. An example might be the flux of internal energy U^0, where $U^0 = n\varepsilon$ and, as before, we will represent the gradient in energy by a Taylor expansion,

$$n\bar{\varepsilon}_z = n\bar{\varepsilon}_{z=0} + \left(\frac{\partial n\bar{\varepsilon}}{\partial z}\right)_{z=0} z$$

By exactly the same arguments used in deriving Eq. (9.45), we conclude that the flux of energy, F, is

$$F = -\frac{1}{3}\bar{c}\lambda\left(\frac{\partial n\bar{\varepsilon}}{\partial z}\right) \qquad \text{J/m}^2 \cdot \text{s} \tag{9.46}$$

The flux of such molecular properties may be readily related to important macroscopic fluxes.

Transport Coefficients (Simple Treatment)

Heat, mass, and momentum transfer in matter are called *transport phenomena*. The macroscopic balance relationships governing each of the three transport phenomena are similar. They differ only in the physical meaning attached to the symbols. For example, one-dimensional steady-state heat conduction described in cartesian coordinates is

$$q = -k\frac{dT}{dz} \tag{9.47}$$

where q is heat flux in units of J/s · m^2 and k is a phenomenological coefficient called the thermal conductivity. If the thermal gradient is in units of K/m, then k must have units of W/m · K. Equation (9.47) is called Fourier's law of heat conduction.

Interestingly, Count Rumford, such a familiar and important name from thermodynamics, had asserted somewhat before 1800 that air could not conduct heat at all. This was generally accepted for the ensuing 70 years. But from our molecular point of view, we might imagine a gas between two parallel surfaces at T(hot) and T(cold). Molecules colliding with the hot plate will rebound with more energy. This excess energy will be transferred to other molecules by intermolecular collision. Finally, after innumerable collisions, some higher-energy molecules will collide with the cold plate, they will depart with less energy, and heat will have been transferred from the hot to the cold plate.

The temperature of the gas will vary in some manner between the two plates. Let us suppose that, at every position z between the plates, the local distribution is maxwellian. The classical average energy will be $\frac{1}{2}kT$ for each of s squared energy terms, as we have seen earlier,

$$\bar{\varepsilon} = s(\tfrac{1}{2}kT) \qquad \text{J/molecule}$$

The thermodynamic molar internal energy per unit volume is just $n\bar{\varepsilon}$, and with $\varepsilon_0 = U_0^0 = 0$, the average energy may also be written

$$\bar{\varepsilon} = c_v T = (\tfrac{1}{2}sk)T$$

where the heat capacity is in units of J/molecule · K.

The macroscopic gradient $d(n\bar{\varepsilon})/dz$ is desired, for we wish to compare this molecular result with the macroscopic Fourier equation,

$$\frac{d(n\bar{\varepsilon})}{dz} = n\frac{d\bar{\varepsilon}}{dz} = n\left(\frac{1}{2}sk\right)\frac{dT}{dz} = C_V^0\frac{dT}{dz}$$

where C_V^0 is the molar heat capacity in J/mol · K. We also have here taken the molecular number density to be constant between the hot and cold plates. Then with F of Eq. (9.46) here identified with the flux of heat,

$$q = -\frac{1}{3}\bar{c}\lambda\left(\frac{1}{2}nsk\frac{dT}{dz}\right)$$

By comparison with Fourier's law, Eq. (9.47), it is clear that the thermal conductivity k becomes

$$k = \tfrac{1}{6}\bar{c}\lambda skn \tag{9.48}$$

where k on the right-hand side is Boltzmann's constant. Let us first restrict attention to structureless or nonatomic species; s then is 3, and the thermal conductivity becomes

$$k = \left(\frac{1}{6}\right)\left(\frac{8kT}{\pi m}\right)^{1/2}\left(\frac{1}{\sqrt{2}n\pi\sigma^2}\right)(s)(k)(n)$$

$$= \frac{1}{\sigma^2}\left(\frac{k^3T}{\pi^3 m}\right)^{1/2} \tag{9.49}$$

where, with consistent SI units, k has the expected units of W/m · K. In practical units, with σ in nm, this becomes

$$k = 2.260 \times 10^{-4}\,\frac{(T/MW)^{1/2}}{\sigma^2} \tag{9.50}$$

Thermal conductivity is thus predicted to be independent of pressure and vary with the ratio of temperature to molecular weight to the one-half power. Interestingly, the thrust of a rocket engine also varies as the ratio of T/MW to the one-half power. Data on many gases suggest that the pressure independence of k is reasonable usually up to about 10 atm. But the effect of temperature is much greater than is predicted here. The problem arises from our unrealistic assumption of hard-sphere molecules. We will introduce more reasonable intermolecular models in a later section.

This theory also poorly describes the thermal conductivity of polyatomic gases. This is not surprising in view of the complex intramolecular modes of all such molecules that have been totally ignored in this simplified treatment. This complexity presents a plethora of opportunities for energy transfer on collision. We will return to this point in a later section.

Returning to the original derivation of a flux, let us suppose that, rather than average energy, it is momentum that is moving through the gas. Then just as before Eq. (9.45) now is rewritten

$$F = \frac{1}{3}\bar{c}\,\lambda\left(\frac{\partial nm\bar{c}}{\partial z}\right) \tag{9.51}$$

where this flux has units of momentum per square meter per second or, equivalently, units of force per unit area.

The phenomenological law known as Newton's law of viscosity is best visualized by imagining a fixed plate and a parallel plate moving at a fixed velocity with a fluid in between. Such a system appears in Fig. 9.9. The fluid will have a velocity distribution impressed on it that varies from zero at the fixed plate to the velocity of the moving plate. The force per unit area necessary to move the upper plate is proportional to the velocity gradient between the plates, that is,

$$F_x = -\,\eta\left(\frac{\partial u_x}{\partial z}\right) \tag{9.52}$$

This is Newton's law of viscosity, and the proportionality constant is a

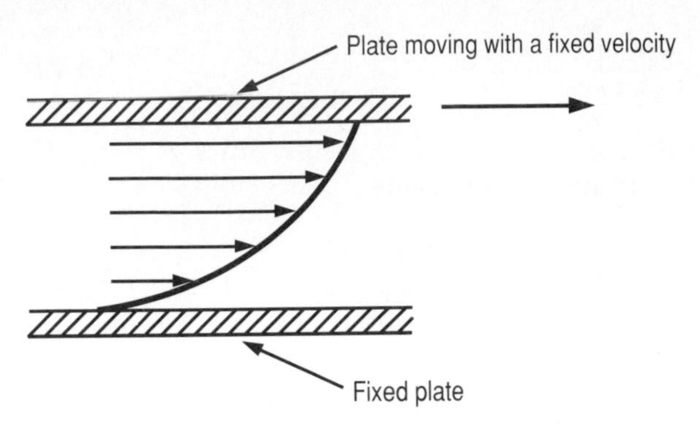

Figure 9.9 Schematic illustration of Newton's law of viscosity.

phenomenological coefficient characteristic of the fluid, called the *viscosity*. Equation (9.52) is the analog of Fourier's law, Eq. (9.47).

In the molecular expression for flux, Eq. (9.45), we can write

$$F_x = -\frac{1}{3}\bar{c}\lambda\left(\frac{\partial nm\bar{c}}{\partial z}\right) = -\frac{1}{3}\bar{c}\lambda nm\left(\frac{\partial \bar{c}}{\partial z}\right)$$

where we have again taken the number density of molecules between the plates to be constant. If we also equate the macroscopic velocity gradient to the gradient in the mean molecular speed, the molecular equivalent to Eq. (9.52) then allows the evaluation of the viscosity as

$$\eta = \frac{1}{3}\bar{c}\lambda nm \tag{9.53}$$

which becomes

$$\eta = \frac{1}{3}\left(\frac{8kT}{\pi m}\right)^{1/2}\left(\frac{1}{\sqrt{2}n\pi\sigma^2}\right)(n)(m) = \frac{2}{3}\left(\frac{kTm}{\pi}\right)^{1/2}\frac{1}{\pi\sigma^2} \tag{9.54}$$

In practical units, this may be rewritten as

$$\eta = 1.812 \times 10^{-8}\frac{(TM)^{1/2}}{\sigma^2} \tag{9.55}$$

where T is in K, M is the molecular weight, σ is the collision radius in nm, and η is the viscosity in kg/m · s.

The momentum flux perpendicular to the direction of flow has been compared to a faster train slowly passing a slower train moving in the same direction on adjacent tracks. As the trains pass, passengers on each throw packages into the other train. The net effect is to slow the faster train and speed the slower train.

The macroscopic diffusion of matter is described by Fick's law,

$$j_{Az} = -D_{AB}\left(\frac{\partial n_A}{\partial z}\right) \tag{9.56}$$

where n_A is the concentration of A in molecules/m^3, z is the distance along which the gradient exists in m, j_{Az} is the flux of molecules of A in the z direction in molecules/m$^2 \cdot$ s, and D_{AB} is a phenomenological coefficient called the diffusivity of A through B, which must obviously then have units of m^2/s. The laws of Fourier, Newton, and Fick are similar. As before, the minus signifies that the mass motion is counter to the concentration gradient. That is, just as heat flows down a temperature gradient, mass diffuses down a concentration gradient. Diffusion is pictorially evident from considering the evaporation of liquid A with the vapor moving upward through a column of B as depicted in Fig. 9.10.

It is easier to first imagine the diffusion of A through itself. We expect that it will be important that the masses and sizes of the diffusing species and the background species be the same. To experimentally measure a diffusion process, the wanderer must be labeled. Self-diffusion may be reasonably noted from isotopic labeling studies. Remember that this argument does not describe mass movement due to a

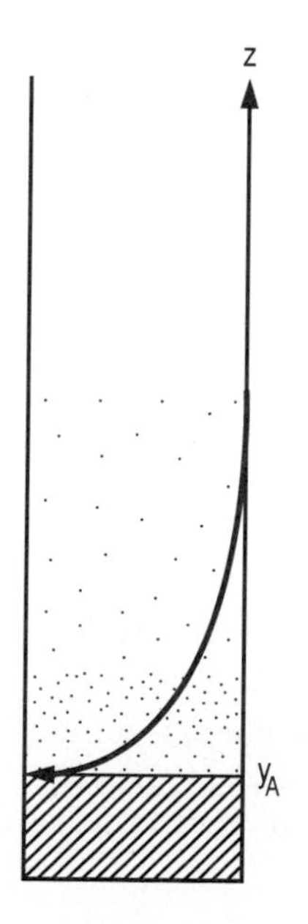

Figure 9.10 Gas from an evaporating liquid diffusing through a second gas. *(From R. B. Bird, W. E. Stewart, and E. N. Lightfoot, Transport Phenomena, Wiley, New York, 1960. Used by permission.)*

pressure gradient. This latter is called viscous flow, and it is not the phenomenon described by Fick's law.

We have earlier derived an expression for the net flow of molecules under a small concentration gradient as

$$N(\text{net}) = j_{Az} = -\frac{\bar{c}\lambda}{3}\left(\frac{\partial n_A}{\partial z}\right) \tag{9.45}$$

Clearly then the self-diffusion coefficient is

$$D_{AA} = \frac{\bar{c}\lambda}{3} \tag{9.57}$$

or

$$D_{AA} = \left(\frac{8kT}{\pi m}\right)^{1/2}\left(\frac{1}{\sqrt{2}n\pi\sigma^2}\right)\left(\frac{1}{3}\right)$$

which, on simplification, becomes

$$D_{AA} = \frac{2}{3}\left(\frac{k^3}{\pi^3 m}\right)^{1/2}\frac{T^{3/2}}{p\sigma^2} \tag{9.58}$$

In practical units with σ in nm, p in kPa, and T in K, this expression for the self-diffusion coefficient in m^2/s becomes

$$D_{AA} = 1.51 \times 10^{-7}\frac{(T^3/M)^{1/2}}{p\sigma^2} \tag{9.59}$$

An important number in convective heat transfer calculations is the Prandtl number, defined as $C_p\eta/k$. It is instructive to calculate N_{Pr} from the expressions derived above and compare its value with experimental numbers for several gases. The value of N_{Pr} is

$$N_{\text{Pr}} = \frac{C_p\eta}{k} = \frac{[(\frac{1}{2}sk + \frac{2}{2}k)/m]\,(\frac{1}{3}\bar{c}\lambda nm)}{\frac{1}{6}\bar{c}\lambda skn}$$

where the additive k converts the theoretical C_V to C_p and where division of the sum by m converts J/K into specific heat, J/K · kg. On simplification,

$$N_{\text{Pr}} = \frac{s+2}{s} \tag{9.60}$$

For a diatomic molecule, $s = 7$ and $N_{\text{Pr}} = 1.3$. For a linear triatomic molecule, $s = 13$ and $N_{\text{Pr}} = 1.15$. In reality N_{Pr} is temperature-dependent. Its experimental values at room temperature for N_2 and CO_2

are 0.71 and 0.76, respectively. Nonetheless, this reasonable agreement from such a simple theory must be taken as heartening.

We must now, however, consider the evaluation of the transport coefficients from a more realistic theory.

A More Rigorous Kinetic Theory

The Boltzmann equation is the starting point for more rigorous developments in kinetic theory. Even here, the assumptions are severe. So there really is no truly rigorous kinetic theory. The Boltzmann equation and its solution presents mathematical problems far beyond anything that might be judged reasonable for purposes of this book. Analogously perhaps, we need not study the quantum mechanics of the solid phase in order to use a mass spectrometer even though this device depends for its function on solid-state electronic components. Central to the function of the mass spectrometer is certain pulsing circuitry. This circuitry is better understood if one understands transistors, which in turn demands an understanding of solid-state physics.

We want to use the best kinetic theory even if we elect not to take the long mathematical excursion to reproduce its origins. We can, however, readily review the gist of these tedious arguments. To begin, it is helpful to recall the concept of a phase-space. Imagine first that we describe a molecule with s position coordinates and the corresponding s momentum coordinates. This molecule is completely described by a single point in a $2s$-dimensional phase-space. Similarly a collection of 10^{23} molecules will be described by a swarm of an equal number of points in this phase-space. As the coordinates and momenta change because of collisions, the history is represented by the motion of these points. The points could even be represented by a distribution function called the density in phase-space. Liouville's theorem (which can be rigorously proven) states that this density in phase-space at any point does not change with time. The density function is somewhat like an incompressible fluid.

Now to simplify, we see that n monatomic species are completely described by a swarm of n points in a six-dimensional phase-space. The distribution function, $f(q, p, t)$, describes the density of points in phase-space at all time. At equilibrium, this function becomes the Maxwell-Boltzmann distribution function. The Boltzmann equation describes the evolution of this distribution function due to collisions. The Boltzmann equation has the form of a continuity equation.

Solving the Boltzmann equation is quite impossible, but one of the most well-developed approximation techniques is that of Chapman and Enskog. It applies when the distribution function is near equilibrium. It involves an expansion of the distribution function in powers of a perturbation parameter which produces the departure from equilibrium. The Boltzmann equation takes the molecules to be freely moving except at the moment of a binary collision, but even so, it is intractable. The time of this collision is small compared to the time of free flight between collisions. So the assumptions, even before the mathematical approximations, restrict the rigorous kinetic theory to low-density species.

Other approaches to the nonequilibrium problem exist,[2] but certainly the book *Mathematical Theory of Non-Uniform Gases* by Chapman and Cowling[3] is the definitive treatise on kinetic theory. However, Chapman himself has said reading his book is "like chewing glass." It is only for the true experts. Chapman and Enskog, working in England and Sweden, respectively, obtained almost simultaneously and independently their complicated mathematical descriptions of kinetic theory. They both are given equal credit.

The Chapman and Enskog (C-E) approach considers the distribution of atomic velocities to be only slightly displaced from maxwellian. In fact it is only under these conditions that the usual definitions of the transport coefficients apply. The C-E theory is classical, and hence no quantum effects appear, but these are insignificant anyway except for H_2 and He at very low temperatures. The C-E theory is also strictly applicable only to elastic collisions of molecules with spherically symmetric interaction potentials. Remember that an elastic collision was defined as an encounter in which translational energy, as well as the magnitude and direction of momentum, are conserved. A second class of collision occurs if the translational energy of one or both of the collision partners is converted in part into internal energy of one of the species leaving the encounter. Such collisions are said to be inelastic. A third class of collisions is characterized by a rearrangement of the atoms. Such collisions are said to be reactive. In inelastic collisions, it matters not whether the energy is absorbed or ejected from an electronic, vibrational, or rotational mode. In inelastic collisions, kinetic energy is not conserved, although mass and momentum are conserved. Because of this, the coefficients of viscosity and diffusion are not too sensitive to the occurrence of inelastic collisions and the C-E theory may be successfully employed, provided only that the collision partners are not too nonspherical. The thermal conductivity coefficient, however, depends significantly on inelastic collisions (see later).

So the so-called rigorous theory has many serious approximations. Nevertheless, the ideas are useful.

The development of the so-called rigorous kinetic theory begins with the work of Boltzmann, Enskog, Maxwell, and Chapman between about 1870 to 1920. The state of either a pure gas or a mixture is completely specified for purposes of kinetic theory if the distribution function of molecular velocities and positions is known throughout the gas. Although these theories are much too mathematically complex for our purposes, the methods finally yield expressions for the transport properties that involve integrals of the type

$$\Omega^{l,s}(T) = \frac{1}{2\sqrt{\pi}} \int_0^\infty e^{-\mu g^2/2kT} \left(\frac{\mu g^2}{2kT}\right)^{(2s+3)/2} Q^l(g)\, dg \tag{9.61}$$

The familiar symbols have their usual meaning, and g is the relative velocity of approach of the two colliders at large separation.

The dynamics of the collision enter into the transport coefficient calculations through the cross sections $Q^{(l)}(g)$, the evaluation of which requires knowledge of the scattering angle as a function of initial relative speed g and the impact parameter, which is the distance of closest approach if the effect of intermolecular forces is ignored. The scattering angle in turn depends directly on the intermolecular potential function. Let us consider each of these quantities in turn. As we review this sequence for some simple cases, it will become obvious that the "rigorous kinetic theory" of Chapman and Enskog requires only mathematical manipulation to move from any intermolecular potential to a numerical value for, say, the viscosity.

Scattering Angle

The scattering angle is an important concept in kinetic theory. It is essential to express this scattering angle in terms of the intermolecular interaction potential. For spherically symmetric force fields, the centers of mass of the two colliding molecules lie in a plane before, during, and after the collision. Consider a collision in the center-of-mass coordinate system as depicted in Fig. 9.11. The impact parameter b is the distance of closest approach of the centers of mass of the molecules if there is no force field. The scattering angle χ is just the angle of deflection of the two colliders as a result of the encounter. As we saw earlier, the equations of motion for the actual binary collision are the same as would be written for a single particle of mass μ, the reduced mass, being scattered by a spherical potential $\phi(r)$ (see Fig. 9.11b). Although it seems obscure, the scattering angle can be measured by using molecular beam techniques (see later).

Without going into the detail, it is intuitively clear that the equations of motion with energy conserved and with momentum in both magnitude and direction also conserved, it should be possible to develop a rigorous expression for the scattering angle χ (see later). This

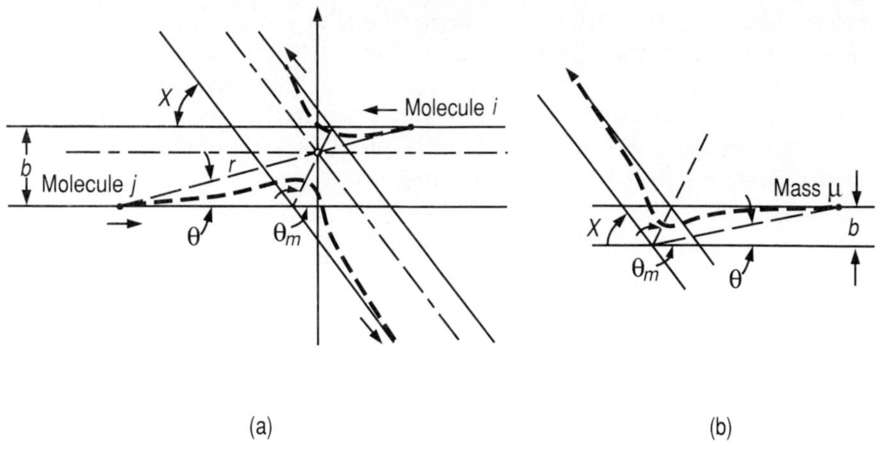

Figure 9.11 (*a*) A real binary collision. (*b*) The equivalent one-body representation.

is analogous to calculating the angle of motion of a ball on the pool table after it is struck by the cue ball. The result is

$$\chi = \pi - 2b \int_{r_m}^{\infty} \frac{dr/r^2}{[1 - \phi(r)/(\frac{1}{2}\mu g^2) - b^2/r^2]^{1/2}} \tag{9.62}$$

The scattering angle is a function of the impact parameter b, the intermolecular potential $\phi(r)$, and the relative velocity at large separation, g. The lower limit of integration, r_m, is the distance of minimum approach, as may be seen in Fig. 9.11*b*. The value of r_m, the minimum distance of approach and the lower limit in the definite integral, is obtained by equating $dr/d\theta$ to zero. This relationship permits the calculation of a scattering angle for any arbitrary impact parameter and initial relative velocity. For realistic intermolecular potentials, the value of r_m decreases rapidly with increasing g. It is therefore temperature-dependent, and it is important in evaluating the transport properties. As an example, let us evaluate the scattering angle for the rigid-sphere model. This is the model that would describe the scattering of balls on a pool table. A graphical representation of the nature of this model as compared to real-gas behavior is familiar but is presented again in Fig. 9.12.

Stated analytically, the rigid sphere model corresponds to

$$\phi(r) = \infty \qquad \text{for } r \leq \sigma$$

$$\phi(r) = 0 \qquad \text{for } r > \sigma$$

The lower limit r_m in our scattering angle formula, Eq. (9.62), is clearly

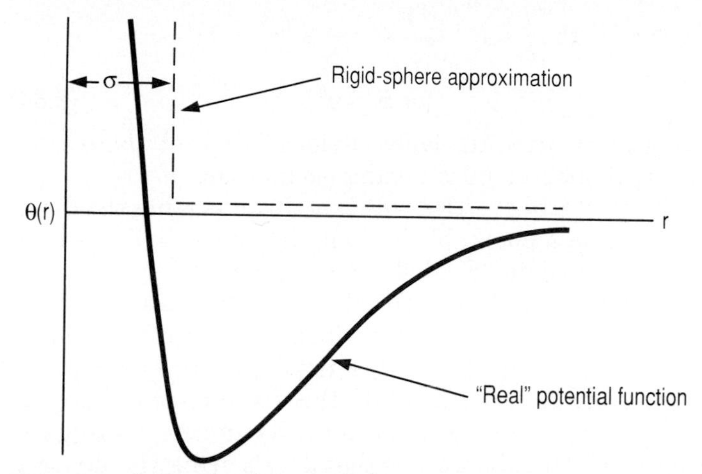

Figure 9.12 Intermolecular potentials.

$$r_m = \sigma \qquad \text{if } b \le \sigma$$

$$r_m = b \qquad \text{if } b > \sigma$$

Hence, for $b \le \sigma$, the scattering angle is

$$\chi = \pi - 2b \int_\sigma^\infty \frac{dr/r^2}{[1 - b^2/r^2]^{1/2}}$$

$$= \pi - 2b \left[\frac{1}{b} \sec^{-1} \frac{r}{b}\right]_\sigma^\infty = \pi - 2\left(\frac{\pi}{2} - \cos^{-1}\frac{b}{\sigma}\right)$$

or

$$\chi = 2 \cos^{-1}\left(\frac{b}{\sigma}\right) \qquad \text{for } b \le \sigma \tag{9.63}$$

and hence

$$\cos \tfrac{1}{2}\chi = \frac{b}{\sigma} \equiv b^*$$

Recalling the half-angle formula from trigonometry, $\cos \frac{1}{2}\theta = \pm [\frac{1}{2}(1 + \cos \theta)]^{1/2}$, and recognizing that χ must vary between 0 and π, it is clear that we may also write

$$\cos \chi = 2b^{*2} - 1 = \frac{2b^2}{\sigma^2} - 1$$

For $b > \sigma$, we have

$$\chi = \pi - 2b\left[\frac{1}{b}\sec^{-1}\frac{r}{b}\right]_b^\infty = \pi - 2\left(\frac{\pi}{2} - 0\right)$$

or

$$\chi = 0 \qquad \text{for } b > \sigma \tag{9.64}$$

This is in keeping with the intuitively obvious fact that there is no scattering if the rigid spheres miss hitting each other.

These results for scattering angles will exactly determine the direction of motion of balls on a pool table but only if the cue-ball has been struck in a way as to have no "English," i.e., no spin.

For all physically realistic intermolecular potential functions, the scattering angle must be obtained by laborious numerical integration of Eq. (9.62).† The scattering angle is the important property in determining the transport properties, while the distance of closest approach, r_m, in a collision is the important property in determining the chemical reaction rate. This distance is found by setting the denominator of the integral of Eq. (9.62) equal to zero.

Cross Sections

The cross section is a proportionality constant that arises in expressing the rate of occurrence of particular molecular scattering events. Since there are many events of interest, there are many cross sections of interest. Exactly the same collision occurs with different cross sections for different events; that is, the cross section for elastic scattering is not the same as that for reactive scattering. This will be discussed in detail in the following chapter on chemical dynamics, but to here make the concept reasonable, consider a beam of molecules entering a target gas of density n_B. We have already seen that the chance of a collision after a distance of travel of s into the target gas is

$$F(s) = e^{-s/\lambda} \tag{9.42}$$

The chance of path length s for the projectile molecules, $F(s)$, also yields the intensity of the beam as a function of its penetration, for

$$I(s) = F(s)n_A v_A$$

where $I(s)$ has units of $\mathrm{m}^{-2}\cdot\mathrm{s}^{-1}$, n_A is the initial number density of molecules of A in the beam, and v_A is their velocity. The fractional change in intensity on passing a distance ds through the target gas is

$$\frac{I(s+ds) - I(s)}{I(s)} = \frac{[e^{-s/\lambda} + e^{-s/\lambda}(-1/\lambda)\,ds] - (e^{-s/\lambda})}{e^{-s/\lambda}} = -\frac{ds}{\lambda}$$

†For a summary of results with several intermolecular potential functions see Chap. 8 of Hirschfelder et al.[2]

or
$$\frac{dI}{I\,ds} = -\frac{1}{\lambda} \tag{9.65}$$

and, on integrating,

$$I(s) = I(0) \exp(-s/\lambda) \tag{9.66}$$

$I(0)$ is the initial flux of the beam as it enters the target gas at $s = 0$. This beam attenuation is analogous to the Beer-Lambert law for the attenuation of light by an absorbing medium. In that case, λ^{-1} is replaced by μ, the absorption coefficient (in units of wave numbers, cm^{-1}) for the particular wavelength in use.

Figure 9.13 shows experimental beam attenuation data for CsCl passing through a fixed length l of either Ar or CH_2F_2. The mean free path is $\lambda = kT/\sqrt{2}\pi p\sigma^2$, Eq. (9.40), and $\ln I(l)/I(0)$ has the expected inverse dependence on pressure. If the attenuation is proportional to ds/λ as given by Eq. (9.65), we also expect attenuation to be proportional to the number of targets, i.e., $n_B\,ds$, or

$$\frac{ds}{\lambda} \propto n_B\,ds$$

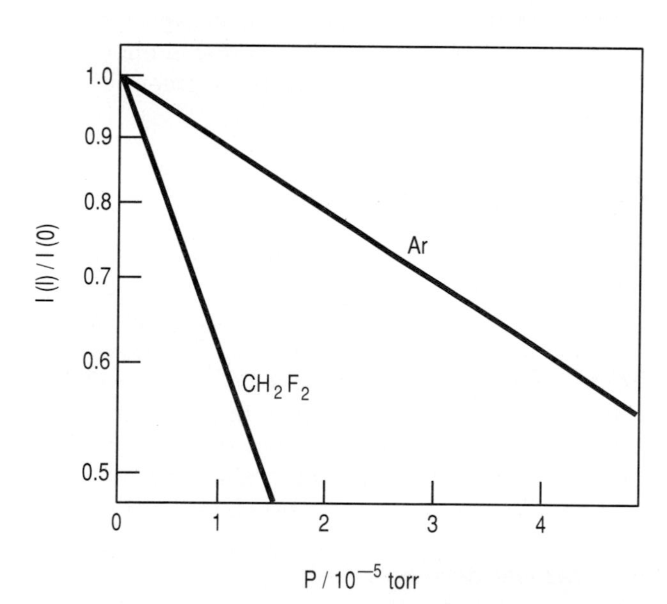

Figure 9.13 Typical attenuation data for the scattering of a thermal beam of CsCl by Ar and by the polar molecule CH_2F_2. The log of the transmission decreased linearly with target gas pressure p and thus n_B. *(Adapted from R. D. Levine and R. B. Bernstein, Molecular Reaction Dynamics, Oxford University Press, New York, 1974. Used by permission.)*

The proportionality constant is the size of each target, so

$$\frac{ds}{\lambda} = Qn_B\, ds$$

or

$$\lambda = (Qn_B)^{-1} \tag{9.67}$$

where Q is here defined as the cross section for all events that attenuate the beam. These could be elastic events, inelastic events with translational energy into vibration, inelastic events with translational energy into rotation, or any sort of reaction whatsoever. One can define many sorts of cross sections. With Eq. (9.66), attenuation may be written in terms of a total cross section, $Q \equiv \Sigma q_i$, as

$$I(s) = I(0) \exp(-sn_B Q) \tag{9.68}$$

In Fig. 9.13, the attenuation for scattering against CH_2F_2 is greater than that against Ar, and thus the total cross section for CH_2F_2 is greater than that for Ar.

We could significantly increase the energy of the incoming CsCl beam and again observe the attenuation to determine the effect of velocity or collision energy on the cross section. Molecules are not rigid spheres. They are soft and deformable, and, as might be expected, we observe a decrease in cross section with increasing collision energy.

The cross sections that are important in the C-E development of the transport properties are written

$$Q^{(l)}(g) = 2\pi \int_0^\infty (1 - \cos^l \chi)\, b\, db \tag{9.69}$$

where l will assume different positive integer values depending on which transport coefficient is being calculated. These cross sections look artificial or ad hoc in this summary discussion, but they arise naturally, even if in a mathematically complex manner. For example, the cross section for diffusivity is $l = 1$, which makes $Q^{(l)}(g) = \pi\sigma^2$ for the special case of the rigid-sphere model (see later). In the Chapman-Enskog approach, the cross section for diffusion is not the same as the cross section for thermal conductivity or viscosity. The cross section for chemical reaction in a collision, which is simply related to the rate of reaction (see later) is still different. For the case of the rigid-sphere gas, the scattering angle was just derived to be

$$\chi = 2 \cos^{-1}\left(\frac{b}{\sigma}\right)$$

Then the impact parameter is

$$b = \sigma \cos\left(\frac{\chi}{2}\right)$$

and then

$$db = -\frac{\sigma}{2}\sin\left(\frac{\chi}{2}\right)d\chi$$

and $b\,db$, needed in the expression for the cross section, is

$$b\,db = -\frac{1}{2}\sigma^2 \sin\left(\frac{\chi}{2}\right)\cos\left(\frac{\chi}{2}\right)d\chi = -\frac{1}{4}\sigma^2 \sin\chi\,d\chi$$

The general expression for the cross sections becomes

$$Q^{(l)} = 2\pi \int_0^\infty (1 - \cos^l \chi)\,b\,db = -\frac{1}{2}\pi\sigma^2 \int_\pi^0 (1 - \cos^l \chi)\sin\chi\,d\chi$$

where integration from $b = 0$ to $b = \infty$ is equivalent to collisions from head-on ($b = 0$) to a grazing ($b = \sigma$) for hard spheres. The cross section is here independent of the relative energy of approach, g. This is not true in general. With $b = 0$, a head-on collision, the scattering angle is π and at $b = \sigma$, a grazing collision, the scattering angle is 0. The idea of the dependence of cross section on scattering angle is thus clear. The cross section when $b > \sigma$ is zero, while for $b \leq \sigma$,

$$Q^{(l)}(g) = 2\pi \int_0^\sigma \left\{1 - \left[2\left(\frac{b}{\sigma}\right)^2 - 1\right]^l\right\} b\,db$$

where we have again used the trigonometric identity $\cos 2\theta = 2\cos^2\theta - 1$. On integration, one obtains

$$Q^{(l)} = \pi\sigma^2\left\{1 - \frac{1}{2}\left[\frac{1 + (-1)^l}{l + 1}\right]\right\} \tag{9.70}$$

for any positive integer value of l, including zero. Notice that the total cross sections, like the scattering angle, are independent of the initial relative velocity g for the special case of the rigid-sphere model. In general, the cross sections will depend on this relative velocity at large separations.

The cross sections all have units of area, and each may be thought of as an effective cross section for collision but one that also includes the effects of relative velocity, impact parameter, and the details of the intermolecular force fields. A slow collision, i.e., g small, will result in greater scattering by a large intermolecular force field. In a fast collision at the same impact parameter, there will be less time for the force field to act, and as a result not as much scattering will occur. The former case will appear as a larger cross section than will the latter.

The dynamics of encounters between particles is needed for discussions of chemical kinetics just as it is in these discussions of transport properties. Later, we will define a cross section for chemical reaction similar to the above definitions. We will also show that the cross sections are simply related to the perhaps more familiar specific rate coefficient. As a matter of fact, the one is the Laplace transform of the other (see following chapter).

Omega Integrals

The several omega integrals that appear in the Chapman-Enskog theory for the transport properties of gases may be each readily evaluated for rigid-sphere molecules. For example, we have just found the scattering angles to be

$$\begin{aligned} \chi(b,g) &= 2\cos^{-1}\left(\frac{b}{\sigma}\right) \qquad b \le \sigma \\ \chi(b,g) &= 0 \qquad\qquad\qquad\quad b > \sigma \end{aligned} \tag{9.71}$$

for molecules of fixed rigid-sphere diameter σ. The scattering angle depends only on the impact parameter b. Nevertheless, we write $\chi(b, g)$ in Eq. (9.71) to remind us that, for all realistic models containing intermolecular attraction, the scattering angle will depend on the distant relative velocity g as well.

We have also just seen how the scattering angle can be integrated over all impact parameters to obtain the cross sections. The result for the rigid-sphere model was

$$Q^l(g) = \pi\sigma^2\left[1 - \frac{1}{2}\left(\frac{1 + (-1)^l}{l + 1}\right)\right] \tag{9.72}$$

Here again we write $Q^l(g)$ even though the cross sections for the special case of rigid-sphere molecules are independent of the distant relative velocity g.

Finally, we integrate over all g to obtain the omega integrals:

$$\Omega^{l,s}(T) = \frac{1}{2\sqrt{\pi}}\int_0^\infty e^{-\mu g^2/2kT}\left(\frac{\mu g^2}{2kT}\right)^{(2s+3)/2} Q^l(g)\,dg \tag{9.61}$$

This general expression is readily specified to the rigid-sphere model where the cross sections are independent of g. The resulting integral is recognized as the gamma function, and the final result is

$$\Omega^{l,s}(T) = \left(\frac{kT}{2\pi\mu}\right)^{1/2}[Q^l(g)]\,\frac{(s+1)!}{2} \tag{9.73}$$

In the evaluation of the scattering angle, the cross sections, and finally the omega integrals, the rigid sphere is, of course, a particularly simple case. More realistic intermolecular models are handled in a conceptually equivalent way. The mathematical detail is, however, always much more tedious.

Among many other attributes of their treatise, Hirschfelder et al.[2] have presented many tables and charts which make the practical calculation of a transport coefficient a relatively simple matter. This ease of usage exists for several common intermolecular models. In this respect, then, their treatise becomes of direct and significant engineering importance. Following these authors, we will define several reduced quantities which will simplify the numerical calculation of a transport property. In defining these reduced variables, we restrict attention to any of the two-parameter models of the intermolecular potential, which all have a characteristic energy ε and a characteristic radius σ. Then,

$$r^* = \frac{r}{\sigma} = \text{reduced intermolecular distance}$$

$$b^* = \frac{b}{\sigma} = \text{reduced impact parameter}$$

$$\phi^* = \frac{\phi}{\varepsilon} = \text{reduced intermolecular potential energy}$$

$$T^* = \frac{kT}{\varepsilon} = \text{reduced temperature}$$

$$g^{*2} = \frac{\frac{1}{2}\mu g^2}{\varepsilon} = \text{reduced relative kinetic energy}$$

It will also be convenient to reduce the cross-reactions and the omega integrals themselves by dividing their value from some arbitrary potential function by the corresponding value obtained for the rigid-sphere model. This is analogous to the earlier reduced second virial coefficient which was defined as the second virial coefficient for the Lennard-Jones potential (or any other) divided by the value for rigid spheres, i.e., $B^* \equiv B/b_0$.

With these reductions, the three basic starting formulas for transport coefficient calculations may be rewritten for the scattering angles,

$$\chi(g^*,b^*) = \pi - 2b^* \int_{r_m^*}^{\infty} \frac{dr^*/r^{*2}}{[1 - b^{*2}/r^{*2} - \phi^*(r^*)/g^{*2}]^{1/2}} \tag{9.74}$$

for the cross sections,

$$Q^{l*}(g^*) = \frac{2}{\left[1 - \frac{1}{2}\left(\frac{1 + (-1)^l}{l + 1}\right)\right]} \int_0^\infty (1 - \cos^l \chi) b^* \, db^* \qquad (9.75)$$

and for the omega integrals,

$$\Omega^{(l,s)*}(T^*) = \frac{2}{(s + 1)! T^{*s+2}} \int_0^\infty e^{-g^{*2}/T^*} g^{*2s+3} Q^{l*}(g^*) \, dg^* \qquad (9.76)$$

The * on the symbols Q and Ω indicates reduction by the corresponding rigid-sphere values. For example, for the Lennard-Jones potential,

$$\Omega^{(l,s)*} = \frac{\Omega^{(l,s)}(\text{Lennard-Jones})}{\Omega^{(l,s)}(\text{rigid sphere})}$$

Rewriting these fundamental relations in this way also emphasizes again the idea of corresponding states. This idea has been extended empirically to dense gases and liquids to yield successful techniques for estimating transport coefficients. This is good news, for extension of our "rigorous" kinetic theory to dense gases and liquids seems mathematically impossible. And besides, the intermolecular potential functions are not known nearly well enough to make such a calculation meaningful even if the mathematics was possible.

In terms of these quantities, the coefficient of viscosity for a pure gas may be written

$$\eta = \frac{5\,(\pi mkT)^{1/2}}{16\pi\sigma^2\Omega^{(2,2)*}} \qquad (9.77)$$

for the thermal conductivity

$$\lambda = \frac{25(\pi mkT)^{1/2}\, C_V^0}{32\pi\sigma^2\Omega^{(2,2)*}M} \qquad (9.78)$$

and for the self-diffusion coefficient

$$D = \frac{3(\pi mkT)^{1/2}}{8\pi\sigma^2\Omega^{(1,1)*}\rho} \qquad (9.79)$$

Here M is the molecular weight, ρ is the density in kg/m^3, and C_V^0 is the heat capacity at constant volume in J/mol · K. These are practical relationships from a first approximation solution from the Chapman-Enskog expansion of the distribution function in phase-space. A second, and more accurate, approximation to, say, the viscosity, involves additional integrals, $\Omega^{(1,s)}$ and $\Omega^{(2,s)}$, where s = 2, 3, and 4. However, calculations with the rigid-sphere potential indicate that the first approximation (involving only $\Omega^{(2,2)*}(T^*)$) will yield a viscosity that is within 98 percent of the "true" value. It is also comforting that the error in using only the first approximation is greater, in general, for the rigid-sphere model than it is for more realistic molecular models.

The cross sections that are calculated using the Lennard-Jones potential are smaller than those for rigid-spheres when the relative energy is high. This is due to the softer repulsion of r^{-12} as opposed to the hard repulsion of $r^{-\infty}$ for rigid spheres. The Lennard-Jones cross sections are several times larger than the rigid-sphere values for low energy collisions because of the attractive forces. This cross section, varying with relative energy g, on integration over all energies of relative approach appears as a temperature dependence of the collision integrals. This dependence is evident in detail from Table 9.2.

This Lennard-Jones model is the most realistic model that has been widely developed for ease of practical calculations. The table of values of the omega integrals have been evaluated by numerical integration of the relations developed above for the Lennard-Jones potential,

$$\phi(r) = 4\varepsilon\left[\left(\frac{\sigma}{r}\right)^{12} - \left(\frac{\sigma}{r}\right)^{6}\right]$$

The Chapman-Enskog expression for viscosity is not too unlike the expression for the viscosity that was first derived from a more naive point of view. For the rigid sphere,

$$Q^{(2,2)} = \tfrac{2}{3}\,\pi\sigma^2$$

and

$$\Omega^{(2,2)*} = 1$$

$$\eta = \frac{5(\pi mkT)^{1/2}}{16\pi\sigma^2} \tag{9.80}$$

which may be compared with

$$\eta = \frac{2(\pi mkT)^{1/2}}{3\pi^2\sigma^2} \tag{9.54}$$

from the naive theory. Or, for rigid spheres, viscosity from the rigorous theory is 0.47 times that from the naive theory. For any model and in practical units, the viscosity is

$$\eta = \frac{(2.669 \times 10^{-8})\sqrt{MT}}{\sigma^2\Omega^{(2,2)*}} \tag{9.81}$$

where σ is in nm, M is the molecular weight, and η is in kg/m · s.

TABLE 9.2 The Collision Integrals $\Omega^{(1,1)*}$ and $\Omega^{(2,2)*}$ Based on the Lennard-Jones Potential

kT/ε	$\Omega^{(2,2)*}$ (for viscosity and thermal conductivity)	$\Omega^{(1,1)*}$ (for mass diffusivity)	kT/ε	$\Omega^{(2,2)*}$ (for viscosity and thermal conductivity)	$\Omega^{(1,1)*}$ (for mass diffusivity)
0.30	2.785	2.662	2.50	1.093	0.9996
0.35	2.628	2.476	2.60	1.081	0.9878
0.40	2.492	2.318	2.70	1.069	0.9770
0.45	2.368	2.184	2.80	1.058	0.9672
0.50	2.257	2.066	2.90	1.048	0.9576
0.55	2.156	1.966	3.00	1.039	0.9490
0.60	2.065	1.877	3.10	1.030	0.9406
0.65	1.982	1.798	3.20	1.022	0.9328
0.70	1.908	1.729	3.30	1.014	0.9256
0.75	1.841	1.667	3.40	1.007	0.9186
0.80	1.780	1.612	3.50	0.9999	0.9120
0.85	1.725	1.562	3.60	0.9932	0.9058
0.90	1.675	1.517	3.70	0.9870	0.8998
0.95	1.629	1.476	3.80	0.9811	0.8942
1.00	1.587	1.439	3.90	0.9755	0.8888
1.05	1.549	1.406	4.00	0.9700	0.8836
1.10	1.514	1.375	4.10	0.9649	0.8788
1.15	1.482	1.346	4.20	0.9600	0.8740
1.20	1.452	1.320	4.30	0.9553	0.8694
1.25	1.424	1.296	4.40	0.9507	0.8652
1.30	1.399	1.273	4.50	0.9464	0.8610
1.35	1.375	1.253	4.60	0.9422	0.8568
1.40	1.353	1.233	4.70	0.9382	0.8530
1.45	1.333	1.215	4.80	0.9343	0.8492
1.50	1.314	1.198	4.90	0.9305	0.8456
1.55	1.296	1.182	5.0	0.9269	0.8422
1.60	1.279	1.167	6.0	0.8963	0.8124
1.65	1.264	1.153	7.0	0.8727	0.7896
1.70	1.248	1.140	8.0	0.8538	0.7712
1.75	1.234	1.128	9.0	0.8379	0.7556
1.80	1.221	1.116	10.0	0.8242	0.7424
1.85	1.209	1.105	20.0	0.7432	0.6640
1.90	1.197	1.094	30.0	0.7005	0.6232
1.95	1.186	1.084	40.0	0.6718	0.5960
2.00	1.175	1.075	50.0	0.6504	0.5756
2.10	1.156	1.057	60.0	0.6335	0.5596
2.20	1.138	1.041	70.0	0.6194	0.5464
2.30	1.122	1.026	80.0	0.6076	0.5352
2.40	1.107	1.012	90.0	0.5973	0.5256
			100.0	0.5882	0.5170

SOURCE: From R. B. Bird, W. E. Stewart, and E. N. Lightfoot, *Transport Phenomena*, Wiley, New York, 1966, p. 746

Example Problem 9.4 Calculate the viscosity of argon at $T^* = 10$. Using Eq. (9.77) with $\varepsilon/k = 124$ K, $\sigma = 0.3418$ nm from Table 9.3, and $\Omega^{(2,2)*} = 0.8242$ from Table 9.2,

$$\eta = \frac{(2.669 \times 10^{-8})\sqrt{40(10)(124)}}{(0.3418)^2(0.8242)}$$

$$= 6173 \times 10^{-8} \text{ kg/m} \cdot \text{s}$$

This may be compared with 6058×10^{-8} kg/m · s from experimental measurement.

Viscosity may be calculated using many intermolecular potentials. The results from some of these appear in Fig. 9.14. Results from Example Problem 9.4 also appear on Fig. 9.14, where at 1240 K, we read $\eta/\sqrt{T} = 175$ or $\eta = 6173 \times 10^{-8}$ kg/m · s. It is interesting to note the completely false prediction of $\eta/\sqrt{T}$ as independent of temperature with a value of 127.4×10^{-8} for the rigid-sphere model. From Fig. 9.14, we see that this prediction is almost at right angles to reality.

These calculations have utilized σ and ε/k evaluated from viscosity measurements. How is the problem solved in reverse to go from an experimental viscosity to obtain the constants in, say, the Lennard-Jones intermolecular potential function? Since there are two constants in the potential, two experimental viscosity measurements at two temperatures are required. The ratio is formed

$$\left(\frac{\eta_1}{\eta_2}\right)_{\text{exp}} = \left(\frac{T_1}{T_2}\right)^{1/2} \frac{\Omega^{(2,2)*}(T_2^*)}{\Omega^{(2,2)*}(T_1^*)} \tag{9.82}$$

For each of a series of assumed values of ε/k, one calculates T_1^* and T_2^*, and then $\Omega^{(2,2)*}(T_1^*)$ and $\Omega^{(2,2)*}(T_2^*)$. This is repeated until an ε/k is selected that makes Eq. (9.82) an equality. With ε/k determined, σ may be obtained from Eq. (9.77) at either T_1 or T_2. Lennard-Jones parameters determined in this way for several simple gases appear in Table 9.3.

Several observations are relevant. The parameters are nonunique. Data at different temperatures and differing quality yield differing parameters. It is possible to force-fit the Lennard-Jones model to data on species that are certainly not spherically symmetric, e.g., a polar molecule like ethanol or a plate-shaped molecule like benzene. Comparisons of the model with experiment for several simple species is evident in Table 9.4. Figure 9.15 graphically shows the quality of fit and emphasizes again the correlating power of corresponding states. Each of these observations has its analog in concerns for the second virial coefficient from the Lennard-Jones model.

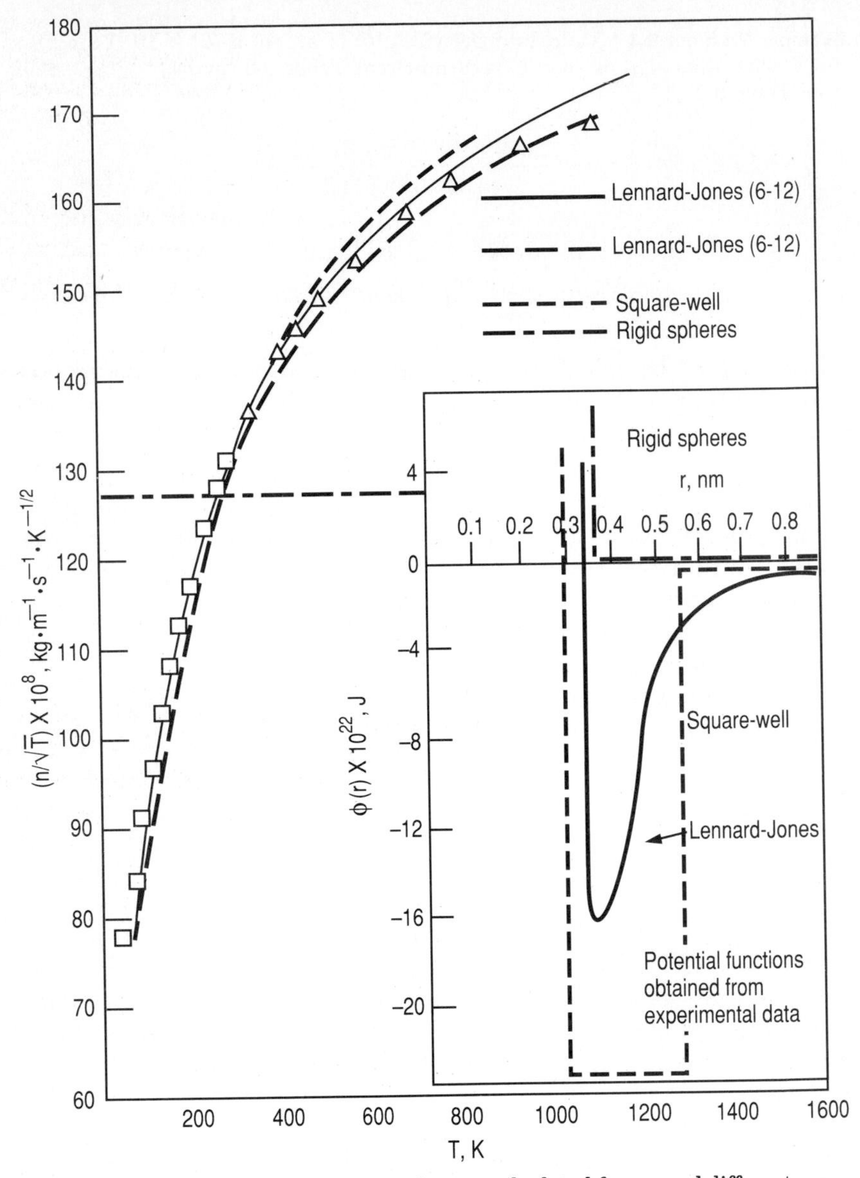

Figure 9.14 The coefficient of viscosity of argon calculated for several different molecular models. Two different curve fits of the Lennard-Jones potential are shown. (*Adapted from Hirschfelder et al.*[2] *Used by permission.*)

The viscosity of mixtures can be calculated as well, but the theoretical mixing rules are complex. Empirical schemes are better.

A popular technique of correlating and estimating viscosity employs the Sutherland potential. Recall from the earlier discussion of real gases that the Sutherland potential is an inverse integer attraction

TABLE 9.3 Lennard-Jones Force Constants for Several Species from Experimental Viscosity Measurements

Substance	Molecular weight M	Lennard-Jones parameters σ, mm	ε/k, K
	Light Elements		
H_2	2.016	0.2915	38.0
He	4.003	0.2576	10.2
	Noble Gases		
Ne	20.183	0.2789	35.7
Ar	39.944	0.3418	124.
Kr	83.80	0.361	190.
Xe	131.3	0.4055	229.
	Simple Polyatomic Substances		
Air	28.97	0.3617	97.0
N_2	28.02	0.3681	91.5
O_2	32.00	0.3433	113.
O_3	48.00	—	—
CO	28.01	0.3590	110.
CO_2	44.01	0.3996	190.
NO	30.01	0.3470	119.
N_2O	44.02	0.3879	220.
SO_2	64.07	0.4290	252.
F_2	38.00	0.3653	112.
Cl_2	70.91	0.4115	357.
Br_2	159.83	0.4268	520.
I_2	253.82	0.4982	550.
	Hydrocarbons		
CH_4	16.04	0.3822	137.
C_2H_2	26.04	0.4221	185.
C_2H_4	28.05	0.4232	205.
C_2H_6	30.07	0.4418	230.
C_3H_6	42.08	—	—
C_3H_8	44.09	0.5061	254.
n-C_4H_{10}	58.12	—	—
i-C_4H_{10}	58.12	0.5341	313.
n-C_5H_{12}	72.15	0.5769	345.
n-C_6H_{14}	86.17	0.5909	413.
n-C_7H_{16}	100.20	—	—
n-C_8H_{18}	114.22	0.7451	320.
n-C_9H_{20}	128.25	—	—
Cyclohexane	84.16	0.6093	324.
C_6H_6	78.11	0.5270	440.
	Other Organic Compounds		
CH_4	16.04	0.3822	137.
CH_3Cl	50.49	0.3375	855.
CH_2Cl_2	84.94	0.4759	406.
$CHCl_3$	119.39	0.5430	327.
CCl_4	153.84	0.5881	327.
C_2N_2	52.04	0.438	339.
COS	60.08	0.413	335.
CS_2	76.14	0.4438	488.

SOURCE: Adapted from R. B. Bird, W. E. Stewart, and E. N. Lightfoot, *Transport Phenomena*, Wiley, New York, 1960.

TABLE 9.4 Comparison of Experimental and Calculated Viscosities Using the Lennard-Jones (6-12) Potential

Values in $\eta \times 10^8$ kg/m · s

T, °K	A		Ne		N_2		CH_4		O_2		CO_2	
	Exp.	Calc.	Exp.	Calc.	Exp.	Calc.	Exp.	Calc.	Exp.	Calc.	Exp.	Calc.
80	688	649	1198	1212								
100	839	814	1435	1451	698	687	403	393	768	757		
120	993	979	1646	1665	826	820	478	472	917	910		
140	1146	1142	1841	1867	948	947	560	553	1061	1059		
160	1298	1300	2026	2054	1068	1070	629	630	1202	1203		
180	1447	1454	2204	2231	1183	1186	703	707	1341	1342		
200	1594	1601	2376	2396	1295	1296	778	780	1476	1474	1015	1014
220	1739	1744	2544	2558	1403	1402	850	852	1604	1602	1112	1114
240	1878	1882	2708	2713	1505	1503	919	921	1728	1726	1209	1212
260	2014	2014	2867	2862	1603	1600	986	987	1845	1845	1303	1308
280	2145	2143	3021	3008	1696	1693	1053	1052	1958	1959	1400	1402
300	2270	2269	3173	3149	1786	1785	1116	1116	2071	2070	1495	1495
400		2839		3812		2202		1405		2578		1923
500		3347		4383		2570		1661		3031		2309
800	4621	4641	5918	5945	3493	3528		2312	4115	4183	3391	3285
1000	5302	5391	6800	6872	4011	4068		2687	4720	4853	3935	3839
1200	5947	6083			4452	4554		3034		5457	4453	4348
1500	6778	6983			5050	5268		3498		6264	5139	5052

SOURCE: Adapted from Hirschfelder et al.,[2] p. 562.

and an infinite repulsion (see Fig. 9.16). The Sutherland potential also leads rigorously to the macroscopic van der Waals equation of state where the constants a and b may be expressed in terms of the Sutherland parameters σ and ε. Even with such a simple potential, the evaluation of scattering angles and cross sections is a major numerical task. Development of the mean free path for a Sutherland gas yields an expression of the form

$$\lambda_s = \frac{\lambda_{rs}}{1 + S/T}$$

where λ_{rs} is the mean free path for rigid-sphere molecules that we developed earlier and S is a positive constant with a different value for each gas. Since viscosity and diffusivity depend on λ, the following form may be obtained:

$$\eta(\text{Sutherland}) = \frac{\eta_{rs}}{1 + S_\eta/T} \tag{9.83}$$

and

$$D(\text{Sutherland}) = \frac{D_{rs}}{1 + S_D/T} \tag{9.84}$$

These are called the Sutherland equations, and they have been widely

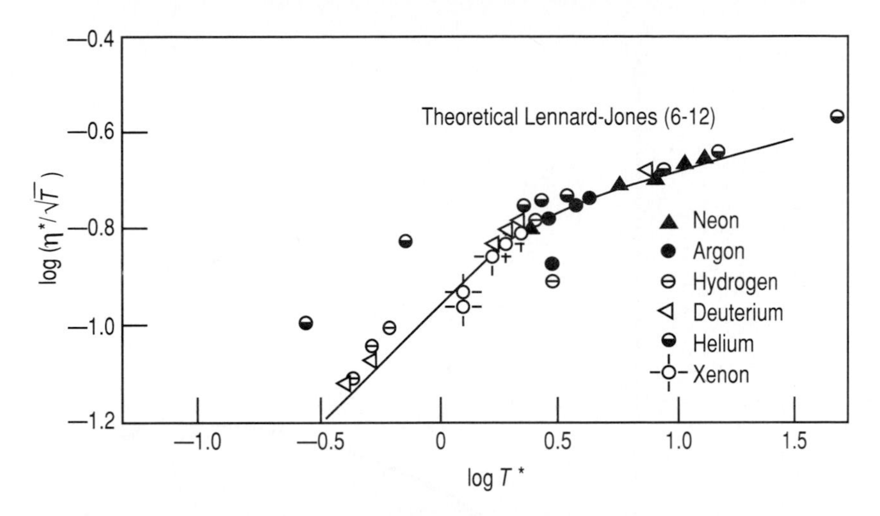

Figure 9.15 Comparison of the theoretical curve of log $(\eta^*/\sqrt{T^*})$ with experimental data (log means logarithm to the base 10). The experimental values are reduced according to the relation $\eta^* \equiv \eta\sigma^2/\sqrt{m\varepsilon}$. *(Adapted from Hirschfelder et al.*[2] *Used by permission.)*

used to fit experimental data. The numerical predictions from the Sutherland model are not good, but the theory has yielded a useful form for empirical use. Thermal conductivity also depends on λ, but inelastic collisions are here so amplified in importance that the details of elastic collisions pale in comparison, and expressions like Eqs. (9.83) and (9.84) are not used.

Thermal Conductivity

The thermal conductivity for monatomic gases, that is, gases with no internal degrees of freedom, may be calculated by using

$$k = \frac{8.322 \times 10^{-4}\sqrt{T/M}}{\sigma^2\Omega^{(2,2)*}(T^*)} \tag{9.85}$$

where σ is in nm, k is in W/m · K, and M is the molecular weight. For the elastic collisions of Chapman-Enskog theory, and strictly for monatomic gases, Eqs. (9.78) and (9.80) show the thermal conductivity to be proportional to the product of viscosity and heat capacity,

$$k = \frac{15}{4}\frac{R}{M}\eta = \frac{5}{2}\left(\frac{3}{2}\frac{R}{M}\right)(\eta) = \frac{5}{2}C_V^0\,\eta \tag{9.86}$$

Some comparisons with experiment for several noble gases are presented in Table 9.5. Calculations with the rigid-sphere potential indi-

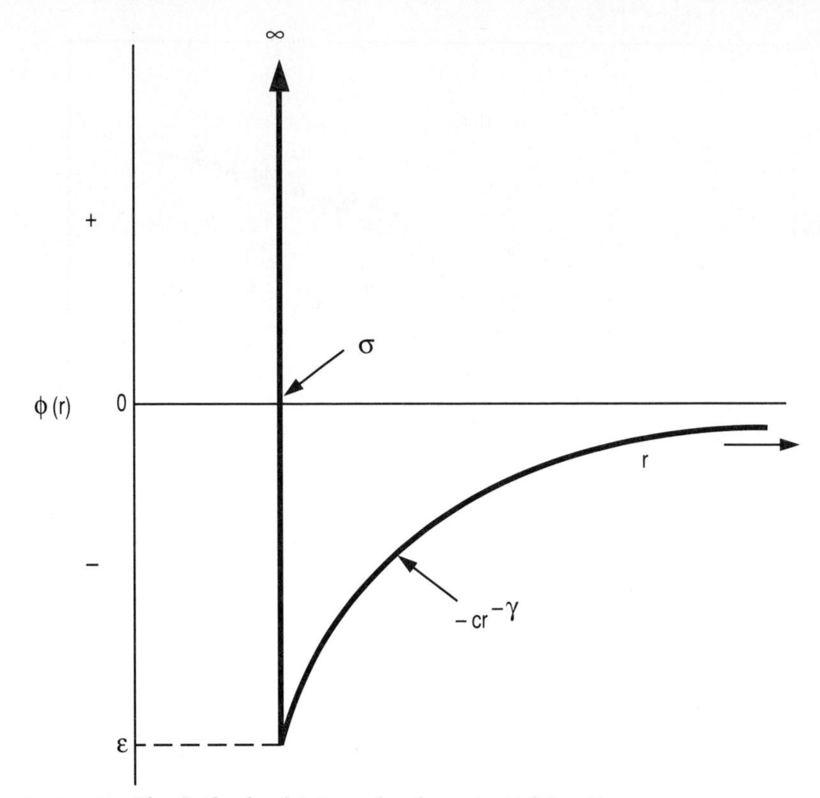

Figure 9.16 The Sutherland intermolecular potential function.

TABLE 9.5 Thermal Conductivity Calculated by Using Lennard-Jones Parameters from Viscosity Data versus Experimental Values

	He		Ne		A		Kr	
T, K	Calc.	Exp.	Calc.	Exp.	Calc.	Exp.	Calc.	Exp.
194.7	2817	2706	878	876	292	293	148	152
273.2	3507	3406	1105	1110	392	394	206	208
373.2	4296	4165	1356	1359	504	506	273	272
579.1	5711	5504	1791	1789	696	685	390	388

SOURCE: Adapted from Hirschfelder et al.,[2] p. 573.

cate that this first-approximation solution involving only $\Omega^{(2,2)}(T^*)$ is within 98 percent of the true convergence value of the Chapman-Enskog theory. Polyatomic gases are much more complicated, for they offer the potential of inelastic collisions in which vibrational, rotational, and translational energies may change.

Certainly the fastest molecules transport the most momentum and the most kinetic energy. The internal energy of the fast molecules is, however, not necessarily greater than the internal energy of slower-moving molecules. Energy is not transported at the same rate by both translational and internal modes. Eucken reasoned that in the proportionality of Eq. (9.86), one should split C_V into its translational and internal modes. For translation, he proposed using the C-E proportionality constant of 2.5, but for the internal modes he let that proportionality constant be 1. Then,

$$k = [2.5C_V^0(\text{trans}) + C_V^0(\text{int})]\eta$$

or, forming a ratio, we get

$$\frac{k(\text{polyatomic})}{k(\text{C-E})} = \frac{2.5\, C_V^0(\text{trans}) + C_V^0(\text{int})}{2.5C_V^0(\text{trans})} = \frac{4}{15}\frac{C_V^0}{R} + \frac{3}{5} \qquad (9.87)$$

This is called the *Eucken correction*. It is widely used as a multiplier with polyatomic gases to correct predictions of thermal conductivity using Chapman-Enskog theory. Detailed treatments of inelastic collisions are much more intricate, and they are not yet practically useful.

The quality of agreement that can be obtained with still very simple polyatomic gases is evident in Table 9.6, where the value of the Eucken correction is also shown. Energy exchange involving internal modes is clearly a very significant part of the thermal conductivity of polyatomic gases. Note that for a monatomic gas, $C_V^0 = 3/2R$, the Eucken correction becomes 1.

The thermal conductivity of mixtures can be calculated as well, but the mixing rules are complex. Empirical schemes are better.

Example Problem 9.5 We have calculated the Prandtl number for N_2 using the naive theory. It is useful to calculate this quantity again using the Lennard-Jones collision integrals. For N_2, $\varepsilon/k = 91.5$ K and $\sigma = 0.3681$ nm (from viscosity data), and at 533.2 K (260°C),

TABLE 9.6 Thermal Conductivity of Polyatomic Gases Calculated by Using Lennard-Jones Parameters from Viscosity Data (with the Eucken Correction)

		λ		
Gas	T, K	Calc.	Exp.	$\frac{4}{15}\frac{C_V^0}{R} + \frac{3}{5}$
O_2	200	436	438	1.268
	300	615	635	1.278
CO_2	200	235	227	1.370
	300	386	398	1.527
CH_4	200	493	522	1.407
	300	741	819	1.479

SOURCE: Adapted from Hirschfelder et al.,[2] p. 574.

$$\eta = \frac{2.669 \times 10^{-8}\sqrt{28(533.2)}}{(0.3681)^2(0.9015)} = 2.670 \times 10^{-5} \text{ kg/m} \cdot \text{s}$$

where $\Omega^{(2,2)*}(T^*) = 0.9015$ at $T^* = 533.2/91.5 = 5.83$, and

$$k = \frac{8.322 \times 10^{-4}\sqrt{533.2/28}}{(0.3681)^2(0.9015)} = 0.02973 \text{ W/m} \cdot \text{K}$$

The heat capacity of N_2 is $C_V^0 = 0.7621$ kJ/kg · K at 260°C. The Eucken correction to the thermal conductivity is

$$\frac{4}{15}\frac{C_V^0}{R} + \frac{3}{5} = \frac{4}{15}\left[\frac{0.7621(28)}{8.314}\right] + \frac{3}{5} = 1.284$$

and

$$k = 0.02973(1.284) = 0.038 \text{ W/m} \cdot \text{K}$$

The Prandtl number is then

$$N_{\text{Pr}} = \frac{C_p^0\eta}{k} = \frac{1.059(2.670 \times 10^{-5})(10^3)}{0.038} = 0.74$$

which may be compared with the earlier value of 1.3 from the naive theory and an experimental value of 0.73. A number of such comparisons appear in Table 9.7 wherein the Eucken relation for the Prandtl number of polyatomic gases at low densities has been used,

$$N_{\text{Pr}} = \frac{C_p^0\eta}{k} = \frac{C_p^0}{C_p^0 + 1.25R}$$

Diffusion Coefficient

The diffusion coefficient of a pair of gases may be calculated from

$$D_{1,2} = \frac{3}{16}\frac{\sqrt{2\pi k^3T^3/\mu}}{p\pi\sigma^2[\Omega_{1,2}^{(1,1)*}(T_{1,2}^*)]} \tag{9.88}$$

or, in practical units,

$$D_{1,2} = \frac{0.2663 \times 10^{-6}\sqrt{T^3(M_1 + M_2)/2M_1M_2}}{p\sigma_{1,2}^2[\Omega_{1,2}^{(1,1)*}(T_{1,2}^*)]} \tag{9.89}$$

where p is the pressure in kPa, $T_{1,2}^*$ is $T/(\varepsilon_{1,2}/k)$ with T in K, σ is in nm, and D_{12} will be in m²/s. The calculation of $[\Omega_{1,2}^{(1,1)*}(T_{1,2}^*)]$ involves Lennard-Jones parameters for the dissimilar molecular interaction $\sigma_{1,2}$ and $\varepsilon_{1,2}/k$. These may be determined from measurements of binary diffusivity as a function of temperature, or from crossed-beam experiments, or from second virial measurements on the mixture as well as each pure species, or by other methods. The dissimilar

TABLE 9.7 Predicted and Observed Values of $C_p^0\eta/k$ for Gases at Atmospheric Pressure

Gas	T, K	$C_p^0\eta / k$ (theoretical)	$C_p^0\eta / k$ (from observed values of C_p, η, and k)
Ne	273.2	0.667	0.66
Ar	273.2	0.667	0.67
H_2	90.6	0.68	0.68
	273.2	0.73	0.70
	673.2	0.74	0.65
N_2	273.2	0.74	0.73
O_2	273.2	0.74	0.74
Air	273.2	0.74	0.73
CO	273.2	0.74	0.76
NO	273.2	0.74	0.77
Cl_2	273.2	0.76	0.76
H_2O	373.2	0.77	0.94
	673.2	0.78	0.90
CO_2	273.2	0.78	0.78
SO_2	273.2	0.79	0.86
NH_3	273.2	0.77	0.85
C_2H_4	273.2	0.80	0.80
C_2H_6	273.2	0.83	0.77
$CHCl_3$	273.2	0.86	0.78
CCl_4	273.2	0.89	0.81

SOURCE: Adapted from R. B. Bird, W. E. Stewart, and E. N. Lightfoot, *Transport Phenomena*, Wiley, New York, 1966, p. 256.

coefficients may also be estimated from the empirical combining rules,

$$\sigma_{1,2} = \tfrac{1}{2}(\sigma_1 + \sigma_2)$$
$$\varepsilon_{1,2} = \sqrt{\varepsilon_1\varepsilon_2} \tag{9.90}$$

With so little available data, use of these rules is common. The nature of agreement with experiment that has been obtained is evident in Table 9.8. Calculations with the rigid-sphere model indicate that the above first approximation involving only

$$[\Omega_{1,2}^{(1,1)*}(T_{1,2}^*)]$$

is within 88 percent of the convergence value from the Chapman-Enskog theory. To this first approximation, only 1,2 interactions influence the diffusivity; 1,1 and 2,2 interactions are distinctly secondary in importance. The C-E theory suggests that measurements of diffusion would yield excellent insight into the force law between dissimilar molecules.

TABLE 9.8 Binary Diffusion Coefficients Calculated by Using the Lennard-Jones Potential with Force Constants for Pure Species from Viscosity Data (p = 101.325 kPa)

				$D_{1,2}$, $m^2/s \times 10^4$	
Gas pair	$\sigma_{1,2}$, nm†	$\varepsilon_{1,2}/k$, K†	T, K	Calc.	Exp.
A-CO_2	0.3707	153	293.2	0.136	0.14
N_2-O_2	0.3557	102	293.2	0.199	0.22
N_2-C_2H_4	0.3957	137	298.2	0.156	0.163
N_2-nC_4H_{10}	0.4339	194	298.2	0.0986	0.0960
N_2-cisbutene-2	0.4467	188	298.2	0.0947	0.095
H_2-CO_2	0.3482	79.5	298.2	0.634	0.646
H_2-C_2H_4	0.3600	82.6	298.2	0.595	0.602
H_2-cisbutene-2	0.4111	113	298.2	0.413	0.378
CO_2-CH_4	0.3939	161	273.2	0.138	0.153

†Calculated by using the combining laws, Eq. (9.90).
SOURCE: Adapted from Hirschfelder et al.,[2] p. 579.

Example Problem 9.6 Calculate the binary diffusion coefficient for CO_2 and CH_4 at 273.2 K.

	ε/k	σ	M
CO_2	190	0.3996	44
CH_4	137	0.3882	16
CO_2:CH_4	161	0.3939	

Then, $T^*_{1,2} = 273.2/161 = 1.697$ and $\Omega^{(1,1)*}_{1,2}(T^*_{1,2}) = 1.141$

and
$$D_{1,2} = \frac{0.2663 \times 10^{-6}\sqrt{(273.2)^3(44 + 16)/2(44)(16)}}{(101.3)(0.3939)^2(1.141)}$$
$$= 13.8 \times 10^{-6}\ \mathrm{m^2 \cdot s^{-1}}$$

As was true with the other transport coefficients, it is again clear that Lennard-Jones force constants may be selected that well-model macroscopic behavior even though the Lennard-Jones potential function cannot describe the actual physical interaction of the molecule. The molecular theory here provides a form or a scaffold for an empirical curve fit. This is not uncommon. For example, the two constants in the van Laar model of real liquid solutions may be derived in terms of the two constants in the van der Waals equation of state. In practice, one forgets that tie and considers the two van Laar constants to be empirical parameters to be evaluated by fitting experimental data on activity coefficients. Alternatively, one might imagine that the transport properties are just not very sensitive to the detailed physics of the bimolecular collisions even though it is these collisions that are the root cause of the existence of the transport properties in the first place.

The coefficient of self-diffusion calculated from the Lennard-Jones potential agrees very well with experiment for many simple gases. This agreement is evident in Fig. 9.17, where the data are plotted such as to show one universal curve and emphasize again the power of the idea of corresponding states. The force constants were determined from viscosity data. These were then used to calculate diffusivity, and the agreement of the latter with experimental diffusivity measurements is excellent. Self-diffusion is measured using isotopes.

As was the case for second virial coefficients, it is possible to calculate the viscosity using a number of models of the intermolecular potential. The calculated viscosity of CO_2 is compared with experiment over a wide range of temperature in Fig. 9.18. The inset shows the character of the rigid-sphere, square-well, Sutherland, and Lennard-

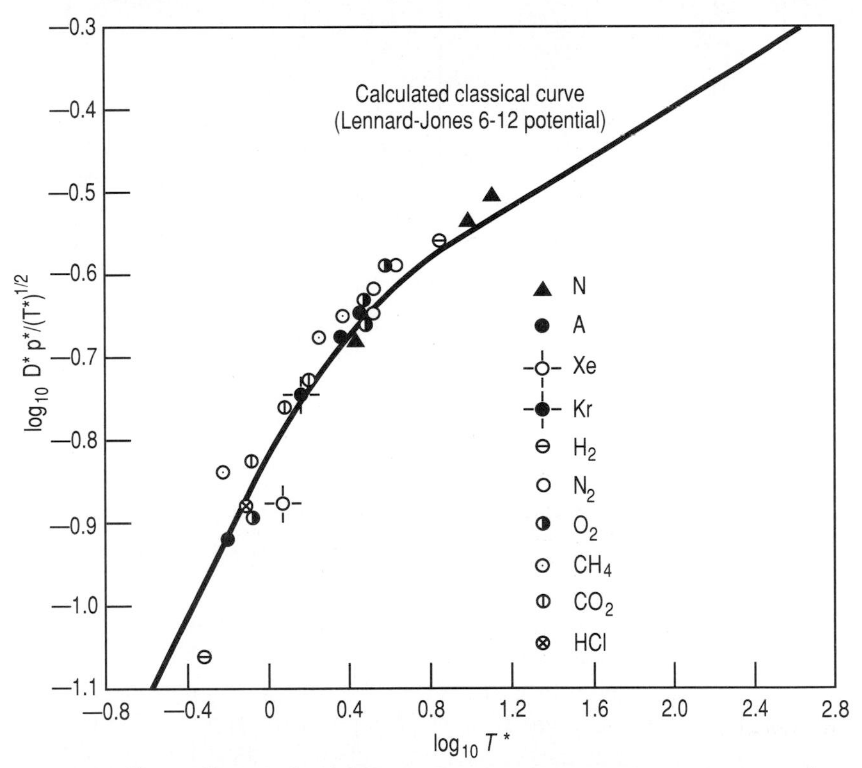

Figure 9.17 The coefficient of self-diffusion in reduced units; $T^* \equiv T/\varepsilon/k$, $p^* \equiv p\sigma^3/\varepsilon$, and $D^* \equiv (D/\sigma)\sqrt{m/\varepsilon}$. The solid curve is the calculated curve: $D^*p^*T^{*-3/2} = 0.2115(1/\Omega^{(1,1)*})$. The plotted points are the experimental values in reduced units. (*Adapted from Hirschfelder et al.*[2] *Used by permission.*)

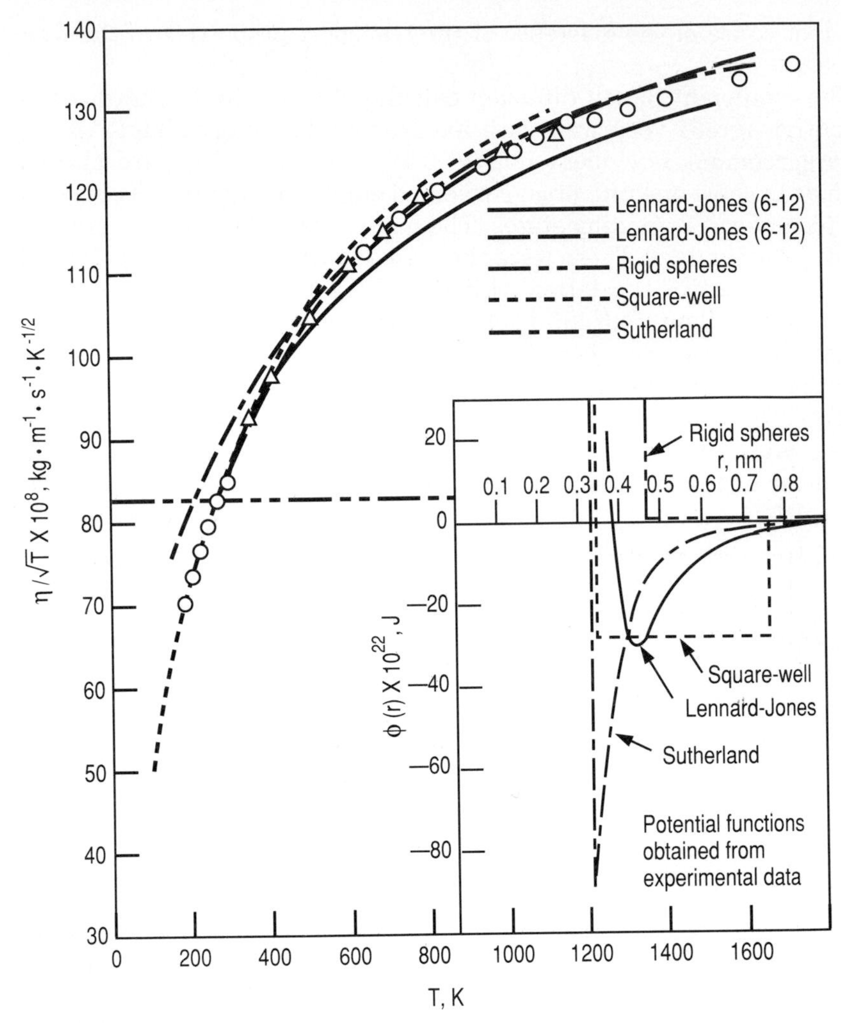

Figure 9.18 The coefficient of viscosity of carbon dioxide as calculated for several molecular models. Two different curve fits of the Lennard-Jones potential are shown. (*Adapted from Hirschfelder et al.*[2] *Used by permission.*)

Jones models. Although there is some variation, each does a rather good job of fitting the experimental data except for the rigid-sphere model, which is completely erroneous. The coefficient of self-diffusion of CO_2 over a much narrower range of temperature appears in Fig. 9.19. The quality of fit is not as good as with viscosity data for any of these molecular models. Note the complete fallacy of the rigid-sphere model.

The use of more realistic models to describe the transport properties of polar molecules is troubled by the mathematical complexity of these more realistic potentials. The Stockmeyer potential

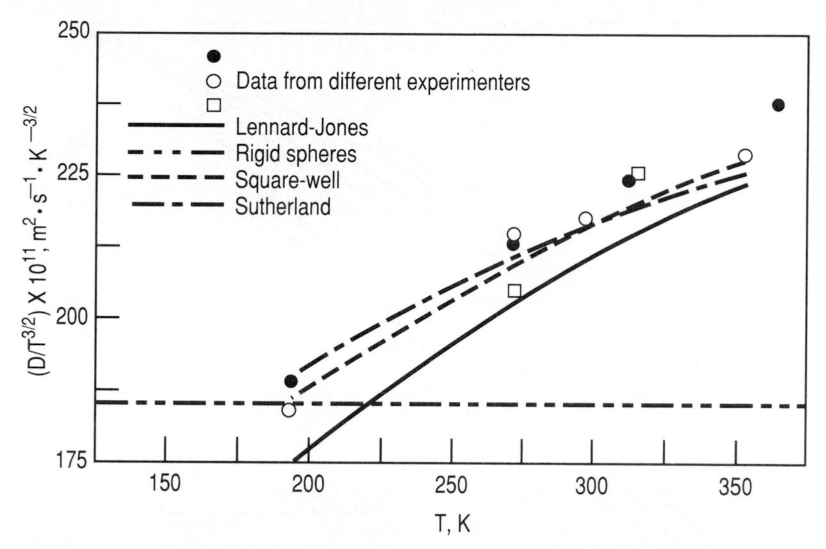

Figure 9.19 The coefficient of self-diffusion for carbon dioxide as calculated for several molecular models with D in m^2/s and T in K. *(Adapted from Hirschfelder et al.[2] Used by permission.)*

$$\phi(r, \theta, \phi) = 4\varepsilon\left[\left(\frac{\sigma}{r}\right)^{12} - \left(\frac{\sigma}{r}\right)^{6}\right] - \frac{\mu^2}{r^3} f(\theta, \phi) \tag{9.91}$$

was used to describe the second virial coefficients of steam. At higher temperatures, collisions are of higher energy; the attractive part of the potential is less important, for the colliding molecules are near each other for a shorter length of time. One may then get a reasonable approximation by considering only the strongest attractions from head-on collisions of the dipolar molecules. The angle dependence then vanishes, and the potential when opposite charges are geometrically opposite is just

$$\phi(r) = 4\varepsilon\left[\left(\frac{\sigma}{r}\right)^{12} - \left(\frac{\sigma}{r}\right)^{6}\right] - \frac{2\mu^2}{4\pi\varepsilon_0 r^3} \tag{9.92}$$

where μ is the dipole moment. Equation (9.92) is called the Krieger potential. The force between dipoles in head-on collisions was discussed in Chap. 5. The results of calculations of viscosity for several polar molecules appear in Fig. 9.20, where a reduced dipole moment defined as

$$\delta^* = \frac{1}{2}\frac{\mu^2}{\varepsilon\sigma^3}$$

is used to characterize each species. The dipole moments of H_2S, SO_2, CH_3OH, and NH_3 are 0.931, 1.611, 1.680, and 1.437 respectively. Fig-

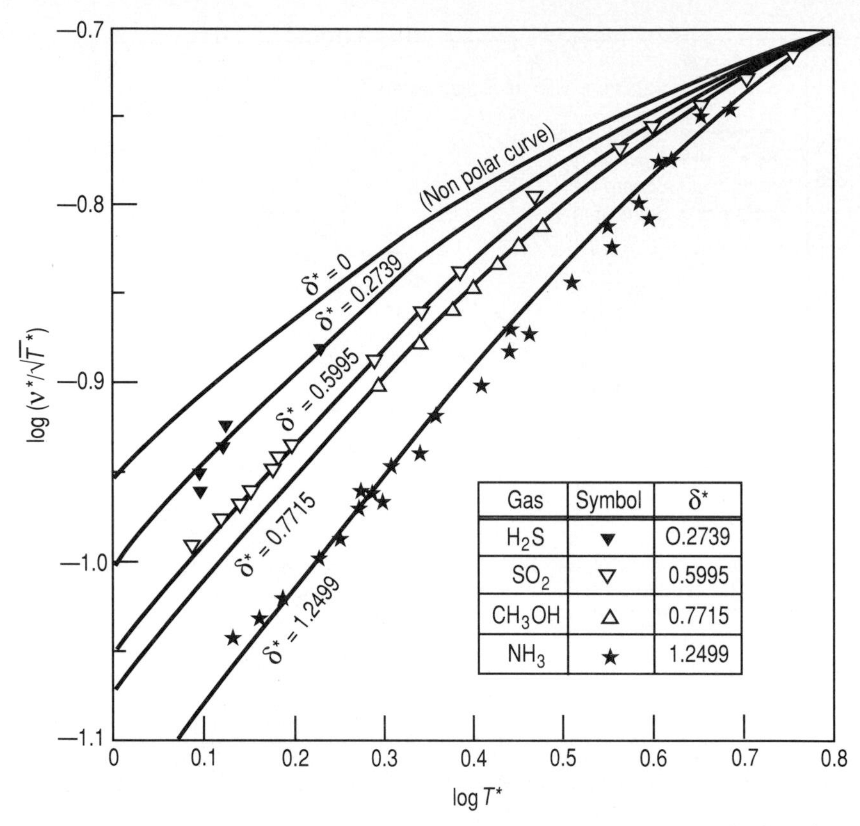

Figure 9.20 Comparison of experimental data with the calculated results based on the Krieger potential function given in Eq. (9.92). The deviations from nonpolar behavior are greater for larger values of δ^*, in keeping with the principle of corresponding states for polar molecules. The reduced quantities are $T^* \equiv T/\varepsilon/k$, $\eta^* \equiv \eta\sigma^2/\sqrt{m\varepsilon}$, and $\delta^* \equiv \mu^2/2\varepsilon\sigma^3$. Logarithm is to the base 10. (*Adapted from Hirschfelder et al.*[2] *Used by permission.*)

ure 9.20 also shows the simple Lennard-Jones curve for a nonpolar species, i.e., $\delta^* = 0$. This curve for $\delta^* = 0$ is analogous then to the single curve earlier presented in Fig. 9.15. Experimental data are well-reproduced by this technique. The curves portray again the significance of the idea of corresponding states.

Summary

Clearly, the Chapman-Enskog kinetic theory of dilute gases is accurate, and (very importantly) it has also been reduced to a form that is convenient for numerical estimates of the transport coefficients. Such estimates are convenient for engineering work. The results of this theory are not applicable to dense gases or to liquids. For such systems there is currently no practically useful kinetic theory, and estimates of properties are better based on empirical relationships.

References

1. E. W., McDaniel, *Collision Phenomena in Ionized Gases*, Wiley, New York, 1964.
2. J. O. Hirschfelder, C. F. Curtiss, and R. B. Bird, *Molecular Theory of Gases and Liquids*, Wiley, New York, 1954.
3. S. Chapman and T. G. Cowling, *The Mathematical Theory of Non-Uniform Gases*, Cambridge Press, 1939.

Further Reading

1. J. O. Hirschfelder, C. F. Curtiss, and R. B. Bird, *Molecular Theory of Gases and Liquids*, Wiley, New York, 1954.
2. E. W. McDaniel, *Collision Phenomena in Ionized Gases*, Wiley, New York, 1964.
3. R. D. Present, *Kinetic Theory of Gases*, McGraw-Hill, New York, 1958.

acteristic broadening and flattening with increased T. Curve b is the reaction cross section as a function of ε. It begins at some threshold value of ε_a and increases with ε in some manner. Curve b is independent of temperature, but the population in the high-energy tail, curve a, is strongly temperature-dependent. Curve c, which is cross-hatched underneath, is just the product of a and b. This cross-hatched area is the specific rate coefficient k of Eq. (10.38), and it sums all collisions that result in reaction. If we multiply numerator and denominator of the ratio of integrals in Eq. (10.40) by

$$\left(\frac{E}{RT}\right)^{s/2-1} \frac{1}{(s/2-1)!}$$

we see that this ratio is just the average energy, contained within s squared terms, of all collision pairs that react. That is, the ratio is the number of molecules of energy E that react, multiplied by E, summed over all E, and divided by the total number of molecules that react. The ratio is then very much like Eq. (9.11) for calculating any average. This average of all molecules that react is the average energy of all molecules within the shaded area of Fig. 10.3. Equation (10.40) can be written more concisely as

$$E_A = E(\text{ave}) - \tfrac{3}{2}RT \qquad (10.41)$$

The Arrhenius activation energy is the difference between the average energy of all molecules that react and the average energy of all molecules.† This activation energy is also represented schematically on Fig. 10.3.

State-to-State Kinetics

The usual way of studying chemical kinetics is to observe the effect of varying concentrations of reactants and of varying temperature. But both of these quantities are statistical, and hence the results of such studies are of little value in testing and formulating theories of chemical kinetics which must arise from a detailed point of view. That is, we must be concerned with collisions of reactants in specific quantum states. We must be concerned with the detailed dynamics of molecular encounters and the subsequent atomic rearrangements that may occur. By thus observing the reactivity of molecules in precisely defined energy states, we may determine the "fine structure" of the reaction rate. This is primarily being pursued experimentally by using molecular-beam techniques and photolytic techniques.

We have defined the overall reaction cross section. It is similarly

---

†This is the Tolman expression for the activation energy.[4]

possible to define a detailed cross section where we are concerned with the collision of molecules of specified energy states to yield products also in specified states:

$$A_i + B_j \rightarrow C_k + D_l$$

There is significant experimental variation, for depending on the character of the potential energy as a function of the geometry of all of the molecules, either translational energy or vibrational energy is experimentally found to be more effective in producing a larger reaction cross section (see later). The subscript i, j, k, or l is a shorthand notation for all quantum numbers that are necessary to specify the precise internal state of each molecule. We expect the rate of reaction to be different for molecules in different internal energy states, and hence the total rate becomes

$$-\frac{dn_A}{dt} = \sum_{i,j} \sum_{k,l} k_{i,j}^{k,l}\, n_{Ai} n_{Bj} \tag{10.42}$$

Here the molecules of A in each quantum state i are considered to be different species. Such a fragmentation of the rate into its component parts is quite well-defined and natural. In any event, the detailed cross section is related to the detailed specific rate coefficient in the same manner as portrayed in Eq. (10.38) for the overall reaction cross section,

$$k_{i,j}^{k,l}(T) = (\pi\mu)^{-1/2} \left(\frac{2}{kT}\right)^{3/2} \int_0^\infty q_{i,j}^{k,l}(\varepsilon)\varepsilon e^{-\varepsilon/kT}\, d\varepsilon \tag{10.43}$$

where the integration is over all translational energies of i relative to j. We recognize from this expression that $k_{i,j}^{k,l}(T)$ is simply proportional to the Laplace transform of the product of $q_{i,j}^{k,l}(\varepsilon)$ and ε. This is an informative insight, for it shows the equivalence of the *specific rate coefficient* language of the chemist and the *reaction cross section* language of the chemical physicist.

We have already assumed a maxwellian translational velocity distribution. If we further assume that the internal energy states of A and B are thermally equilibrated, we can calculate the number densities of A_i and B_j from their internal molecular partition functions, e.g.,

$$n_{A_i} = \left(\frac{n_A}{f_A(\text{int})}\right) \exp\left(\frac{-\varepsilon_{A_i}}{kT}\right)$$

This merely makes the usually good assumption that the rate of internal energy exchange is greater than the rate of chemical reaction so that the Maxwell-Boltzmann distribution is maintained. The phenomenological rate constant for the original bimolecular reaction, Eq. (10.43), may then be written

$$k(T) = \frac{(2/kT)^{3/2}}{(\pi\mu)^{1/2} f_A(\text{int}) f_B(\text{int})} \sum_{i,j} \sum_{k,l} \exp\left[\frac{-(\varepsilon_{A_i} + \varepsilon_{B_j})}{kT}\right] \times \int_0^\infty q_{i,j}^{k,l}(\varepsilon)\varepsilon \exp\left(\frac{-\varepsilon}{kT}\right) d\varepsilon \quad (10.44)$$

Although Eq. (10.44) appears formidable, it really is not, for the double summation is of limited extent. For example, few vibrational states are occupied except at very high temperatures.

This is the most general collision expression that may be written while still assuming a thermodynamic or Maxwell-Boltzmann system. It may be specified further and in various ways, and it will yield results corresponding to the various collision theories of chemical kinetics that have already been discussed. For example, for the simplest assumptions of (1) constant reaction cross section of $q = \pi\sigma^2$, above some threshold value ε_a of the total relative translational energy ($s = 3$) rather than the translational energy along the line of centers ($s = 2$), and (2) the internal energy configurations of reactants and products have no effect on the kinetics, Eq. (10.44) becomes,

$$k(T) = \left(\frac{2}{kT}\right)^{3/2} (\pi\mu)^{-1/2} (\pi\sigma^2) \int_{\varepsilon_a}^{\infty} \varepsilon e^{-\varepsilon/kT} \, d\varepsilon$$

that is, Eq. (10.38), but with a particular assumption about the variation of $q(\varepsilon)$ with ε. The integral is related to Eq. (9.20) with $s = 4$. The integral may be rigorously evaluated to yield

$$k(T) = \left(\frac{2}{kT}\right)^{3/2} (\pi\mu)^{-1/2}(\pi\sigma^2)\,(kT)^2\Gamma(2)e^{-\varepsilon_a/kT}\left[\left(\frac{\varepsilon_a}{kT}\right) + 1\right]$$

or

$$k(T) = \sigma^2 \left(\frac{8\pi kT}{\mu}\right)^{1/2} \exp\left(\frac{-\varepsilon_a}{kT}\right)\left[\frac{\varepsilon_a}{kT} + 1\right] \quad (10.45)$$

If one elects to use only the first term in the series that arises from successive integration by parts, one retains only ε_a/kT in the bracketed term, and one recovers the simplest collision relationship, Eq. (10.17). Equation (10.45) may be equally well viewed as just the fraction of all 1, 2 pairs that have relative translational energy of at least ε_a in 4 squared terms [Eq. (9.20) with $s = 4$] multiplied by the total collision frequency $z_{1,2}$, given by Eq. (9.37).

The energy levels are often known, but the cross sections are rarely known, and so we frequently revert to some form of approximation. The best hope for experimental determination of the state-specific cross sections $q_{i,j}^{k,l}(\varepsilon)$ seems to be from laser photolytic experiments or from molecular-beam experiments.

It is interesting that the integral in the general expression for the rate coefficient, Eq. (10.44), is the Laplace transform of the product of the cross section and the energy. Recall that the Laplace transform $g(p)$ of a function $f(t)$ is defined

$$g(p) = \int_0^\infty f(t)e^{-pt}\,dt$$

In our rate constant expression, $f(t)$ is $\varepsilon q_{i,j}^{k,l}(\varepsilon)$ and p is $(kT)^{-1}$. These integral transforms have been tabulated for many functions[5] $f(t)$, so that the development of expressions for the cross sections from a rate constant expression and vice versa may often be readily accomplished.

As an example of this interrelationship, consider the case of reaction between molecules whose internal structure is unimportant; $q_{i,j}^{k,l}(\varepsilon)$ is just $q(\varepsilon)$, and the specific rate coefficient, Eq. (10.44), becomes

$$k = (\pi\mu)^{-1/2}\left(\frac{2}{kT}\right)^{3/2}\int_0^\infty q(\varepsilon)\varepsilon e^{-\varepsilon/kT}\,d\varepsilon \qquad (10.46)$$

Reaction will occur only if the relative kinetic energy of the collision is greater than some threshold value ε_a. Let us make the reasonable first-order approximation of a cross section of zero for $\varepsilon < \varepsilon_a$ and of $\pi\sigma^2$ (the area of the effective disk presented by a spherical molecule to an impinging molecule) for $\varepsilon \geq \varepsilon_a$. Above ε_a, the cross section is constant. We can write the Laplace transforms of the piecewise continuous function

$$f(t) = \begin{cases} 0 & \text{for } \varepsilon < \varepsilon_a \\ \varepsilon\pi\sigma^2 & \text{for } \varepsilon \geq \varepsilon_a \end{cases}$$

The transform then becomes,

$$g(p) = \int_0^\infty f(t)\,e^{-\varepsilon/kT}\,d\varepsilon = \int_0^{\varepsilon_a}(0)e^{-\varepsilon/kT}\,d\varepsilon + \int_{\varepsilon_a}^\infty(\pi\sigma^2\varepsilon)\,e^{-\varepsilon/kT}\,d\varepsilon$$

or $$g(p) = \pi\sigma^2\,e^{-\varepsilon_a/kT}\left(\int_0^\infty xe^{-x/kT}\,dx + \int_0^\infty \varepsilon_a e^{-x/kT}\,dx\right)$$

where we have substituted $x = \varepsilon - \varepsilon_a$ to put the integral into the standard Laplace form involving integration from 0 to ∞. The two integrals are found from tables[6] to be

$$g(p) = \pi\sigma^2 e^{-\varepsilon_a/kT}\,[(kT)^2 + \varepsilon_a(kT)]$$

$$\dot{\phi} = \frac{d\phi}{dt} = \frac{bg}{r^2}$$

that is, Eq. (10.52), and

$$\dot{r} = \frac{dr}{dt} = \pm\left[g^2 - \frac{2v(r)}{\mu} - \frac{b^2g^2}{r^2}\right]^{1/2}$$

that is, Eq. (10.51), and together we obtain,

$$\frac{dr}{d\phi} = \pm\frac{r^2}{bg}\left[g^2 - \frac{2v(r)}{\mu} - \frac{b^2g^2}{r^2}\right]^{1/2} = \pm\frac{r^2}{b}\left[1 - \frac{2v(r)}{\mu g^2} - \frac{b^2}{r^2}\right]^{1/2} \tag{10.54}$$

The negative sign yields the incoming branch of the trajectory, and the positive sign yields the symmetric outgoing branch of the trajectory (see Fig. 10.5). The distance of minimum approach, r_m, i.e., the value of r that makes $dr/d\phi = 0$, is the largest positive value of r that makes the rightmost bracketed term of Eq. (10.54) equal to zero,

$$1 - \frac{b^2}{r_m^2} - \frac{2v(r_m)}{\mu g^2} = 0 \tag{10.55}$$

A real solution to Eq. (10.55) does not always exist. For attractive potentials of the form $v(r) = -cr^{-n}$ with $n \geq 2$, there are values of b and g for which there is no solution. Rather, the moving particle orbits the target—in a sort of sticky collision—and then spirals back out. On the other hand, solutions to find r_m for some attractive potentials do exist and are well-behaved. Solutions always exist for repulsive potentials, however. If b is large enough for a given g, an orbit of the sort shown in Fig. 10.6*a* occurs, and this is certainly in keeping with one's intuition. At the same g, a smaller value of b less than a critical b_0, a spiraling orbit of the sort shown in Fig. 10.6*b* will

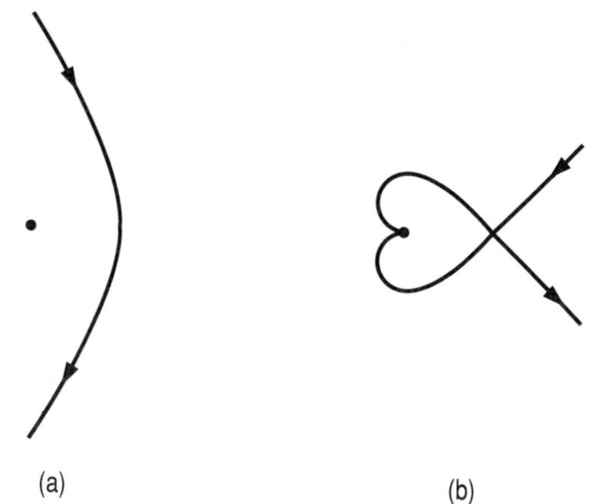

Figure 10.6 Types of orbit of a particle moving in the attractive inverse-fourth-power potential field. (*a*) Large angular momentum with a nonspiraling orbit; (*b*) small angular momentum with a spiraling orbit.

occur. Orbits may be categorized as spiraling or nonspiraling, and this distinction is frequently seen in the literature of kinetics and of transport phenomena. To explain ion-molecule reaction,[8] we assume that no reaction occurs in type a trajectories, but reaction is certain for type b trajectories. This "certainty" of course assumes that such spiraling interactions, but with only inelastic energy transfer, are few compared to ion-molecule reactive collisions. If all this be so, the cross section for reaction is clearly

$$q(g) = \pi b_0^2(g)$$

where the critical impact parameter b_0 is a function of the distant relative velocity g. We can find b_0, the impact parameter at the transition from a to b behavior in Fig. 10.6 that produces a spiraling, but perfectly balanced, circular orbit, by setting the discriminant of the square root of Eq. (10.55) equal to zero. This yields

$$b_0 = \left[\frac{2(zC)^2\alpha}{\mu g^2}\right]^{1/4}$$

and the cross section becomes

$$q(g) = (\pi)\left[\frac{2(zC)^2\alpha}{\mu g^2}\right]^{1/2}$$

The specific rate coefficient is just $gq(g)$ [see Eq. (10.34)], or

$$k = \pi\left[\frac{2(zC)^2\alpha}{4\pi\varepsilon_0\mu}\right]^{1/2} \tag{10.57}$$

All this is in exact agreement with experimental observations using the mass spectrometer that show the cross section for reaction (i.e., tantamount to the specific rate coefficient) as varying inversely with the square root of the ion repeller voltage and independent of temperature. The primary ions are accelerated by this repeller voltage E_r to translational energy $E_r = \frac{1}{2}m(p^+)g^2$. Therefore a cross section varying inversely with the square root of repeller voltage is equivalent to a cross section varying inversely with the translational energy. The sense of the agreement is exactly correct, and the nature of the numerical agreement that may be expected is evident in Table 10.3.

Not surprisingly, these data suggest that the above model based on encounters of spherical species is not correct for collisions where orientation effects are important. For example, both HCl and HBr have permanent dipole moments, and the ion-molecule intermolecular potential is, in that case, not determined by the polarizability alone. We would expect the cross section for polar neutrals to be greater because

TABLE 10.3 Theoretical and Experimental Rate Constants for Several Ion-Molecule Reactions

Reaction	Polarizability of the neutral molecule, $\alpha \times 10^{30}$ m^3	Reduced mass μ, g	$k \times 10^9$, cm^3/molecule · s Experimental	Theoretical
$Ar^+ + H_2 \rightarrow ArH^+ + H$	0.7894	1.919	1.68	1.1
$Ar^+ + HD \rightarrow \begin{cases} AH^+ + D \\ AD^+ + H \end{cases}$	0.7829	2.810	1.43	0.9
$Ar^+ + D_2 \rightarrow ArD^+ + D$	0.7749	3.661	1.35	0.8
$Kr^+ + H_2 \rightarrow KrH^+ + H$	0.7894	1.969	0.48	1.0
$Kr^+ + D_2 \rightarrow KrD^+ + D$	0.7749	3.845	0.30	0.7
$Ne^+ + H_2 \rightarrow NeH^+ + H$	0.7894	1.832	0.27	1.1
$N_2{}^+ + D_2 \rightarrow N_2D^+ + D$	0.7749	3.523	1.72	0.8
$CO^+ + D_2 \rightarrow COD^+ + D$	0.7749	3.523	1.63	0.8
$O_2 + H_2{}^+ \rightarrow O_2H^+ + H$	1.60	1.897	7.56	1.5
$O_2 + D_2{}^+ \rightarrow O_2D^+ + D$	1.60	3.579	3.56	1.1
$D_2 + D_2{}^+ \rightarrow D_3{}^+ + D$	0.7749	2.015	1.43	1.0
$H^{35}Cl^+ + H^{35}Cl \rightarrow H_2Cl^+ + Cl$	2.63	17.994	0.43	0.6
$H^{81}Br^+ + H^{81}Br \rightarrow H_2Br^+ + Br$	3.61	41.001	0.22	0.5

SOURCE: Adapted from Gioumousis and Stevenson.[8]

of the greater attractive forces, but we observe the opposite. Ion-molecule reactions involving hydrocarbons reveal cross sections that are linear functions of the reciprocal voltage and for which a small temperature dependence is observed. The present model fails to describe these processes as well.

Example Problem 10.1 Calculate the specific rate coefficient for the reaction of Ar^+ with H_2 at 2000 K.

The specific rate coefficient is given by Eq. (10.57),

$$k = \pi \left[\frac{2(zC)^2 \alpha}{(4\pi\varepsilon_0 \mu)} \right]^{1/2}$$

where the vacuum permittivity has been included to convert α in its tabulated units of m^3 into its proper units of J · m^4/C^2. When the ion-molecule reaction occurs in a medium other than a low-pressure gas, one includes the dielectric constant of the medium, D, with the $4\pi\varepsilon_0$. The fact that the reaction occurs at 2000 K is immaterial, for unlike Arrhenius-type behavior, this ion-molecule reaction is not temperature-dependent. Then,

$$k = \pi \left[\frac{2(1.602 \times 10^{-19})^2 (0.7894 \times 10^{-30})(6.023 \times 10^{23})}{(1.919 \times 10^{-3})(4\pi \times 8.854 \times 10^{-12})} \right]^{1/2}$$

$$= 1.1 \times 10^{-15} \text{ m}^3/\text{molecule} \cdot \text{s}$$

Avogadro's number appears to convert the reduced mass in kg/mol to kg/molecule. The units may be verified:

$$k = \left[\frac{(\mathrm{C})^2(\mathrm{m}^3)}{(\mathrm{kg})(\mathrm{C}^2 \cdot \mathrm{J}^{-1} \cdot \mathrm{m}^{-1})}\right]^{1/2} \Rightarrow \left[\frac{\mathrm{m}^6}{\mathrm{s}^2}\right]^{1/2} \Rightarrow \frac{\mathrm{m}^3}{\mathrm{s}}$$

Ion-molecule reactions occur in flames where temperatures may be 1000 to over 4000 K. They also occur in chemical vapor deposition (CVD), in radiation chemistry, and in plasma processing. Reactions may be endothermic, e.g.,

$$H_3O^+ + CH_2O \rightarrow CH_3O^+ + H_2O \qquad \Delta H^0 = 41 \text{ kcal/mol}$$

or exothermic, e.g.,

$$CH_3O^+ + O \rightarrow H_3O^+ + CO \qquad \Delta H^0 = -157 \text{ kcal/mol}$$

Reactions may involve positive as well as negative ions. An example of ion concentration as a function of distance above a 15-cm-diameter flat-flame burner appears in Fig. 10.7 Flames form a good milieu or "solvent" in which to add reagents to study ion-molecule reaction kinetics.

Termolecular Collision Frequency†

There are a very few elementary reaction processes of known third-order kinetic behavior. Three-body collisions are rare compared to two and the chance of a perhaps required geometry, vibrational phase angle, etc. is also rare. We then expect reactive termolecular processes to be unusual. For chemical reaction purposes, we may reasonably define a ternary collision among hard spheres as occurring when the third sphere is within some small distance δ of a binary collision, that is, when a binary collision complex is struck by a third species. An estimate of the termolecular collision frequency can be made in the following way.

Consider a collision complex $M_1 \cdot M_2$ as existing when the two molecules M_1 and M_2 are within a separation σ_a, i.e., the complex is said to exist when $\sigma_{1,2} < \sigma < \sigma_a$. We would expect an equilibrium concentration of these complexes,

$$(M_1 \cdot M_2) = K(M_1)(M_2)$$

One molecule of M_1 is "complexed" with another when the second lies within the volume $\frac{4}{3}\pi(\sigma_a^3 - \sigma_{1,2}^3)$. There are n_2 molecules of M_2 per cubic meter, and then there are $n_2(\frac{4}{3}\pi)(\sigma_a^3 - \sigma_{1,2}^3)$ molecules of M_2 within the spherical shell about each M_1, or a total of $n_1 n_2(\frac{4}{3}\pi)(\sigma_a^3 - \sigma_{1,2}^3)$ complexes per cubic meter of gaseous mixture. An additional factor of ½

†Adapted from Benson.[9]

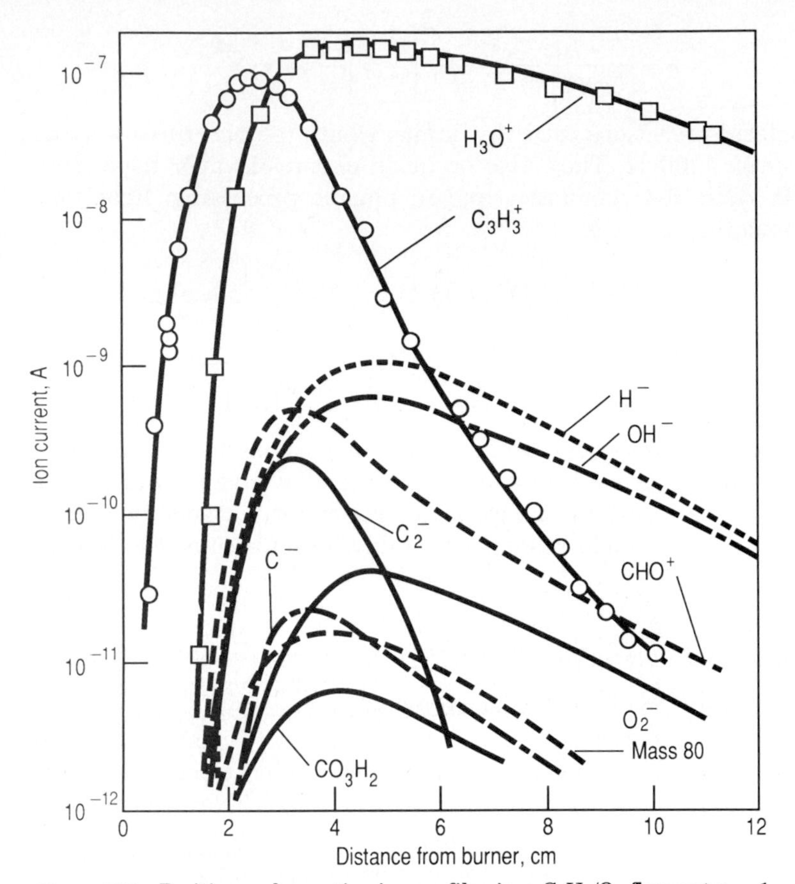

Figure 10.7 Positive and negative ion profiles in a C_2H_2/O_2 flame at $p = 1$ torr and a stoichiometric feed. (*Adapted from H. F. Calcote et al., 10th Symposium on Combustion, 1964. Used by permission.*)

would be included to avoid the double count if the complexes were merely a dimer of one species. Therefore, the equilibrium constant is given by $K = (4/3\pi)(\sigma_a^3 - \sigma_{1,2}^3)$. But the rate of termolecular collisions is just the rate at which these complexes will be struck by a third molecule, which is, from our previous developments, given by

$$z_{1,2,3} = [K(n_1)(n_2)](n_3)\pi(\sigma_a + \sigma_3)^2 \left(\frac{8kT}{\pi\mu}\right)^{1/2}$$

The bracketed term, $K(n_1)(n_2)$, is just the concentration of M_1M_2 complexes, and

$$z_{1,2,3} = n_1 n_2 n_3 \left(\frac{4}{3}\pi\right)(\sigma_a^3 - \sigma_{1,2}^3)\pi(\sigma_a + \sigma_3)^2 \left(\frac{8kT}{\pi\mu}\right)^{1/2} \tag{10.58}$$

where the reduced mass of the $M_1 \cdot M_2$ and M_3 collision is

$$\mu = \frac{m_{1,2} m_3}{m_{1,2} + m_3}$$

To get some feel for the magnitude of $z_{1,2,3}$, imagine that we have an equimolar mixture of all three components at standard temperature and pressure (STP) or $m_1 = m_2 = m_3 = 10^{19}$ molecules/cm^3. Also, take $\sigma_a = 0.4$ nm, $\sigma_{1,2} = 0.3$ nm, $\sigma_3 = 0.2$ nm, and the velocity $(8kT/\pi\mu)^{1/2} = 400$ m/s. If these numbers are typical, we can assert that the termolecular frequency $z_{1,2,3} \cong 7 \times 10^{23}\ \text{s}^{-1} \cdot \text{cm}^{-3}$. Clearly, the bimolecular frequency is five to six orders of magnitude greater than the termolecular frequency [see Eq. (9.38) and following]. Third-order molecular events are unlikely. Since $n = p/kT$, the termolecular collision frequency varies as p^3.

I atom recombination is an interesting example of a termolecular process:

$$\text{I} + \text{I} + M \rightarrow \text{I}_2 + M$$

where the third body M is referred to as an energy-transfer catalyst. The character of M has a profound effect on the rate constant, as is evident in Fig. 10.8.[10] Here we see a remarkable correlation between the rate of atom recombination and the boiling point of the third body. The range of rates is about 240, which is far larger than the range of hard-sphere collision diameters of the atoms and the third bodies. Catalysts in Fig. 10.8 include M's as diverse as He, benzene, water, carbon tetrachloride, propane, and ethyl iodide. This correlation suggests that the forces that bind a molecule to itself to form a liquid, that is, the M-M forces, are similar to the forces that bind M to I to form a reactive encounter. The collisions are more sticky the greater is this force, and there are more binary complexes per cubic meter than predicted by the rigid-sphere picture that is described here. There are then more ternary encounters. The boiling point is, of course, a measure of the M-M attractive forces, and then it correlates with the reaction rates.

Quantum Theory of Rate Processes

The popular collision models of chemical reaction that were developed above are classical arguments, i.e., nonquantum arguments. To be sure, there was a passing reference to the idea of quantized vibrational energy levels when expressions for the degeneracy, Eqs. (10.20) to (10.24), were proposed as well as when we discussed the detailed

$$CH_4 + O \rightarrow CH_3 + OH$$

where all of the experimental data between $300 \le T \le 2200$ K are reasonably fit by

$$k = AT^m e^{-\Delta H/RT}$$

with $A = 1.94 \times 10^{-17}$ $cm^{-3} \cdot s^{-1}$, $m = 2.075$, and $\Delta H = 7.63$ kcal/mol.[11] Using CH_3F as a reasonable isoelectronic model of the transition state, [CH_3-HO], it is readily possible to develop activation energies and preexponential factors that yield reasonable agreement with experiment as shown in Table 10.5 and graphically in Fig. 10.11. This development may equally well postulate a bent or a linear transition state with results for k that differ by a factor of 2 or more.

Eyring Theory of Absolute Reaction Rates

A prospective closely related to that just described is widely known as the *Eyring theory of absolute reaction rates* although there is nothing about it that is absolute. As before, one imagines a transition state in equilibrium with the sea of ordinary molecules. But we also inject an efficiency factor for passing out of the jug of Fig. 10.9 to form products rather than being reflected back. This is called the transmission coefficient κ, which for every reaction will have a value between 0 and 1. Then,

$$k = \nu\kappa e^{-\varepsilon_0/kT} \frac{f_s^\dagger}{f_s} \tag{10.62}$$

The complex pathway out of the jug of Fig. 10.9 is called the reaction coordinate (more later). The two molecules of a bimolecular reaction move through the energy valley called *reactants* along a trail called the *reaction coordinate* that leads up and over an energy saddle and down into an adjoining valley called *products*. A misalignment with

TABLE 10.5 Arrhenius Parameters for the Reaction O + CH_4 as Predicted using a Bent (B) or a Linear (L) Transition State

		T = 300 K	T = 1000 K	T = 2000 K
$\log A_T$	L	10.01	11.12	11.68
	B	10.69	11.63	12.05
ΔH_T, kcal/mol	L	8.12	8.32	7.98
	B	9.06	8.78	7.50
k_2, $cm^{-3} \cdot s^{-1}$	L		1.2×10^{-12}	3.9×10^{-11}
	B	7.4×10^{-18}	3.1×10^{-12}	1.0×10^{-10}
	exp	7.4×10^{-18}	7.0×10^{-13}	2.0×10^{-11}

SOURCE: D. M. Golden, *J. Phys. Chem.*, *83*, 108 (1979).

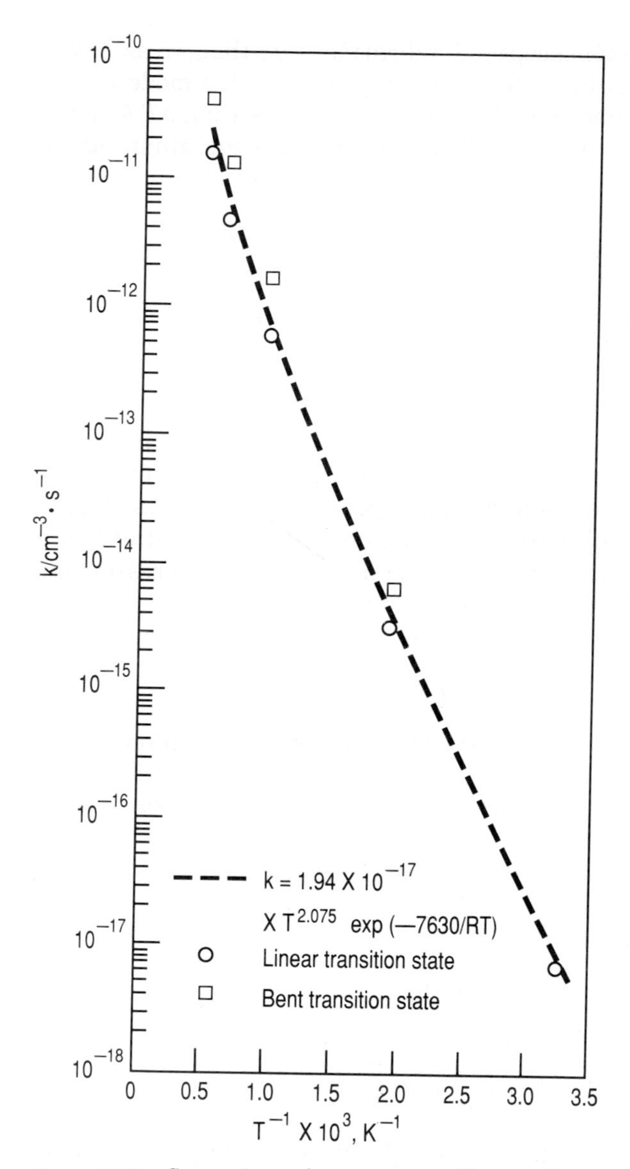

Figure 10.11 Comparison of experiment with transition-state theory estimates for $O + CH_4 \rightarrow HO + CH_3$. (*Adapted from D. M. Golden, J. Phys. Chem., 83, 108 (1979). Used by permission.*)

the trail at the saddle results in reflection rather than passage, and κ then is less than 1. All this is again an abstract picture of chemical reaction, but it is also easily visualized.

If we imagine all components of the partition functions for $f_s^\dagger$ and f_s

TABLE 10.6 Loss of Rotational Freedom n in Bimolecular Reactions

Reactants	Transition State	n
atom + linear	linear	0
linear + linear	linear	2
linear + linear	nonlinear	1
linear + nonlinear	nonlinear	2
nonlinear + nonlinear	nonlinear	3

which corresponds to a linear collision complex ($n = 2$).[12] But experimental data on the similar dissociation reaction

$$O_2 + Ar \leftrightarrows 2O + Ar$$

are well-represented by

$$k = (3.6 \times 10^{19})T^{-1}e^{-118{,}000/RT}$$

with k in cm^3/mol · s and at temperatures between 3300 and 7500 K.[12] Table 10.6 suggests the temperature dependence of the preexponential factor for this reaction should be $T^{-0.5}$. The difference could be due to the neglect of the temperature dependence of the vibrational partition functions in Eq. (10.67). More likely it is due to difficulties in obtaining good rate data where inaccuracies of a factor of 2 to 10 are not uncommon. It is essential that one critically evaluate data from the scientific literature before beginning serious calculations.

There is little choice in either setting $\nu = 10^{13}$ s^{-1} in Eq. (10.61) or setting $\nu = kT/h$ in Eq. (10.62). Both arguments involve the partition function of an unobservable transition state about which we can only make reasoned guesses. Both perspectives do provide a self-consistent scheme for speculation about the dependency of reaction rate on structure.

State-to-State Measurements

As we have seen, overall, macroscopic kinetics is the sum total of molecular events involving reactants in specific quantum states forming products in similarly specific quantum states and all arising from collisions with specific geometries and energies. The best measurements of such state-to-state reaction kinetics are obtained from laser experiments and from crossed-molecular-beam experiments. Translational-velocity-selected molecular beams may be obtained with choppers. Rotational state selection has occurred.[13] The experiments are exquisite. However, the impact parameter and the relative geometry of the colliding molecules eludes control. The phase angles of the many vibrators that may be involved are also uncontrollable. Complete description also demands knowledge of the translational energy and internal quantum states of the product molecules as well. This has been approached, for example, by infrared chemiluminescent measurements.[14] The effect of all of these microscopic variables on the elementary act of energy exchange or chemical reaction is called *chemical dynamics*. The remainder of this chapter is devoted to chemical dynamics.

Crossed Molecular Beams in the Study of Chemical Kinetics

Most chemical reactions occur by bimolecular encounters, and it is of fundamental interest to understand such encounters. As we have seen, processes that are observed to be unimolecular will become bimolecular if the pressure is sufficiently reduced so that the dependence of concentration of labile molecules on collision frequency is controlling. In a crossed-molecular-beam experiment, one produces a beam (a fine pencil of molecules all moving in the same direction and perhaps also with the same velocity) of each of two compounds which are directed such as to cross each other at some point in space. Clearly such beams must be produced in a low-pressure environment to minimize background scattering and attendant diminution in beam intensity. Beam experiments permit insight into much of the fine detail of molecular dynamics, for one can observe the effect of single collisions.

An example of a crossed-beam experiment appears in Fig. 10.12. The experiment was designed to study the reaction

$$D + H_2 \rightarrow HD + H$$

The D atoms were made by thermolysis in a hot tube at 2800 K. The atoms emerge from an orifice in the furnace, and they are collimated by a series of orifices or slits, with high-speed pumping of the space between each such orifice. This pencil of D atoms is directed such as to encounter a similar beam of H_2 molecules. In the apparatus of Fig. 10.12, modulation of the atomic beam and use of phase-sensitive detection increased sensitivity even with a background of HD. Rotating the detector about the collision covolume allowed determination of the scattering angle with respect to the incoming high-velocity beam of D. Even molecular velocities of the product HD could be measured by noting the phase shift of the modulated signal. The HD beam shape, its phase shift, and its scattering angle allow determination of internal energy. Interestingly, with no velocity selection at these low pressures, one is observing the reaction of a maxwellian beam of D at 2800 K with a maxwellian beam of H_2 at 100 K. This is very different from the typical reaction in a three-necked boiling flask. Of course, the act of state selection, whether velocity, vibrational, or whatever, negates

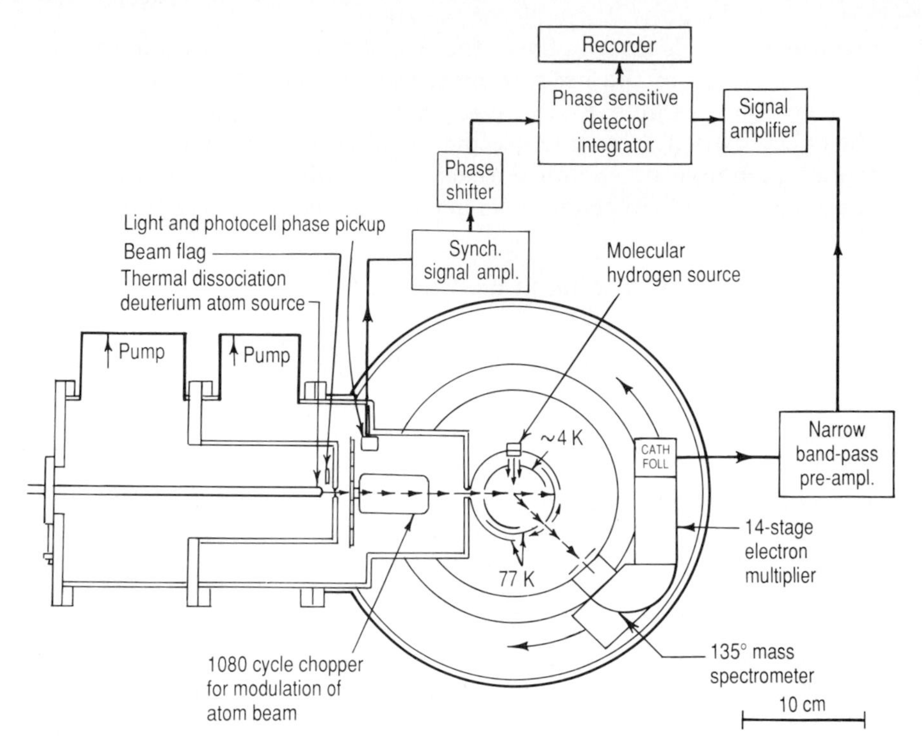

Figure 10.12 A typical crossed-beam apparatus. (Partly schematic, but principal dimensions approximately to scale.) *(From S. Datz and E. H. Taylor, J. Chem. Phys., 39, 1896 (1963). Used by permission.)*

the idea of temperature. The experimental results in this case suggested a short-lived linear collision complex yielding largely unexcited HD.

With rotating slotted disks, one can make a velocity selection to probe the effect of translational energy on reaction. Such a velocity selector appears schematically in Fig. 10.13. Here a beam originates at S and passes through the rapidly rotating slotted disks. Only those species of appropriate speed will travel the distance L and pass through the second slotted disk C_2 within the specific time span set by the speed of rotation of the shaft and the offset of the slots. Changing the speed of rotation will change the molecular speed that is selected and which will arrive at R and continue on into a reactive scattering experiment.

One can conveniently produce a molecular beam aerodynamically using a nozzle expansion scheme like that shown schematically in Fig. 10.14. In a nozzle expansion source, one produces a "reagent" beam of cold molecules essentially free of rotation and perhaps vibrationally cold as well. Rotational, vibrational, and electronic state selection of

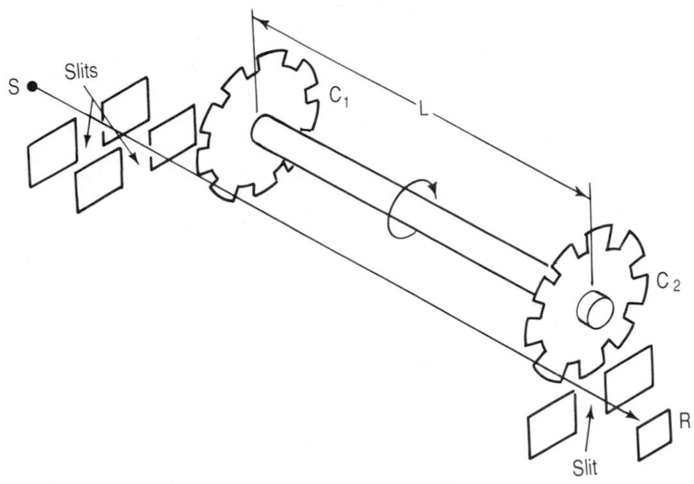

Figure 10.13 Schematic of apparatus for velocity selection from a molecular beam. *(From G. W. Castellan, Physical Chemistry, Addison-Wesley, 1971. Used by permission.)*

the reagent beam is possible by using inhomogeneous electric and magnetic fields. For example, spin-up or spin-down hydrogen atoms may be selected. Molecules in the beam can be oriented in space to study steric effects. Facile interfacing of the beam with lasers is readily possible, allowing studies of rotational, vibrational, and electronic excitation of reagent beam molecules on their chemical reactivity.

If the species and their energy states, both of reactants and products, that recoil from the collision covolume in a crossed-beam experiment may be detected, one may deduce, in principle, all of the information that is needed to completely describe not only the chemical kinetics, but also the transport properties and thermodynamic gas imperfection properties of the several species. Stringent requirements are placed on detection since one must unravel elastic, inelastic, and reactive scattering processes which are all occurring simultaneously.

It is a powerful technique that yields data and then theories for real understanding of reaction kinetics as opposed to mere description of rate phenomena.

Trajectory Calculations

To understand the dynamics of a bimolecular reaction event, we need to know the energy of the colliding system as a function of its configuration. With no external fields, the total energy is constant, but there will be movement of energy between translation, rotation, and

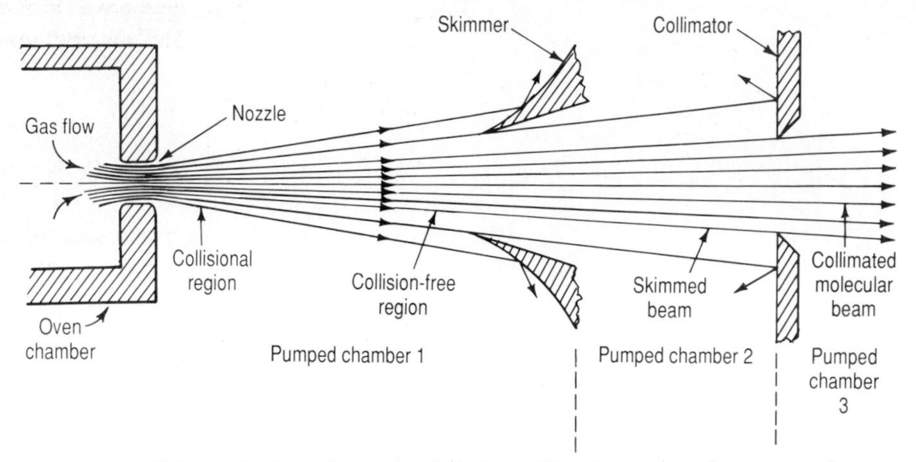

Figure 10.14 Schematic view of a nozzle-skimmer-collimator source of a supersonic molecular beam. While the pressure in the oven chamber may be of the order of several atmospheres, it is typically 10^{-3} torr in the nozzle-skimmer chamber and less than 10^{-6} torr in the collimator chamber. The distance from nozzle to skimmer is of the order of 10^2 nozzle diameters. John Fenn, Chemical Engineering Department, Yale University, was one of the pioneers of this technique. (*Adapted from R. B. Bernstein, Chemical Dynamics via Molecular Beam and Laser Techniques, Oxford University Press, 1982. Used by permission.*)

vibration in both kinetic and potential form as the collision occurs. Visualizing all this for even a simple polyatomic system is impossible—it is just too complicated—but we can picture it in a schematic way by considering a simplified collision of a triatomic system which is confined to be collinear. For example, consider

$$F + H_2 \rightarrow FH + H \tag{10.68}$$

which is an important collision in the chemically pumped HF laser. The potential energy as a function of the two interatomic distances in this collinear collision is schematically depicted in Fig. 10.15 and more specifically in Fig. 10.16. There are two, not three, interatomic distances because for simplicity of analysis we allow only collinear collisions. The potential energy is shown as contours like elevation maps familiar from land survey work. The potential energy is a map that relates the internal energy of colliding molecules to their complete geometry. Reaction is represented by a point of reduced mass μ entering from the right along the reactant valley, going up and over the high mountain pass, and exiting upward along the product valley. If one were to cut the reactant valley at a right angle to the abscissa, the cut edge would look like a Morse curve (Fig. 3.9) that describes the vibration of H_2. Similarly, a cut perpendicular to the ordinate would reveal a Morse-like curve describing the HF potential. The difference in ele-

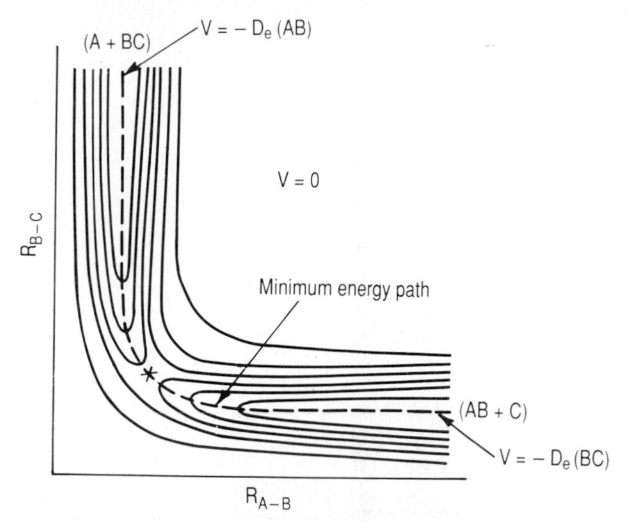

Figure 10.15 A schematic potential contour map for a collinear collision of an atom and a diatomic molecule. The entrance valley has a depth $V = D_e(BC)$, the dissociation energy of BC (as measured from the bottom of the valley) and similarly for the exit valley. At the saddle point (×) the potential energy is above that of either valley but below that of the plateau which corresponds to complete dissociation of all three atoms.

vation from the top of the pass to the reactant valley is the activation energy for reaction. This is much less than the depth of the valley, i.e., the H_2 bond energy of 102 kcal/mol. Similarly, on reversing the motion, we see the activation energy for the reaction of H with HF and the HF bond energy respectively. If the three-particle system has enough energy, it can exit either valley up to the high plateau above, at $V = 0$, which corresponds to complete dissociation into three atoms. The difference in elevation of the two valleys is the thermodynamic exoergic [in the case of reaction (10.68)] or endoergic [in the reverse of reaction (10.68)] character of the reaction.

The configuration at the saddle at the top of the pass has already been called the transition state. There is sometimes a slight depression at the saddle point, and the point traversing the pass can become trapped there to form a long-lived metastable transition state. A very shallow depression would produce a sticky collision (see later).

Even though we cannot draw such a neat picture, we schematically envisage all reactions as occurring similarly regardless of their molecular complexity.

Preparation of such a potential energy surface is computationally within reach for very simple systems. Energy as a function of geometry has been calculated via quantum-mechanical *ab initio* techniques

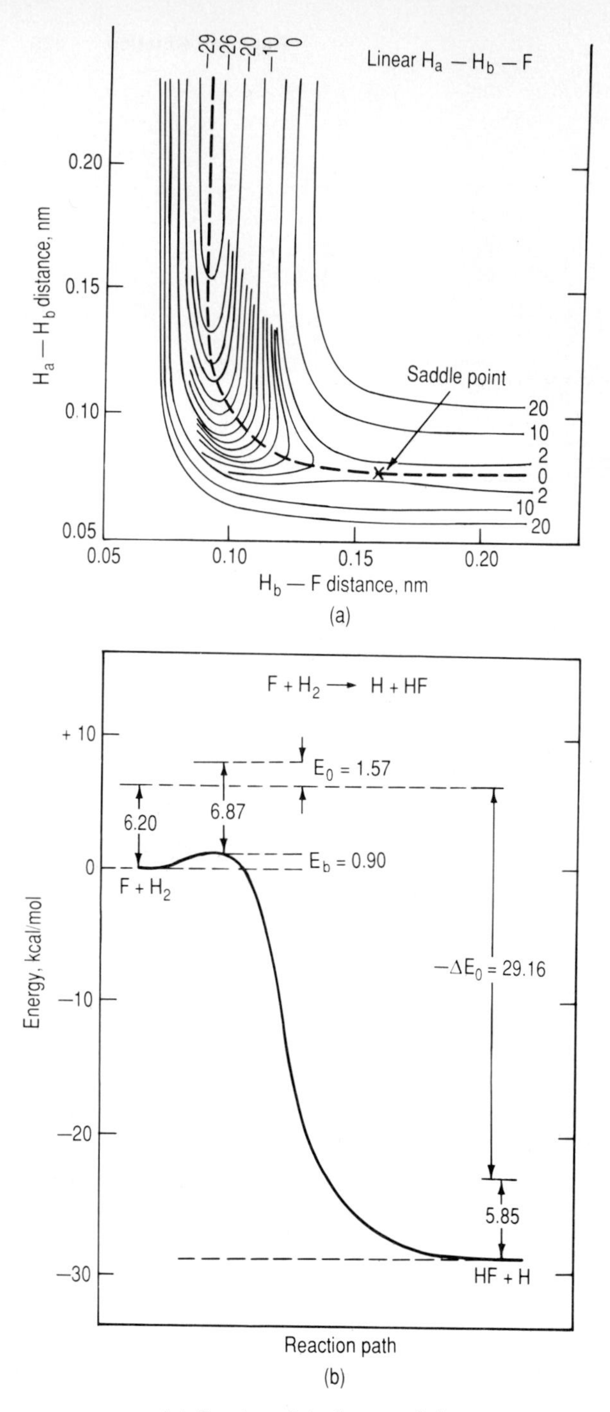

Figure 10.16 (*a*) Semiempirical potential contour map for collinear FH_2. The dashed curve through the saddle point (×) is the minimum-energy path. (*b*) Potential energy profile along the dashed reaction path of the collinear FH_2 system. (*Adapted from Levine and Bernstein.*[3] *Used by permission.*)

as well as by semiempirical quantum schemes (see Chap. 8). Because the quantum calculations are so difficult, such diagrams are usually prepared semiempirically using known diatomic potentials. This is the basis of the so-called London-Eyring-Polanyi-Sato (LEPS) method. Figure 10.16 is an example of such a surface. The lower figure is a schematic of the energy along the reaction coordinate, i.e., the dashed line of the upper figure. The activation energy for the reaction of Eq. (10.68) is 0.9 kcal/mol and the exothermicity is 29 kcal/mol. The activation energy for the reverse reaction is also 29 kcal/mol. The zero-point vibrational energy of H_2 is $\frac{1}{2}h\nu$ = 6.20 kcal/mol and correspondingly for HF and HHF.

The calculation of energy as a function of configuration as represented by Fig. 10.15 allows new insight into the idea of steric hindrance. For example, a LEPS calculation of the energy of the ClHI system as a function of angle of approach, θ, of the Cl atom relative to the center of mass of HI and at a fixed HI separation,

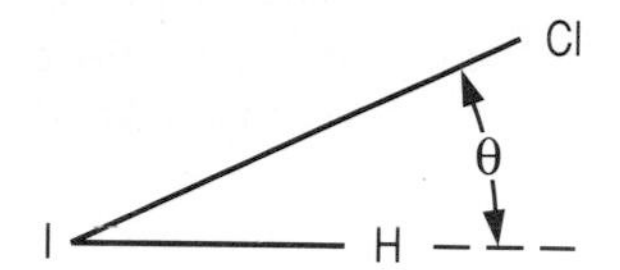

shows a low-energy approach by the chlorine atom in a narrow cone about the H atom. Only such collisions will lead to effective H atom abstraction,

$$HI + Cl \rightarrow HCl + I$$

Otherwise, the large negative cloud that is the I atom effectively shields the small H atom from attack by the Cl. In this simple system, an illustrative, if still semiempirical, calculation is possible. Nevertheless, we imagine this example as a schematic representation of the steric character of all reactions.

With the potential energy described by the contour map for all separations, it is possible to solve the classical-mechanical equations of motion for a collision of initial impact parameter b, initial relative velocity g, and initial position in the vibrational and rotational motion of the reactant. Dynamically, this would be analogous to rolling a marble through the surface of Fig. 10.16. The marble starts far up the valley at large F-H separation. It is at some point up the wall (vibrational position), and it starts down the valley toward the pass at some velocity and direction (initial relative kinetic energy and impact parameter). Newton's laws of mechanics allow one to trace the path through the valleys and pass. The result of such a trajectory calculation appears in Fig. 10.17 for both a reactive collision and a nonreactive collision wherein the marble was merely reflected back up the valley of reactants from which it came.

Experimental reactions, of course, occur at all impact parameters, all relative velocities, and all phase angles of the vibrators. It is then necessary to make many trajectory calculations to enable comparisons of averaged trajectory results with experiment. The initial conditions are randomly selected to begin each trajectory using a so-called Monte Carlo technique. Typically the ratio of effective to total collisions from about 1000 trajectories, i.e., 1000 simulated collisions, will reasonably correspond to the specific rate coefficient. The two heavy dots of Fig. 10.18 have each been determined by using the LEPS surface of Fig. 10.16. The dashed line is from activated complex theory using reasonable assumptions about the partition function of the F-H-H transition state and the LEPS predicted activation energy of 0.9 kcal/mol. Both Monte Carlo trajectory calculations and activated complex theory have yielded reasonable agreement with experiment. Uncertainties in trajectory calculations arise from the use of classical mechanics to describe collision dynamics and the use of potential energy surfaces that are highly approximate.

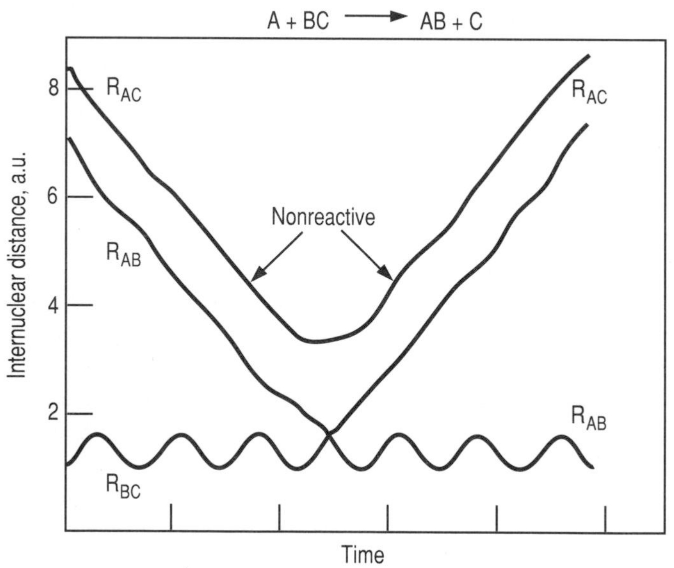

Figure 10.17 Collision trajectory for a reactive collision showing vibration of the reactant BC molecule prior to the encounter with the A atom, the fast switchover of the atoms, and the oscillation of the newly formed AB bond as atom C departs. Also, collision trajectory of a nonreactive collision showing large values of R_{AC} both before and after collision. *(Adapted from Levine and Bernstein.[3] Used by permission.)*

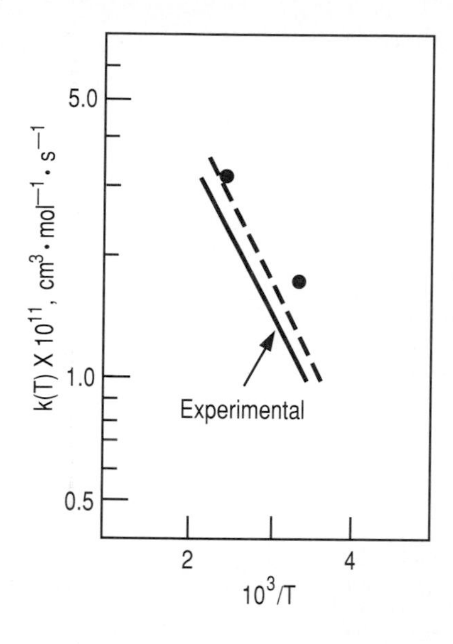

Figure 10.18 Arrhenius plot of bimolecular rate constant for the reaction of F + H_2. The dashed line represents calculations made by activated complex theory, the points are results of classical trajectory computations. The same semiempirical potential surface, that of Fig. 10.16, was used for both. That this surface is reasonable is evident from the comparison with the experimental line. *(Adapted from Levine and Bernstein.[3] Used by permission.)*

Trajectories are calculated by solving the equations of motion of classical mechanics. To place products in specific quantum states, the now continuous rotational and vibrational energies that come from the trajectory calculation must be "quantized" in some way. This is done by assigning the product species to the quantum state nearest, but less than, the classical energy that has been calculated. The difference is assigned to translation.

It is reasonable to wonder whether high translational energy or high internal energy will be more effective in producing reactive collisions. And an important related question is that of where does the exothermicity of the reaction reside in the product species. More specifically, is the energy released in a reaction contained in relative translation of the product species or is it in the vibration of the products? The answer to both questions is evident from the character of the potential energy surface.

Consider a surface such as that appearing in Fig. 10.19 where the exothermicity is released early in the collision. That is, the saddle point occurs at still rather large R_{AB} separations. Trajectory calculations reveal the reaction to follow the heavy line, and the product AB leaves the collision with high vibrational excitation. Such surfaces with an early saddle are said to be attractive surfaces. Conversely, some surfaces have a late saddle that occurs when R_{BC} is already rather large. In this case, trajectory calculations predict the product AB to leave the collision with high translational energy. Such surfaces are said to be repulsive.

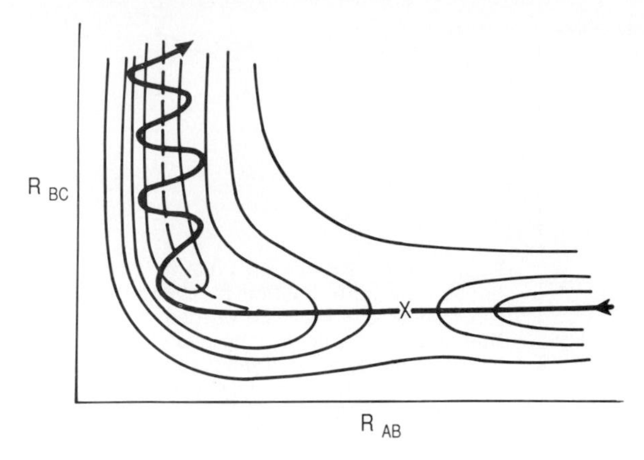

Figure 10.19 Schematic representation of efficient channeling of the reaction exoergicity into product vibrational excitation, for an attractive potential energy surface. The early release of the exoergicity (past the saddle point ×) leads to a deviation of the actual collision trajectory (solid line) from the minimum-energy path (broken line). Had the trajectory followed this reaction path, the exoergicity would have been released as relative translational energy of the products. (*Adapted from Levine and Bernstein.*[3] *Used by permission.*)

Figure 10.20 compares predicted and experimental vibrational excitation from the reaction

$$H + F_2 \rightarrow HF + F$$

that is so important in the HF chemically pumped infrared laser. The LEPS surface was used in the trajectory calculations. It is attractive, the exothermicity appears as vibrational excitation of the product HF, and the excitation is peaked at $v = 6$. This reaction produces a vibrational population inversion which is the necessary condition for any infrared laser (more later). For the laser to operate, the vibrationally hot HF that has been chemically produced must be stimulated down before the inversion is relaxed by collision to the equilibrium Maxwell-Boltzmann distribution. At equilibrium—that is, for a thermodynamic system—the ground-state vibrators are most numerous, and the distribution would appear more as schematically shown by the dashed curve of Fig. 10.20.

Experiment reveals that the vibrators first relax among themselves to form an internal Maxwell-Boltzmann distribution, but characterized by a higher temperature than that of the surrounding translation-rotation "heat bath." This purely vibrational relaxation is called *V-V* transfer. This vibrational temperature can be measured by spectroscopically determining the relative population of each level and then calculating the corresponding temperature. The lower tem-

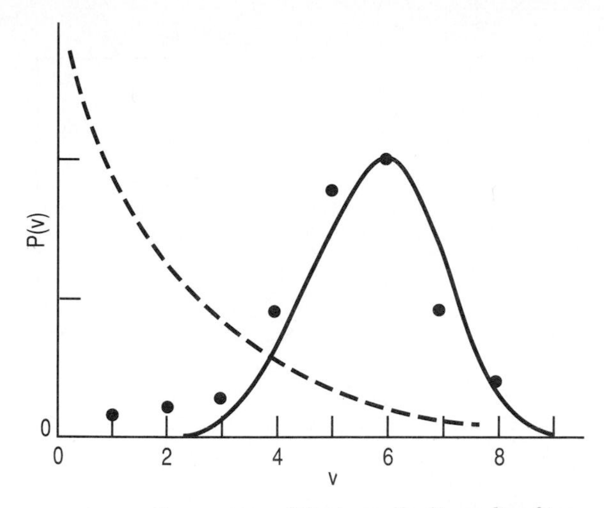

Figure 10.20 Comparison of the (smoothed) results of trajectory computations based on an LEPS surface (solid line) for the location of the energy released in the reaction $H + F_2 \rightarrow HF + F$, with the experimental values (•). The ordinate is the relative probability of formation of HF in the vibrational state v. *(Adapted from Levine and Bernstein.[3] Used by permission.)*

perature of the translation-rotation heat-bath could be measured in the usual way using a thermocouple or other conventional thermometer. The slower relaxation of the vibration into the translation-rotation heat bath then occurs. This is called *V-T* transfer. When all of this is complete—perhaps requiring a total of a few microseconds, depending on the system—the reaction has come to equilibrium and the final temperature is called the *adiabatic flame temperature.*

Reactions that occur on repulsive surfaces cannot produce a chemically pumped laser because the exothermicity will now appear as translation.

Related to the question of the fate of exothermicity in reaction is that of the relative effectiveness of translation versus internal energy in producing reaction in the first place. Recall that, in each of the collision theories, effective collisions depended on their occurrence with an energy in excess of some threshold value ε_a. But there was no concern for where this energy might best reside for collisions to be most effective. Trajectory calculations permit such an analysis, and Fig. 10.21 schematically depicts the situation for both attractive and repulsive surfaces. In each instance, there is sufficient energy in the collision for reaction to occur; that is, the total energy exceeds ε_a. On an attractive surface, high translation leads to reaction while high vibrational excitation leads to reflection of the collision back up the reactant valley. Conversely, for a repulsive surface, high translation leads to repulsion while high vibrational excitation leads to reaction. There

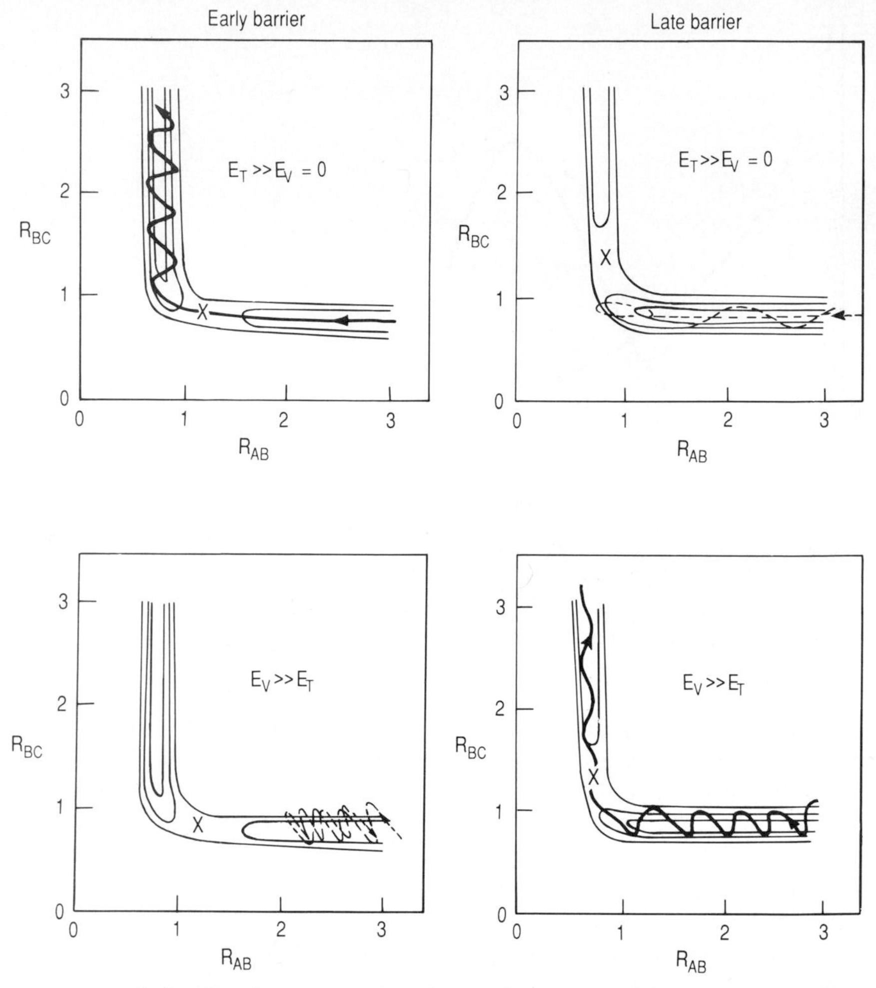

Figure 10.21 Influence of reactant energy for the thermoneutral $A + BC \rightarrow AB + C$ reaction on an LEPS surface with a barrier at ×. (*Adapted from Levine and Bernstein.*[3] *Used by permission.*)

is discrimination or bias in collisions, for it is clear that all forms of energy are not at all equally effective in producing reaction.

Relaxation Times

At equilibrium, the distribution of molecules among each energy mode is uniquely specified as maxwellian. It is only for such maxwellian systems that temperature is defined and for which thermodynamics exists. Some sudden energy change, resulting from perhaps a high-speed expansion, or a flame, or an explosion, or an acoustic process, or perhaps from a photolysis, can each result in the displacement of molecules from their equilibrium Boltzmann distribution. Because of collisions, this nonthermodynamic system will relax back to a Maxwell-Boltzmann state. The rate of such relaxation processes is then of practical concern. Several parameters are in common use to describe the rate of such relaxation processes. To illustrate these parameters and to interrelate them, consider the collision

$$\mathrm{HF}(v = 1) + M \rightarrow \mathrm{HF}(v = 0) + M^*$$

If the vibrational quantum in HF is removed as translation of the collider M, this reaction is called a *V-T* transfer process. The rate of the process depends on the dynamics of the bimolecular encounter and the specific character of each species involved. The relaxation rate will be different if M is a ground-state hydrogen fluoride, or a helium, or a benzene, etc. Although the total energy in the collision is conserved, translational energy of the products is much greater than that of the reactants from a *V-T* transfer, and such collisions are inelastic. The vibrational quantum can also appear as rotation of M, and then one speaks of a *V-R* exchange. Of course, these reactions can go the other way, i.e., *T-V* or *R-V* transfer can occur. At equilibrium, concentrations are constant, and detailed balancing demands

$$-\frac{dn_0}{dt} = \frac{dn_1}{dt} = -k_{10}nn_1 + k_{01}nn_0 = 0$$

where n_0 and n_1 are the number densities of HF ($v = 0$) and HF ($v = 1$) respectively, and n is the number density of the buffer or background gas of M that we will, for simplicity, take to be present in excess. Most collisions then are of the nn_1 or nn_0 type, although both HF ($v = 0$) and HF ($v = 1$) of course collide with everything that is present at rates proportional to concentration. The rate coefficients, k_{10} or k_{01}, will be different for each collision partner.

At equilibrium, the net rate up and down is zero, and

$$\frac{k_{01}}{k_{10}} = \frac{n_1}{n_0} = e^{-(\varepsilon_1 - \varepsilon_0)/kT} = e^{-h\nu/kT} = e^{-h\omega c/kT} = K \qquad (10.69)$$

where ν is the frequency (in s^{-1}) or ω is the frequency (in wave numbers, cm^{-1}) of the vibrational quantum of HF. The ratio of k_{10}/k_{01} is a thermodynamic equilibrium constant for the ratio of the two reactants and it must be the same for any collision partner at fixed temperature. This stretches somewhat the usual definition of an equilibrium constant, for we here are concerned with the concentration of the two "species" of HF, one in the $v = 0$ state and the other in the $v = 1$ state.

The average number of collisions that a molecule must suffer for it to lose one quantum of vibrational energy from the $v = 1$ to $v = 0$ state is Z_{10}. The number of collisions to activate the molecule from $v = 0$ into $v = 1$ is Z_{01}. Clearly, transition between any two rotational, vibrational, or electronic states will likewise have its associated value of Z. In our example of HF in a vast background of some buffer M, each molecule of HF suffers collisions with M at a rate [see preceding Chap. 9, Eq. (9.37)] of

$$z = n\sigma^2\left(\frac{8\pi kT}{\mu}\right)^{1/2} \tag{9.37}$$

This is called the gas-kinetic collision rate. The ratio of the collision number to this collision rate, Z_{10}/Z, with units of s, is called the relaxation time,

$$\tau_{10} = Z_{10}/z = Z_{10}/n\sigma^2\left(\frac{8\pi kT}{\mu}\right)^{1/2} \tag{10.70}$$

Since the buffer or relaxing gas has been taken to be present in great excess, its number density n is essentially p/kT, and τ_{10} will vary inversely with pressure. Instead of Z_{10}, some writers use this $p\tau_{10}$ product as the basic measure of relaxation rate.

The reciprocal of the number of collisions to remove the vibrational quantum is the average probability per collision that a transition will occur, for example, $P_{10} = 1/Z_{10}$.

A specific rate coefficient defining relaxation can be developed by considering an equilibrium distribution of, say, HF that has been instantaneously perturbed by moving Δn_1 molecules into the $v = 1$ level from the $v = 0$ level. This nonequilibrium circumstance immediately begins to relax at a rate

$$\frac{dn_1}{dt} = \frac{d(n_1(\text{eq}) + \Delta n_1)}{dt} = \frac{d\,\Delta n_1}{dt}$$

$$= -\,k_{10}(n_1(\text{eq}) + \Delta n_1)(n) + k_{01}(n_0(\text{eq}) - \Delta n_1)(n)$$

where the rate constants have the usual units of m^3/molecule · s. Relaxation back to equilibrium implies $k_{10} > k_{01}$, that is, K from Eq. (10.69) is small, and recalling that at equilibrium, $k_{10}n_1(\text{eq}) = k_{01}n_0(\text{eq})$, we write

$$\frac{d\Delta n_1}{dt} = -\,(k_{10} + k_{01})n\,\Delta n_1 = -\,(k_{10} + Kk_{10})\,n\,\Delta n_1$$

$$\cong -\,k_{10}n\,\Delta n_1 \equiv -\,\Delta n_1/\tau \tag{10.71}$$

which defines τ, the relaxation time, as $(k_{10}n)^{-1}$. The return to equilibrium is clearly exponential in time, for on integrating Eq. (10.71), we get

$$\Delta n_1(t) = \Delta n_1\,(t = 0)\,e^{-t/\tau} \tag{10.72}$$

With $n = p/kT$, Eq. (10.71) becomes

$$k_{10} = \frac{kT}{p\tau} \tag{10.73}$$

Equation (10.73) is the same relationship between the relaxation time τ and the specific rate coefficient k_{10} as defined by Eq. (10.71). In practical units, k_{10} must be in $m^3/mol \cdot s$ rather than in $m^3/molecule \cdot s$. With Avogadro's number and p in kPa, Eq. (10.73) becomes

$$k_{10}(m^3/mol \cdot s) = \frac{8.315 \times 10^{-3}\,T}{p\tau} \tag{10.74}$$

Some writers also use the cross section defined as earlier [see Eq. (10.34)],

$$k_{10}(m^3/mol \cdot s) \equiv v_{10}(m/s)n_A(mol^{-1})q_{10}(m^2) \tag{10.75}$$

The average relative speed in a bimolecular collision is $v_{10} = (8kT/\pi\mu_{10})^{1/2}$ [see Eq. (9.34)], where μ_{10} is the reduced mass of the two colliding species, here HF and the buffer M. In practical units, with μ in g/mol,

$$k_{10}(m^3/mol \cdot s) = 8.703 \times 10^7\,(T/\mu)^{1/2}\,q_{10}(nm^2) \tag{10.76}$$

So the relaxation rate can be expressed as Z_{10}, P_{10}, τ_{10}, $p\tau_{10}$, k_{10}, or q_{10}, and we see how they all are interrelated. Various books and articles use each of these designations.

A summary of energy-transfer processes and the order of magnitude of the associated $p\tau$ product appears in Fig. 10.22. We see enormous differences in rates. R-T transfer is fast, but V-T transfer is usually about four orders of magnitude slower even though V-V transfer is similarly fast. This means that it is perfectly proper to talk about a vibrational temperature different from that of the rotation-translation (R-T) heat bath. This is because $\tau_{R\text{-}R} < \tau_{V\text{-}V}$ or $\tau_{R\text{-}T} < \tau_{V\text{-}T}$ as shown in Fig. 10.22. Typically for a gas at 1 atm, we expect the removal of one quantum of vibrational energy to require about 100 μs. At 0.01 atm, this same relaxation will require 10 ms. In general, small transfers of energy are much more likely than large transfers, and the transfer of vibrational energy to or from a molecule is more likely if the molecule has a low vibration frequency (more later).

A simple physical and qualitative interpretation of these phenom-

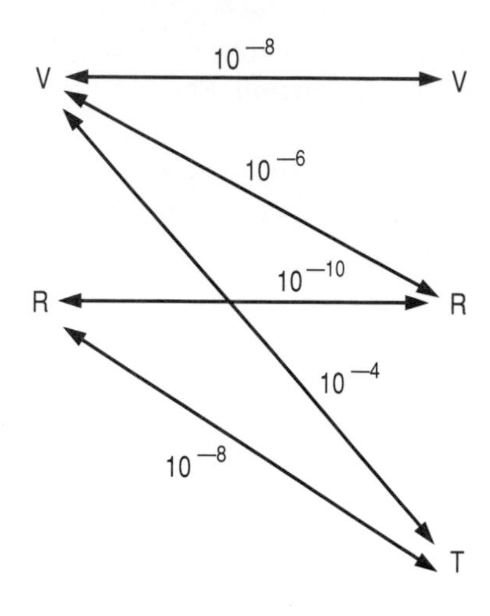

Figure 10.22 Schematic of energy transfer processes and their typical relaxation times with $p\tau$ in atm · s. (*Adapted from Levine and Bernstein.*[3] *Used by permission.*)

ena is given by the so-called Ehrenfest adiabatic principle, which states that when a changing force acts on a quantized periodic motion, no energy transfer will occur when the change in force is small during a period of the motion. Such events are said to be adiabatic. Conversely, if the change in force is large during the period, energy transfer will occur and the event is said to be nonadiabatic. The duration of a collision is some range of the intermolecular force a divided by the average speed $\bar{c}$. The period of a vibration is ν^{-1}, and vibrational energy is likely to be transferred in a collision when $a/\bar{c} < \nu^{-1}$ or when $(8k\mathrm{T}/\pi m)^{1/2} > a\nu$. Energy transfer is favored by collision partners of low mass, low vibration frequencies, highly repulsive intermolecular potentials (i.e., small a), and high temperatures. Consider Cl_2 at 300 K with an average speed of 300 m · s^{-1}, a collision diameter of 0.2 nm, and a frequency of 556 cm^{-1}. Thus $\bar{c}$ = 300 is less than $a\nu$ or $0.2 \times 10^{-9} (556 \times 3 \times 10^{10})$ by an order of magnitude, and we expect a small V-T rate. Experiment with Cl_2 shows a probability of transfer of 2×10^{-5} per collision.

The V-T relaxation rate is observed to be strongly temperature-dependent. To a good approximation, at higher temperatures log Z_{10} varies as $T^{-1/3}$, as is evident from Figs. 10.23, 10.24, and 10.25. For O_2 the $T^{-1/3}$ behavior is well-obeyed over a change in relaxation rate of at least four orders of magnitude. Some polyatomic molecules behave similarly, while some show a maximum in Z, i.e., a minimum in the relaxation rate. Diatomic and polyatomic molecules that are polar and exhibit strong intermolecular attraction seem to be more likely to deviate from the $T^{-1/3}$ behavior (see later).

The Landau-Teller analysis predicts just such a $T^{-1/3}$ dependence even though absolute values are poorly predicted. In this analysis, vi-

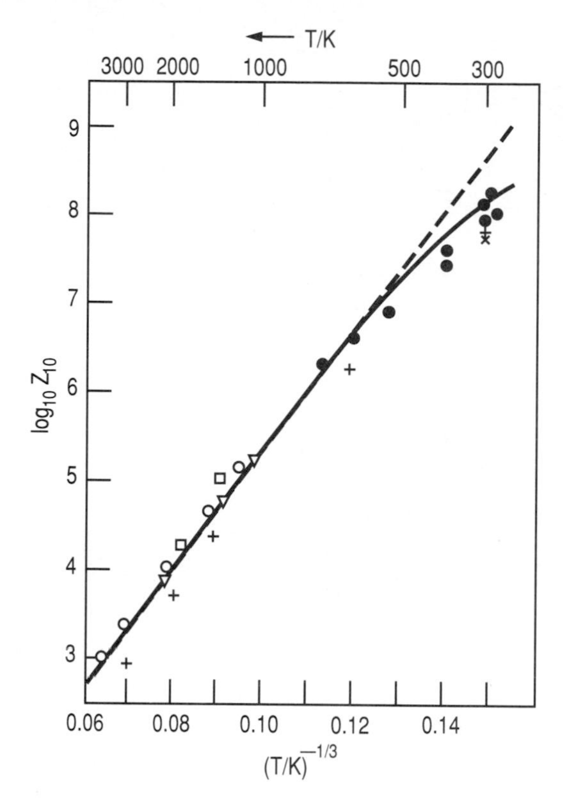

Figure 10.23 Vibrational relaxation rate of O_2 as a function of temperature. *(Adapted from J. D. Lambert, Vibrational and Rotational Relaxation in Gases, Oxford University Press, 1977. Used by permission.)*

sualize a head-on collision of atom A with a diatomic harmonic oscillator B-C as shown here,

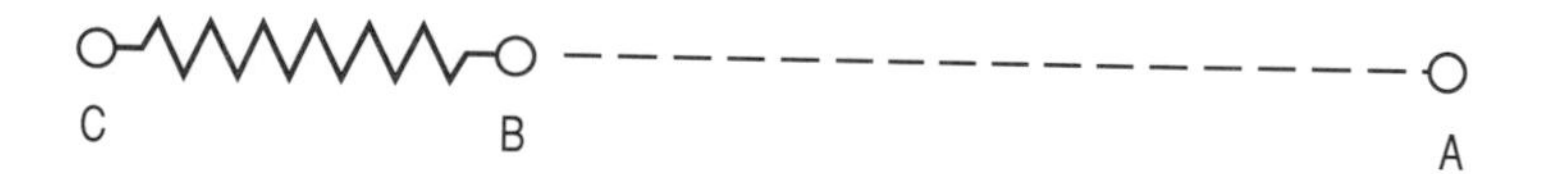

Consider two extreme cases. The interaction between A and B is as between rigid spheres. B receives a sharp head-on blow, it is set into motion toward C, compressing the Hooke's law spring. A recoils straight backward. During the short time of the collision, B behaves as a free particle. The compression of the spring finally stops, and B executes harmonic motion with all of the energy it received from the impact with A. This is called a *nonadiabatic collision*.

Alternatively, consider a slow collision with long-range interaction

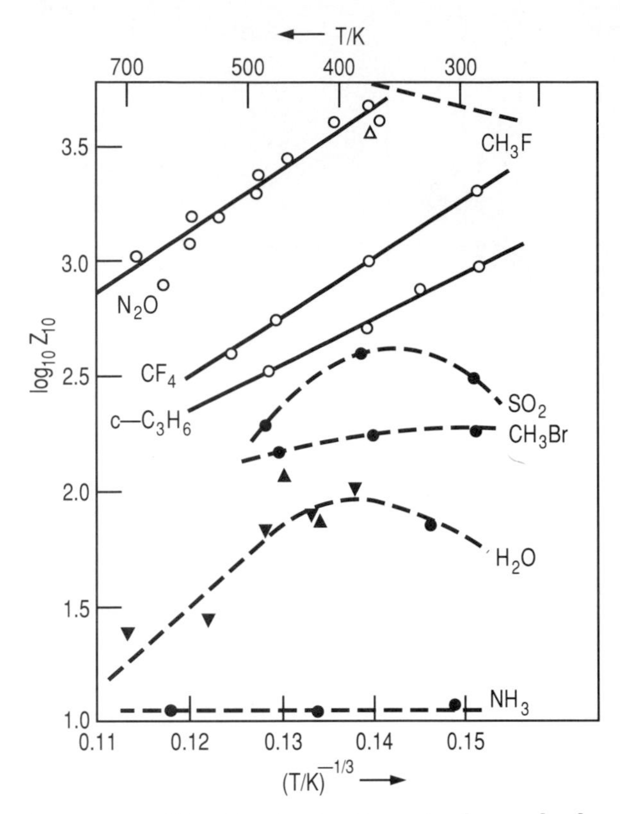

Figure 10.24 Vibrational relaxation rate of several polyatomic molecules. *(Adapted from J. D. Lambert, Vibrational and Rotational Relaxation in Gases, Oxford University Press, 1977. Used by permission.)*

between A and B so that B moves in response to A as A approaches. Finally, this long-range force stops A, which slowly recoils and recovers from B all of the energy B had received as A approached. The kinetic energy of A has been transferred completely to potential energy between BC and AB at the turning point. No energy has been permanently transferred, and this is called an *adiabatic collision*.

The Landau-Teller theory is based on a simple exponential repulsive potential, $\phi(r) = \phi_0 \exp(-\alpha r)$, and it assumes that the probability of energy transfer on collision will depend on the negative exponential of the ratio of the duration of collision to the period of vibration. Energy transfer occurs when this ratio is small, that is, when the character of the collision in the above pictorial is nonadiabatic. Duration of collision is set equal to a length of trajectory, l, over which interaction between A and B occurs, divided by the speed of relative motion of A toward B, labeled w. On the other hand, the

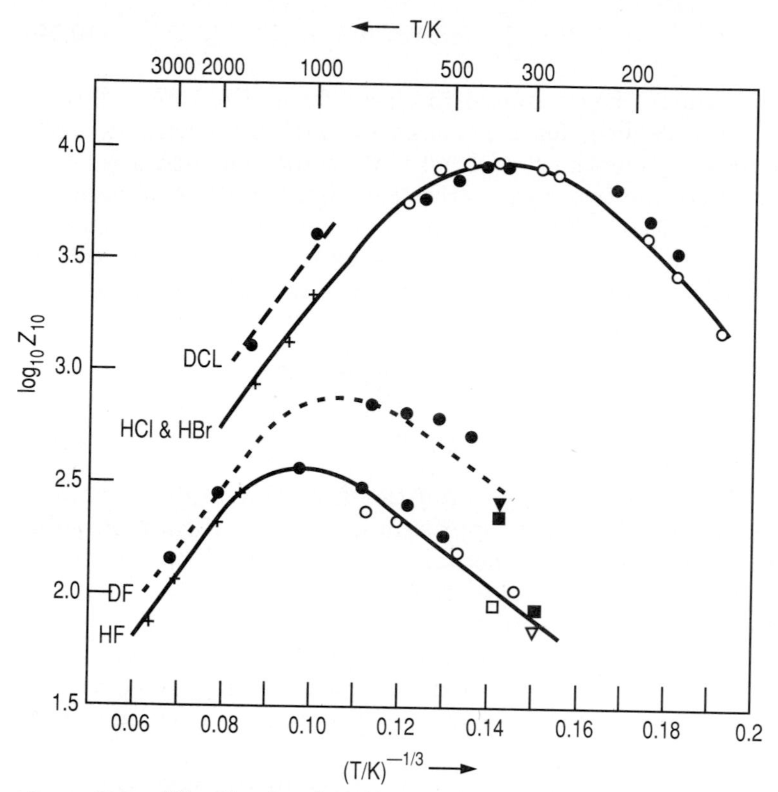

Figure 10.25 Vibrational relaxation rates of several polar diatomic molecules. (*Adapted from J. D. Lambert, Vibrational and Rotational Relaxation in Gases, Oxford University Press, 1977. Used by permission.*)

required period of vibration is inversely proportional to its frequency. Landau and Teller then assert that the probability of exchange of internal and relative translational energy is given by

$$P(w) = P_0 \exp\left(-\frac{2\pi\nu l}{w}\right) \tag{10.77}$$

where P_0 is a proportionality constant.

The relative velocity w has a Maxwell-Boltzmann distribution,

$$\frac{dn(w)}{n} = \left(\frac{\mu}{\pi kT}\right)^{1/2} \exp\left(-\frac{\mu w^2}{2kT}\right) dw \tag{10.78}$$

Note that Eq. (10.78) is the distribution law in one dimension, that is, along the line of centers between atoms A and B [Eq. (9.9), Chap. 9].

The total probability of energy exchange, of course, results from an integration of the product of Eq. (10.77) and Eq. (10.78) over all w,

$$P = \int_0^\infty P(w)\, dn(w) \qquad (10.79)$$

This integral cannot be evaluated in closed form. The integral, however, is sharply peaked, for $P(w)$ rises exponentially with w while $dn(w)$ declines exponentially with w. Only collisions with a relative speed near where the two curves cross can significantly contribute to the value of P in Eq. (10.79). The mathematical detail is unnecessary. Suffice it to realize that, expanding the exponentials about this w where the two curves cross and integrating, one finally obtains,

$$P = a \exp\left\{ -\frac{3}{2}\left[\frac{\mu(2\pi\nu l)^2}{kT}\right]^{1/3}\right\}$$

or

$$P = a \exp(-bT^{-1/3}) \qquad (10.80)$$

Here a and b are collections of quantities from the original approximate model as well as from its approximate mathematical manipulation to obtain Eq. (10.80). The actual values of a and b bear little resemblance to the numbers needed to fit real relaxation data. But the form of Eq. (10.80), and particularly its dependence on $T^{-1/3}$, does well-describe real behavior. It is gratifying that an approximate mathematical manipulation of an approximate model should agree so well with experiment, as is revealed by the typical data that appear in Figs. 10.23, 10.24, and 10.25.

The relaxation of simple molecules is typified by the data on oxygen that appear in Fig. 10.23. The Landau-Teller behavior is evident down to low temperatures where attractive forces that the Landau-Teller theory ignores increasingly come into play and destroy the $T^{-1/3}$ dependence. Not surprisingly, this is even more evident in data on strongly polar molecules like HCl and HF where the strong dipole-dipole attraction actually causes the relaxation rate to fall (see Fig. 10.25). The direction of the change is evident from our impulse model where long-range forces will make for a more adiabatic collision at the low relative velocities at low temperatures. Polar molecules are commonly described using the Stockmayer potential, which shows both a deeper attraction and a steeper repulsion than does the Lennard-Jones potential (see Chap. 5). The maximum attraction between dipoles occurs for head-on collisions where $V(\text{max}) = -2\mu^2/4\pi\varepsilon_0 r_0^3$, and this favorable orientation will be preferred at lower temperatures where $V(\text{max}) > RT$. At these lower temperatures, vibrational relaxation rates will decrease with increasing temperature. At higher temperatures, $V(\text{max}) < RT$, and Landau-Teller behavior is evident. All of this behavior is evident in the experimental data reported in Figs. 10.24 and 10.25. This same behavior is seen for polyatomic molecules, and it is enhanced when intermolecular forces are so strong as to produce dimers such as occur with H_2O or SO_2.

A useful empirical correlation that fits data on *V-T* relaxation of most pure polyatomic gases at 300 K is the so-called Lambert-Salter plot.[15] This is simply a plot of log Z_{10} against the lowest vibrational frequency of the molecule. The gas is pure, so the buffer M is the species itself. Data fall into two classes as is evident in Figs. 10.26 and 10.27 and Tables 10.7 and 10.8, depending on whether the molecule does or does not have constituent hydrogen atoms. Deuterated species invariably have a larger Z_{10} than does the corresponding molecule with ordinary hydrogen. The data of both figures fit

$$P_{10} \equiv \frac{1}{Z_{10}} = \exp(-CE) \tag{10.81}$$

where E is the energy of the lowest of the $3n - 5$ (or 6) total vibrators in cm^{-1}, and then

$$C_{\text{I}} = 8.3 \times 10^{-3}\ \text{cm} \qquad \text{(for hydrogen-containing molecules)}$$

and $C_{\text{II}} = 17 \times 10^{-3}$ cm (for non-hydrogen-containing molecules)

This fits the data to within a maximum deviation of roughly a factor

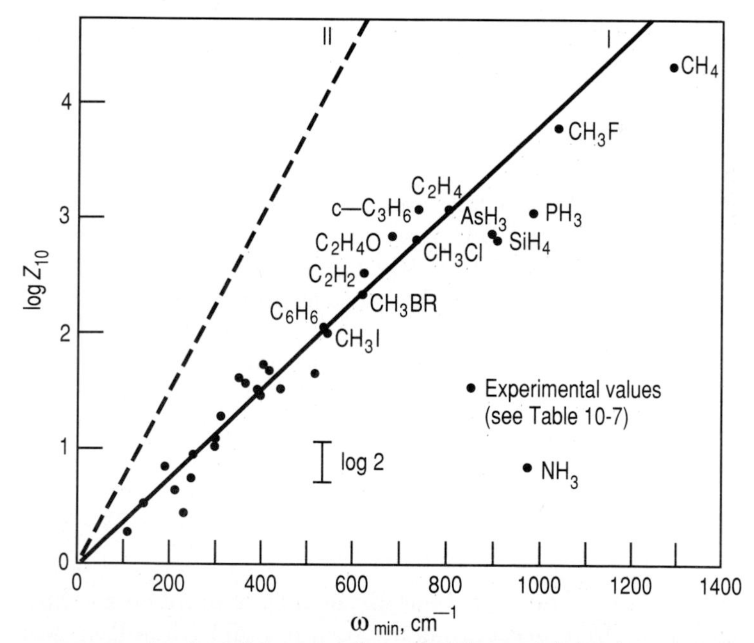

Figure 10.26 Lambert-Salter plot for molecules containing two or more hydrogen atoms. (*Adapted from Lambert.*[15] *Used by permission.*)

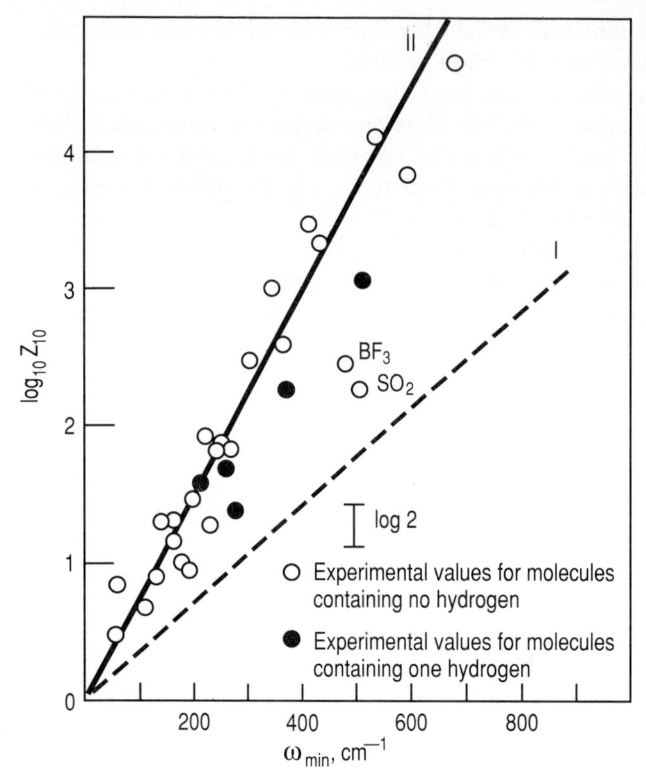

Figure 10.27 Lambert-Salter plot for molecules containing one or no hydrogen atoms. (*Adapted from Lambert.*[15] *Used by permission.*)

of 2, as is evident from Figs. 10.26 and 10.27. In molecules with low frequencies ($\omega \leq 100$ cm^{-1}), *V-T* transfer occurs essentially on every collision. Predominantly repulsive intermolecular interactions conform well, of course, to the normal Landau-Teller plot, and to the Lambert-Salter plot as well.

The behavior of mixtures is also readily correlated. The relaxation time τ was defined as the number of collisions required to produce a given relaxation divided by the gas-kinetic collision rate. For example, the vibrational self-relaxation time for ν (1 → 0) is

$$\tau_{10} = \frac{Z_{10}}{z} = \frac{Z_{10}}{2n\sigma^2(\pi kT/m)^{1/2}} \tag{10.82}$$

where the collision frequency has been divided by 2 to avoid the double count [see Eq. (9.36) of preceding chapter]. If a vibrationally relaxing gas is mixed with another gas, two processes can occur:

TABLE 10.7 Vibrational Relaxation Data at 300 K for Polyatomic Molecules Containing Two or More Hydrogen Atoms

Molecule	ω_{min}, cm^{-1}	Z_{10}
CH_4	1306	15160
CH_3F	1048	4800
c-C_3H_6	740	990
C_2H_4	810	970
PH_3	991	890
C_2H_4O	685	603
AsH_3	905	605
CH_3Cl	732	575
SiH_4	914	570
C_2H_2	612	280
CH_3Br	611	190
CH_2F_2	529	83
CH_3I	533	78
CH_2CHCl	395	48
CH_2CHF	500	41
CH_2CHBr	345	36
CH_2:C:CH_2	354	33
C_2H_5OH	431	29
$CH_3COOC_2H_5$	376	29
CH_2ClF	385	26
CH_3COOCH_3	303	15
C_3H_7OH	243	8
CH_3CH:CH_2	177	6
CH_3CHCl_2	240	5
NH_3	950	5
$C(CH_3)_4$	200	4.2
CH_2ClCH_2Cl	125	3
c-C_5H_{10}	207	2.6
n-C_4H_{10}	102	1.6
n-C_5H_{12}	88	0.88
n-C_6H_{14}	61	0.66

SOURCE: Adapted from Lambert.[15]

$$A(v = 1) + A \rightarrow A + A \qquad \text{with } \tau_{AA}$$

$$A(v = 1) + B \rightarrow A + B \qquad \text{with } \tau_{AB}$$

That is, the vibrationally hot species can collide with a ground-state species of itself, A, or with the added second species, B. Since the rate constants for the two processes add, Eq. (10.73) suggests that the total relaxation time for A is

$$\frac{1}{\tau_A} = \frac{1 - x_B}{\tau_{AA}} + \frac{x_B}{\tau_{AB}}$$

A plot of the reciprocal of the observed total relaxation time τ_A vs.

TABLE 10.8 Vibrational Relaxation Data at 300 K for Polyatomic Molecules Containing One or No Hydrogen Atoms

Molecule	ω_{min}, cm^{-1}	Z_{10}
CO_2	673	53000
COS	527	13000
N_2O	589	7320
CS_2	397	2950
CF_4	437	2070
SF_6	344	1005
CF_3Cl	356	427
BF_3	480	380
CF_3Br	297	316
CCl_4	218	86
$CFCl_3$	248	72
CF_2Cl_2	261	68
C_2Cl_4	237	68
CF_2BrCl	200	30
$SiCl_4$	150	20
CF_2Br_2	165	14
C_2F_4	190	9.6
$GeCl_4$	134	8
C_2F_6	68	7
$SnCl_4$	104	5
CHF_3	507	1230
$CHClF_2$	369	191
$CHCl_3$	260	52
$CHBrCl_2$	215	37
$CHCl_2F$	276	26
$CHBr_2Cl$	168	20

SOURCE: Adapted from Lambert.[15]

mole fraction of B in the A,B mixture will be linear, allowing the convenient determination of τ_{AB}. The rate coefficient or cross section or collision number that one would calculate from this experimental relaxation time τ_{AB} must now, of course, be calculated using the dissimilar collision frequency of molecules of B on a single molecule of A that was derived earlier [see Eq. (9.37)],

$$z_{AB} = 2n_B\left(\frac{\sigma_A + \sigma_B}{2}\right)^2\left(\frac{2\pi kT}{\mu_{AB}}\right)^{1/2}$$

Helium is almost always more effective in relaxing a gas than are self-collisions. However, very polar molecules such as HCl, H_2O, and SO_2 relax faster when pure than when mixed with even He. This shows the powerful effect of the intermolecular attractive forces. Remember that the linear Landau-Teller plot is based on a sharp repulsive interaction only. Polar molecules do not fit this model. Polar molecules can also be extraordinarily rapidly relaxed by the addition of molecules which allow an incipient sort of chemical reaction. For example, pure CO_2, with a minimum vibrational frequency of 673 cm^{-1}, behaves normally and relaxes slowly with $Z_{10} \cong 50{,}000$ at 300 K. On adding water, Z_{10} is reduced to about 25, and on adding HF, $Z_{10} \cong 160$. This striking catalytic effect is evidently due to the formation of

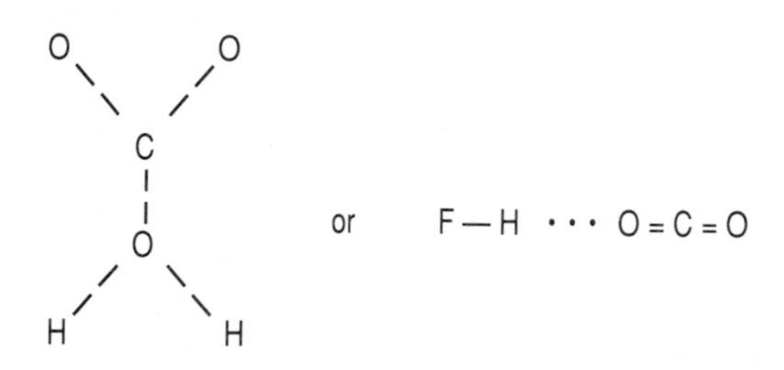

Remembering the rule that energy transfer is faster when the energy that is moving is smaller, we may surmise relative *V-V* relaxation times from the size of the energy imbalance in the collision event. In polyatomic molecules, it is usual that the gaps in energy levels between frequencies are less than the energy level of the lowest mode. The picture that emerges imagines rapid *V-V* transfer maintaining thermodynamic equilibrium among the several vibrators of a molecule as the whole vibrational energy relaxes from the lowest vibrational level of the molecule into the *R-T* heat bath at a much slower rate. For example, the collision

$$\mathrm{HF}(v = 2) + \mathrm{HF}(v = 0) \rightarrow \mathrm{HF}(v = 1) + \mathrm{HF}(v = 1) + Q$$

is very efficient; that is, Z_{21} is small, $p\tau_{21}$ is small, k_{21} is large, etc. Here there is only the small energy due to the anharmonicity of the HF vibrator that appears as translation of the product species. A positive Q represents internal energy transferred to translation and is an exothermic process, while a negative Q represents an endothermic process where internal energy is increased at the expense of translational energy. The rule relating efficiency to translational energy defect is also evident for *R-V* transitions as well, where, for example,

$$I_2(v = 0) + \mathrm{H}_2(J = 2) \rightarrow \mathrm{I}_2(v = 1) + \mathrm{H}_2(J = 0) + Q$$

is reasonably fast, for the energy deficit Q is -139 cm^{-1}. The rule is also evident for electronic transitions where

$$\mathrm{HF}(v = 1) + \mathrm{Br}(^2\mathrm{P}_{3/2}) \rightarrow \mathrm{HF}(v = 0) + \mathrm{Br}(^2\mathrm{P}_{1/2}) + Q$$

is efficient even though the energy deficit is 453 cm^{-1}. CO relaxation is 2 times faster against p-H_2 as against o-H_2,

$$CO(v = 1) + p\text{-}H_2(J = 2) \rightarrow CO(v = 0) + p\text{-}H_2(J = 6) + 88\text{cm}^{-1}$$

and $CO(v = 1) + o\text{-}H_2(J = 3) \rightarrow CO(v = 0) + o\text{-}H_2(J = 7) - 327\text{cm}^{-1}$

It is clear that fast *V-V* relaxation will produce a boltzmannian system among the vibrators themselves long before the energy will leak out of the lowest vibrator into the *R-T* heat bath. A vibrational temperature higher than that of the surrounding *R-T* heat bath can readily occur.

It is interesting that rotational relaxation into translation evidently varies with energy defect in the same way, but at a much faster rate. The rate is so high that rotational populations reach thermodynamic equilibrium almost as fast as does the translational distribution, and consequently, the role of rotation in heating and cooling processes is usually uninteresting. This is, however, not the case for H_2 and the hydrides where the moments of inertia are small and the energy levels large. Also the gaps between successive energy levels become rapidly larger as J increases. Imagine a hot gas with its rotational population peaked at some high J. Initial deexcitation will be slow because of the large energy gaps, but the gas will relax progressively faster as the rotational temperature drops. No selection rule operates for J. Hydrides in high J levels may appear highly metastable. In a jet engine or rocket engine expansion, the rotational temperature may lag behind the translational temperature. Rotational relaxation data on, for example HCl, are well-described by an empirical relationship very reminiscent of Eq. (10.81),

$$P_{J \rightarrow J'} = N \exp(-C\,\Delta E)$$

where ΔE is the energy difference between states J and J' and where N and C are constants for a particular temperature.[15] Recall, however, that the rate of rotational relaxation decreases with increasing temperature while that for vibration increases with temperature.

The rotational population changes can be followed spectroscopically from HCl produced by

$$H + Cl_2 \rightarrow HCl^* + Cl$$

This reaction produces rotationally excited HCl^* peaking at $J = 13$ while J is peaked at 2 to 3 for a thermodynamic distribution at room temperature. One then merely observes the chemiluminescent decay of HCl^* against a variety of target molecules which are present in large excess. There is ample experimental evidence that the lower J levels relax much faster than do the upper J levels.

Monte Carlo trajectory calculations have been used to estimate the rate of vibrational relaxation both in situations where chemical reaction is and is not possible. H atom exchange is a very efficient means of deactivation in

$$HCl(v = 1) + H \rightarrow HCl(v = 0) + H$$

Calculations on an LEPS potential energy surface for this reaction yielded the results shown in Table 10.9, which shows specific rates between two selected levels, $v \rightarrow v'(k_{vv'})$, as well as the overall rate of deactivation of the HCl(k_v). These data, and that at other temperatures, well-fit an expression of the form

$$k = AT^{-n} \exp\left(-\frac{E}{RT}\right) \tag{10.83}$$

and the values of the parameters in this expression also appear in Table 10.9. There are two channels for reaction, to form either HCl or H_2, and the rather different results for both channels also appear in Table

TABLE 10.9 Rate-Coefficient Parameters and Rate Coefficients for the Reactions of H + HCl (v) from Monte Carlo Trajectory Calculations

v	v'	log A	n	E, cal/mole	$k_{vv'}$ (T = 300 K), cm³/mol · s	k_v (T = 300° K), cm³/mole · s
				H + HCl(v) → HCl (v') + H		
1	0	14.79	0.26	901	3.0×10^{13}	3.0×10^{13}
2	1	14.66	0.31	824	1.9×10^{13}	
2	0	14.97	0.36	874	2.7×10^{13}	4.6×10^{13}
3	2	15.22	0.49	953	2.0×10^{13}	
3	1	13.91	0.14	638	1.2×10^{13}	
3	0	14.82	0.32	765	2.8×10^{13}	6.0×10^{13}
6	5	14.88	0.42	822	1.7×10^{13}	
6	4	14.65	0.35	905	1.3×10^{13}	
6	3	13.92	0.16	692	1.0×10^{13}	
6	2	13.68	0.39	839	1.3×10^{12}	
6	1	13.94	0.17	716	1.0×10^{13}	
6	0	14.79	0.36	811	2.0×10^{13}	7.1×10^{13}
				H + HCl(v) → H_2 (v') + Cl		
1	0	12.80	0.05	976	9.0×10^{11}	9.0×10^{11}
2	1	14.04	0.30	1359	1.9×10^{12}	
2	0	12.99	0.28	739	2.5×10^{12}	4.4×10^{12}
3	2	13.97	0.32	881	3.3×10^{12}	
3	1	13.85	0.29	927	2.8×10^{12}	
3	0	13.28	0.26	876	1.0×10^{12}	7.1×10^{12}
6	4	13.16	0.15	552	2.4×10^{12}	
6	3	14.16	0.46	773	2.8×10^{12}	
6	2	13.40	0.20	938	1.6×10^{12}	6.8×10^{12}

SOURCE: Adapted from R. L. Wilkins, *J. Chem. Phys.*, *63*, 534 (1975).

10.9. An experimental value for v (1 → 0) in the channel to form H_2 at 298 K of 4×10^{11} $cm^3/mol \cdot s$ is in reasonable agreement with the trajectory results of 9×10^{11} $cm^3/mol \cdot s$.

Similar data for the reaction

$$Cl + HCl(v) \rightarrow Cl + HCl(v')$$

appear in Table 10.10, and the trajectory results again are well-fitted by an equation of the form of Eq. (10.83). Appropriate parameters for this reaction, again both for $v \rightarrow v'$ processes and for the overall deactivation, appear in Table 10.10. The high value of the endothermicity (effectively an activation energy) of the reaction

$$Cl + HCl \rightarrow Cl_2 + H$$

of 45 kcal/mol closes this as a possible reaction channel for the vibrational energy levels of HCl that were considered in this trajectory study. The data of Table 10.10 are the result of about 20,000 Monte Carlo trajectory calculations. Although they are not separated in Table 10.10, these trajectory calculations indicate that reactive chlorine atom exchange and inelastic scattering occur about equally in the relaxation of vibrationally excited HCl to produce the tabulated values of $k_{vv'}$. The relaxation rate is revealed to be somewhat faster the greater is the vibrational energy in the reactant HCl. This is as expected, since anharmonicity ensures that, for example, $\nu_6 \rightarrow \nu_5$ is lower in energy than is $\nu_1 \rightarrow \nu_0$. The calculations also reveal that the

TABLE 10.10 Rate-Coefficient Parameters and Rate Coefficients for the Reactions of Cl and HCl (v) from Monte Carlo Trajectory Calculations

v	v'	log A	n	E, cal/mol	$k_{vv'}$ (T = 300 K), $cm^3/mol \cdot s$	k_v (T = 300 K), $cm^3/mol \cdot s$
				$Cl + HCl(v) \rightarrow HCl(v') + Cl$		
1	0	13.75	0.20	830	4.5×10^{12}	4.5×10^{12}
2	1	14.36	0.37	976	5.4×10^{12}	
2	0	13.75	0.28	1009	2.1×10^{12}	7.5×10^{12}
3	2	14.74	0.50	1016	5.8×10^{12}	
3	1	14.36	0.47	984	3.0×10^{12}	
3	0	13.75	0.41	852	1.3×10^{12}	10.1×10^{12}
6	5	14.85	0.33	1172	1.5×10^{13}	
6	4	14.25	0.24	1029	8.0×10^{12}	
6	3	14.15	0.26	1050	5.5×10^{12}	
6	2	14.74	0.47	1270	4.5×10^{12}	
6	1	14.52	0.43	1540	2.2×10^{12}	
6	0	15.02	0.61	1478	2.7×10^{12}	37.7×10^{12}

SOURCE: Adapted from R. L. Wilkins, *J. Chem. Phys.*, *63*, 534 (1975).

relaxed vibrational energy mainly appears as rotation of the product HCl. Finally, agreement of the trajectory calculations with experiment is good where trajectory calculations of k_{10}, k_{21}, k_{32} are 4.5×10^{12}, 5.4×10^{12}, 5.8×10^{12}, which should be compared with experimental values of 6.5×10^{12}, 1.9×10^{12}, 4.8×10^{12} cm^3/mol · s.

The rate of vibrational relaxation may be catalytically enhanced by an added collision partner that has a near coincidence of one of its rotational levels with the required vibrational transition. For example, BCl_3 with a minimum frequency of 243 cm^{-1} is more efficiently relaxed by HCl than by self-collision where Z_{10} is 91 and 145, respectively. Evidently the $J = 4 \rightarrow 6$ and $J = 3 \rightarrow 5$ rotational transitions in HCl and the ν $(1 \rightarrow 0)$ transition of BCl_3 are near-resonant. There are also plenty of $J = 3$ and $J = 4$ HCl's available, for at 300 K, these J's lie near the peak of the Boltzmann distribution of rotational levels of HCl. After the *V-R* exchange between BCl_3 and HCl, the excess rotational energy of HCl is rapidly lost to translation, and thermodynamic equilibrium of the BCl_3/HCl mixture ensues.

As in atoms, electronic energy is also more efficiently exchanged when the translational energy defect is small. For example, in the chemical oxygen iodine laser (COIL), electronically excited $O_2(^1\Delta)$ is produced by the reaction of Cl_2 with H_2O_2,

$$Cl_2(g) + H_2O_2(\text{soln}) \rightarrow O_2(^1\Delta)(g) + HCl(\text{soln})$$

$O_2(^1\Delta)$ is unusual in that it is long-lived. It has a collisionless radiative half-life of 45 min. It is 22.54 kcal/mol above the $^3\Sigma$ ground state of O_2. It can be removed from the chlorine/peroxide reactor and pumped into an optical cavity where it efficiently exchanges with iodine,

$$\underset{0}{I(^2P_{3/2})} + \underset{7877}{O_2(^1\Delta)} \rightarrow \underset{7603}{I(^2P_{1/2})} + \underset{0}{O_2(^3\Sigma)} \qquad (10.84)$$

The electronically excited iodine then lases in the infrared at 1.3 μm. The energy defect in Eq. (10.84) is only 274 cm^{-1}. The use of I in COIL, rather than some other species, is because of this near-resonant, and then efficient, energy exchange. This system is unusual, for it produces an infrared laser starting only from readily available compounds in a simple chemical reaction. There is no need for an external source of electric power. All chemical devices allow the production of high-power continuous-wave (cw) beams. A COIL system has already been operated on H_2O_2 and Cl_2 and then frequency-doubled to 657 nm to yield a 0.5-kW cw beam of red light.

Similarly, in the familiar Ne red-line laser, the gaseous medium is a mixture of typically 10 parts He and 1 part Ne at a total pressure of 1 mm Hg that is excited by an electric discharge. The He is excited to the long-lived 2^1S state at 166,272 cm^{-1} by collisions with electrons in the plasma. On collision, He efficiently transfers to Ne,

$$\begin{array}{cccc} He[(1s)(2s)] + & Ne[(2p)^6] \rightarrow & He[(1s)^2] + & Ne[(2p)^5(5s)] \\ 166{,}272 & 0 & 0 & 165{,}915 \end{array}$$

where the energy defect is only 357 cm^{-1} which then lases to the $Ne[(2p)^5(3p)]$ level and emits a red line at 0.633 μm. The 2^3S He is also produced in the discharge; it has energy of 159,850 cm^{-1}, and it then exchanges with ground-state neon to produce the $(2p)^5(4s)$ state at 159,537 cm^{-1} with an energy deficit of only 313 cm^{-1}. This species lases down to the same $(2p)^5(3p)$ state of neon but with the production of infrared photons at 1.15 μm. Whether in inelastic or reactive collisions, electron spin is usually conserved, and in keeping with this expectation, the lifetime of the ^3S He at 10^{-4} s is much longer than that of the ^1S species at 5×10^{-6} s as they both decay to the ^1S ground state.

The CO_2 Laser

Of the hundreds of lasing systems that have been described, the CO_2 laser has seen, by far, the greatest practical application. The device is widely used for welding, drilling, marking, and cutting procedures. The CO_2 laser is a common device on the factory floor. The CO_2 laser emits several dozen lines near 10 μm. Understanding its operation is an exercise in molecular engineering.

As a linear triatomic species, CO_2 has $3n - 5$ or 4 normal vibrational modes, and these are shown schematically in Fig. 10.28. Because the molecule is linear, the bending motion, ν_2, is doubly degenerate. The frequencies are 1388 cm^{-1} (v_1), 2349 cm^{-1} (v_3), and two at 667 cm^{-1} (v_2). Let m, n, and p represent the number of quanta in the symmetric stretching (v_1), bending (v_2), and asymmetric stretching (v_3) modes, respectively. Intramolecular-collision-induced transfer between v_1 and v_2,

$$CO_2(020) + M \rightarrow CO_2(100) + M - 103\ cm^{-1}$$

occurs with an energy deficit of only 103 cm^{-1}, and it is experimentally observed to be fast. Intermolecular transfer is similarly fast,

$$CO_2(100) + CO_2(000) \rightarrow 2CO_2(010) + 54\ cm^{-1}$$

and $$CO_2(020) + CO_2(000) \rightarrow 2CO_2(010) - 49\ cm^{-1}$$

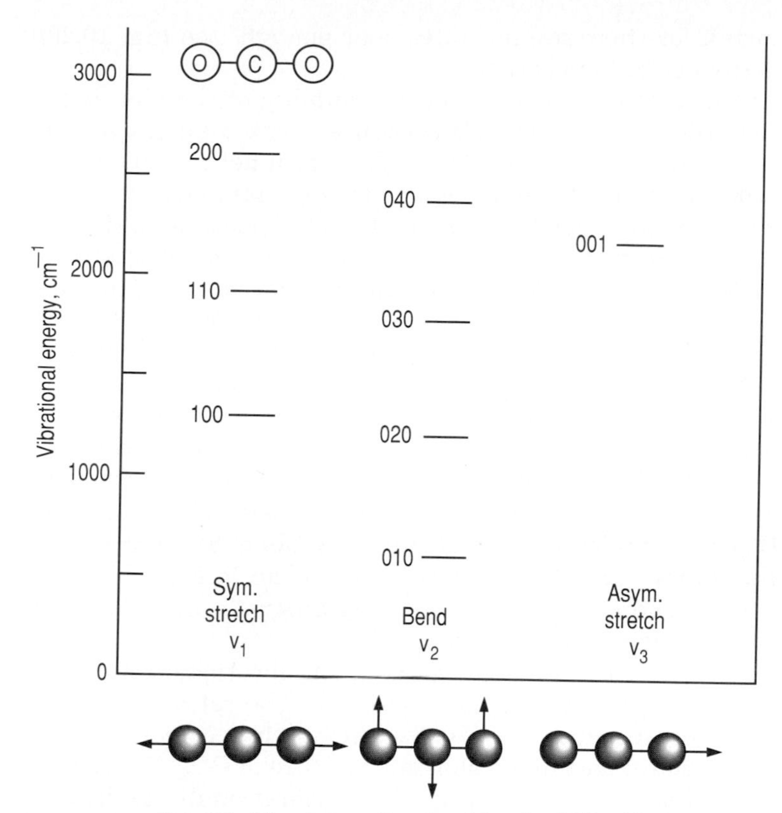

Figure 10.28 Simplified low-lying vibrational levels of CO_2. Their spectroscopic designations are also shown. *(Adapted from Levine and Bernstein.[3] Used by permission.)*

and each occurs with a small energy deficit. Thus the symmetric stretch and bending modes $(m, n, -)$ or $(v_1, v_2, -)$ rapidly reach equilibrium among themselves, or, in other words, these two modes become thermodynamic. Then the energy in these modes much more slowly, about a factor of 10^3 more slowly, leaks into the R-T heat bath from the lowest vibrational energy level of (010) by

$$CO_2(010) + M \rightarrow CO_2(000) + M + 667\ \text{cm}^{-1}$$

This last process is rate-determining in concerns for the rate at which CO_2 achieves overall thermodynamic equilibrium. The energy defect at 667 cm^{-1} is large.

Finally, excess energy placed in the asymmetric stretch mode v_3 is quickly shared among all the molecules of CO_2, so that the asymmetric stretch system comes to its own, but separate, equilibrium. Energy in this mode is not efficiently coupled to the symmetric stretch and bending modes, for there are no states near enough (see Fig. 10.28); i.e., the energy defect is too great.

It is interesting that the physics of the coupling of the vibrational modes of CO_2 allow the side-by-side existence of two different sorts of CO_2. Molecules with excess energy in their symmetric stretch and bending modes, v_1 and v_2, rapidly become maxwellian, i.e., thermodynamic, and this energy slowly leaks into the R-T heat bath. Side-by-side CO_2 molecules with excess energy in their asymmetric stretch mode, v_3, independently and rapidly also becomes maxwellian, and this energy also slowly leaks into the R-T heat bath. We see two different types of CO_2, each thermodynamic, in intimate contact with each other, but with little or no coupling between them.

This sort of separateness is not uncommon. As another example, consider helium that has been readily electronically excited by an electric discharge. The excited electron can be either spin-aligned or spin-opposed to the electron remaining in the 1s level so that the helium is either in a singlet or a triplet state (see Chap. 3). Each kind of helium exhibits its characteristic spectrum, but no intercombination lines are known. There may be then two very different sorts of helium existing side by side in the same "pot."

Returning to CO_2, note that the character of vibrational relaxation in CO_2 makes the familiar CO_2 laser possible. The relevant energy-transfer processes appear in Fig. 10.29, which is just a redrawn version of Fig. 10.28. The lasing medium is a mixture of CO_2, N_2, and He that is subjected to electric discharge. N_2 is vibrationally excited by collisions with electrons,

$$N_2(v = 0) + e \rightarrow N_2(v = 1) + e$$

which in turn excites CO_2 by collision,

$$N_2(v = 1) + CO_2(000) \rightarrow N_2(v = 0) + CO_2(001) - 18 \text{ cm}^{-1}$$

with low energy defect. Stimulated laser emission occurs as shown in Fig. 10.29,

$$CO_2(001) \rightarrow CO_2(100) + h\nu \; (10.6 \; \mu\text{m})$$

and

$$CO_2(001) \rightarrow CO_2(020) + h\nu \; (9.6 \; \mu\text{m})$$

For this scheme to work, however, the CO_2 (001) must not be deactivated by collision before the energy can be stimulated out in the optical cavity. We know that energy in v_3 migrates into v_1 and v_2 slowly because of the large energy defect. The second requirement is the rapid depletion of molecules from the end states at (100) and (020) by

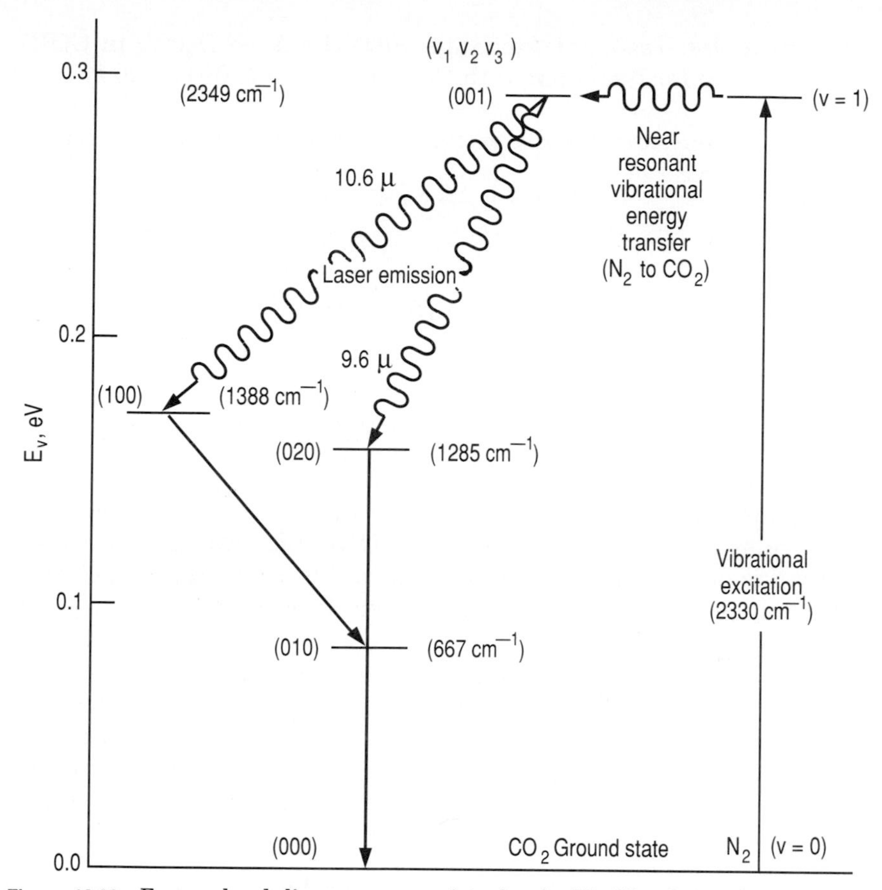

Figure 10.29 Energy level diagram appropriate for the N_2-CO_2 chemical laser (see Fig. 10.28). Vibrationally excited N_2 molecules are the source of the excitation of the 001 vibration of the CO_2. The final states CO_2 (100) and CO_2 (020) are depleted by *V-V* transfer. *(Adapted from Levine and Bernstein.[3] Used by permission.)*

$$CO_2(100) + CO_2(000) \rightarrow 2CO_2(010) + 54 \text{ cm}^{-1}$$

$$CO_2(020) + CO_2(000) \rightarrow 2CO_2(010) - 49 \text{ cm}^{-1}$$

so that the population inversion can be maintained. This collisional deexcitation is fast with energy deficits of only 54 and 49 cm^{-1}. The (010) intermediate level must also be rapidly emptied, and He is added as a particularly good collision partner to promote this 010 → 000 decay. There are many rates, each exerting its influence on the overall macroscopic behavior, and experimentation was necessary to arrive at the optimum proportions of the three gases that compose the lasing medium. Similarly, experimentation was necessary to discover the best additives to the CO_2. However, just as I was selected as a near-resonant exchange with $O_2(^1\Delta) \rightarrow O_2(^3\Sigma)$ in COIL, N_2 was chosen for its deficit with $CO_2(000) \rightarrow CO_2(001)$ of only -19 cm^{-1}.

Interestingly, vibrationally excited DF from the chemical reaction

$$NO + F_2 \rightarrow NOF + F$$

followed by

$$F + D_2 \rightarrow DF^* + D$$

will resonantly exchange with CO_2:

$$CO_2(000) + DF^* \rightarrow CO_2(001) + DF$$

and the subsequent events to produce CO_2 lasing are just as from the electric-discharge-powered device. Only here, CO_2 lasing has been produced by using bottled gases. Like COIL, this laser is particularly interesting because it is a purely chemical laser. No external electric power source is necessary and high-power cw output beams are possible.

The lines available from each vibrational transition of the CO_2 laser are collected in Table 10.11. Recall that the lines arise from the rotational structure of both vibrational levels as is shown schematically in Fig. 10.30. The P branch involves a $\Delta J = -1$ in absorption while the R branch arises from $\Delta J = +1$. The designation $P(4)$ in the upper half of Table 10.11 means $(v = 001, J = 4) \rightarrow (v = 100, J = 5)$.

Exercise Problem 10.2 Calculate the wave number of the $P(4)$ line of the $001 \rightarrow 100$ transition of CO_2 and compare with the value of 957.8 cm^{-1} from Table 10.11. The rotational constant for symmetric CO_2 is 0.39038 cm^{-1}.

A P line involves a ΔJ in absorption of -1, or the lasing $P(4)$ line involves an upper level of $J = 3$ and a vibrational energy of 2349 cm^{-1} at (001). The energy of the upper level is

$$\varepsilon = hc\omega + J(J + 1)Bh$$

$$= (6.626 \times 10^{-34})(3 \times 10^{10})(2349) + 3(3 + 1)(0.39038)(3 \times 10^{10})(6.626 \times 10^{34})$$

$$= 46{,}693 \times 10^{-24} + 93 \times 10^{-24} = 46{,}786 \times 10^{-24}\ \text{J}$$

At the lower level, $J = 4$ and the vibrational energy is 1388 cm^{-1} at (100). The energy of the lower level is

$$\varepsilon = (6.626 \times 10^{-34})(3 \times 10^{10})(1388)$$

$$+ 4(4 + 1)(0.39038)(3 \times 10^{10})(6.626 \times 10^{-34})$$

$$= 27{,}591 \times 10^{-24} + 155 \times 10^{-24} = 27{,}746 \times 10^{-24}\ \text{J}$$

TABLE 10.11 Lines Available from the CO_2 Laser

Branch (number)	Wave number, cm^{-1}	Approximate strength	Branch (number)	Wave number, cm^{-1}	Approximate strength
		001–100 transition			
$P(56)$	907.78	0.3	$R(4)$	964.77	0.3
$P(54)$	910.02	0.3	$R(6)$	966.25	0.65
$P(52)$	912.23	0.3	$R(8)$	967.71	0.75
$P(50)$	914.42	0.3	$R(10)$	969.14	0.8
$P(48)$	916.58	0.55	$R(12)$	970.55	0.85
$P(46)$	918.72	0.55	$R(14)$	971.93	0.95
$P(44)$	920.83	0.6	$R(16)$	973.29	0.95
$P(42)$	922.92	0.7	$R(18)$	974.62	1.0
$P(40)$	924.97	0.75	$R(20)$	975.93	1.0
$P(38)$	927.01	0.85	$R(22)$	977.21	1.0
$P(36)$	929.02	0.85	$R(24)$	978.47	1.0
$P(34)$	931.00	0.95	$R(26)$	979.71	0.95
$P(32)$	932.96	0.95	$R(28)$	980.91	0.9
$P(30)$	934.90	1.0	$R(30)$	982.10	0.9
$P(28)$	936.80	1.0	$R(32)$	983.25	0.85
$P(26)$	938.69	1.0	$R(34)$	984.38	0.85
$P(24)$	940.55	1.0	$R(36)$	985.49	0.8
$P(22)$	942.38	1.0	$R(38)$	986.57	0.75
$P(20)$	944.19	1.0	$R(40)$	987.62	0.75
$P(18)$	945.98	1.0	$R(42)$	988.65	0.6
$P(16)$	947.74	1.0	$R(44)$	989.65	0.45
$P(14)$	949.48	1.0	$R(46)$	990.62	0.35
$P(12)$	951.19	1.0	$R(48)$	991.57	0.25
$P(10)$	952.88	0.95			
$P(8)$	954.55	0.85	$R(50)$	992.49	—
$P(6)$	956.19	0.75			
$P(4)$	957.80	0.55			
		001–020 transition			
$P(50)$	1016.72	0.15	$R(4)$	1067.54	0.25
$P(48)$	1018.90	0.25	$R(6)$	1069.01	0.4
$P(46)$	1021.06	0.4	$R(8)$	1070.46	0.5
$P(44)$	1023.19	0.55	$R(10)$	1071.88	0.6
$P(42)$	1025.30	0.55	$R(12)$	1073.28	0.6
$P(40)$	1027.38	0.65	$R(14)$	1074.65	0.6
$P(38)$	1029.44	0.65	$R(16)$	1075.99	0.6
$P(36)$	1031.48	0.7	$R(18)$	1077.30	0.6
$P(34)$	1033.49	0.7	$R(20)$	1078.59	0.6
$P(32)$	1035.47	0.75	$R(22)$	1079.85	0.55
$P(30)$	1037.43	0.75	$R(24)$	1081.09	0.55
$P(28)$	1039.37	0.75	$R(26)$	1082.30	0.55
$P(26)$	1041.28	0.8	$R(28)$	1083.48	0.5
$P(24)$	1043.16	0.8	$R(30)$	1084.64	0.5
$P(22)$	1045.02	0.8	$R(32)$	1085.77	0.5
$P(20)$	1046.85	0.75	$R(34)$	1086.87	0.5
$P(18)$	1048.66	0.75	$R(36)$	1087.95	0.4
$P(16)$	1050.44	0.7	$R(38)$	1089.00	0.35
$P(14)$	1052.20	0.7	$R(40)$	1090.03	0.35
$P(12)$	1053.92	0.7	$R(42)$	1091.03	0.25
$P(10)$	1055.63	0.7	$R(44)$	1092.01	0.2
$P(8)$	1057.30	0.6			
$P(6)$	1058.95	0.35			
$P(4)$	1060.57	0.15			

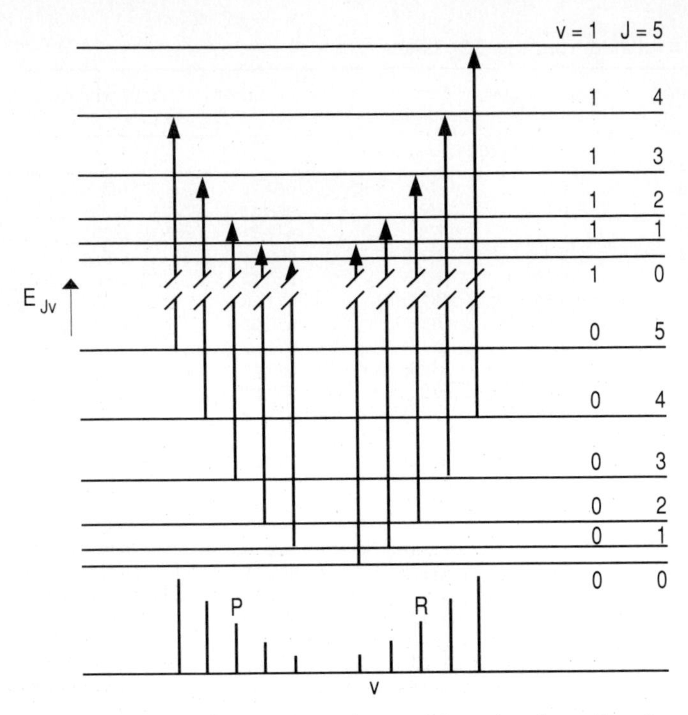

Figure 10.30 Energy levels and transitions in absorption for changes in v and J. (*Adapted from J. D. Graybeal, Molecular Spectroscopy, McGraw-Hill, 1988. Used by permission.*)

The energy difference or the energy of the emitted photon is

$$\Delta\varepsilon = \frac{(46{,}786 - 27{,}746)\,10^{-24}}{(6.626 \times 10^{-34})(3 \times 10^{10})}$$

$$= 957.8 \text{ cm}^{-1}$$

which agrees exactly with the energy appearing in Table 10.11.

The CO_2 laser is the workhorse in practical applications of laser technology for a number of reasons. It has a high overall efficiency of about 30 percent. This is a result of excited CO_2 molecules cascading down from their original levels to pile up in the long-lived (001) level from which lasing occurs. Excited N_2 similarly pile up in the $v = 1$ level, where they efficiently exchange with CO_2 with a deficit of only 19 cm^{-1}. This deficit is made up from the translational energy of the colliding N_2 ($v = 1$) and ground-state CO_2. The added He rapidly relaxes the terminal levels of both lasing processes back to the ground state, and the CO_2 molecules can then be used again. Very large, multikilowatt CO_2 lasers have been built.

Bond-Specific Photochemistry

One of the holy grails of chemistry is the ability to place energy into a molecule exactly where it is needed for some reaction. This specificity is to be contrasted with conventional heating of the reactants which, of course, puts energy throughout the molecule in a distribution determined by the partition function. It is possible to pump a particular mode, but the rapidity of *V-V* intramolecular exchange means that, after perhaps 10 to 100 vibrations, the energy is scrambled over usually all of the vibrational modes of the molecule. Figure 10.22 shows $p\tau$ values in *V-V* exchange of only 10^{-8} atm · s. Molecular dynamics does not retain the memory of the initial conditions. Bond-specific photochemistry remains, however, an active research interest.

A particularly interesting phenomenon is light-induced collisional energy transfer (LICET), which may occur when an excited species A^* and a ground-state species B are irradiated by a laser at the moment of collision. If the photon is selected to be of energy equal to the energy deficit in the collision, it is possible to thereby produce a resonant energy-transfer process. The process may be schematically imagined as shown in Fig. 10.31. We see A^* and B approaching each other where, at some separation r, a photon of just the correct energy pro-

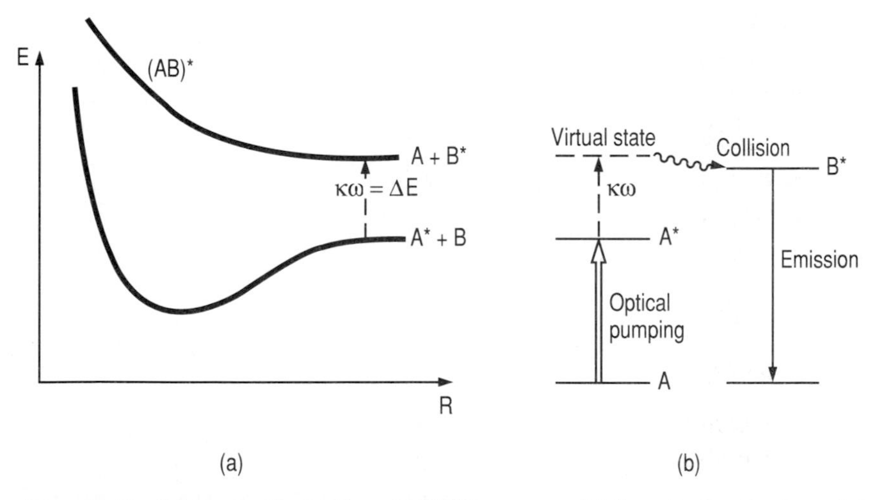

Figure 10.31 Schematic illustration of LICET processes. (*a*) Quasi-molecule viewpoint; (*b*) optical excitation into a virtual state of A with successive collisional energy transfer to B. (*Adapted from W. Demtroder, Laser Spectroscopy, Springer-Verlag, 1982. Used by permission.*)

motes the system to a higher and repulsive electronic energy surface, and $(AB)^*$ separates to form A and B^* rather than A^* and B as would have been the case on the lower-energy surface. Such transfer of electronic excitation occurs with cross-sections greater than gas-kinetic in the irradiation of mixtures of Sr and Ca.[16] The energy relationships for this transfer are shown in Fig. 10.32. In the Sr-to-Ca transfer, the mixed vapors at 800°C and at number densities of Sr of 10^{16} atoms/cm^3 and Ca of 10^{19} cm^{-3} were irradiated by a dye laser at 460.7 nm which produced Sr in its excited (5s, 5p) state. A second so-called transfer photon at 497.7 nm from a second dye laser is absorbed by the Sr:Ca pair at the moment of collision to produce Ca in its excited ($4p^2$) state. After the atoms separate, one sees ground-state Sr and an excited Ca that is evident from its fluoresence at 551.3 nm. Sr does not have an electronic energy level near the ($4p^2$) state of Ca. Just as expected from the idea of minimum energy deficit, maximum transfer occurs when

$$h\nu \text{ (pump)} + h\nu \text{ (transfer)} = E\ (4p^2\ {}^1\text{S state of Ca})$$

With these photons, cross sections as large as 10^{-13} cm^2 were observed. This is greater than gas-kinetic by two to three orders of magnitude.

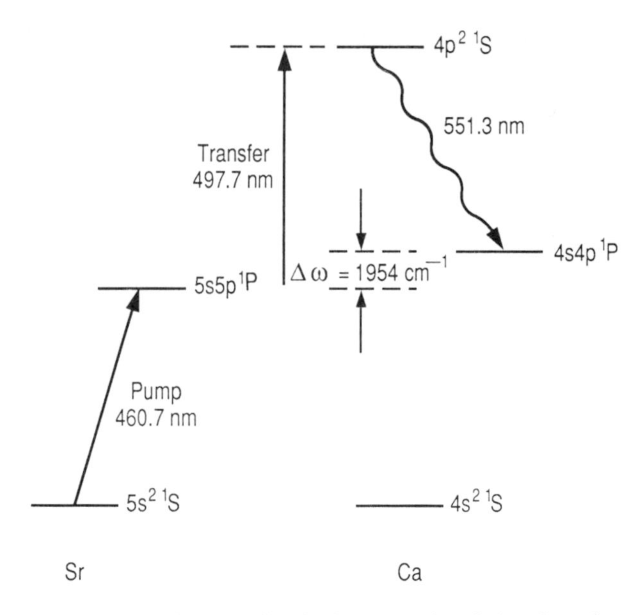

Figure 10.32 Energy level diagram for $Sr^* + Ca + h\nu \rightarrow Sr + Ca^*$ switched collision process. *(Adapted from S. E. Harris et al., Phy. Rev. Letters, 42, 970 (1979). Used by permission.)*

Summary

The goal of molecular engineering as applied to chemical reaction kinetics is to understand reaction rates in terms of bond lengths, vibration frequencies, polarizabilities, and the like. *To understand* means a desire to predict as well as a desire to explain and correlate. Some useful tools toward this understanding include collision theory, trajectory calculations using classical mechanics, the postulation of an activated complex or a transition state, the idea of reaction mechanism, and the like. Each of these concepts or theories results from some molecular insight, and each has value although none is completely satisfactory. Without these insights of molecular engineering, we have only empirical phenomenology of the sort that appears in traditional texts on reactor engineering. All of these are reaction- and system-specific, and they have no value toward understanding or correlating or predicting.

Just as the equilibrium properties of nonideal gases and liquids elude predictive molecular understanding, so also do the kinetic and transport properties. Both failures are primarily a result of the now-dominant role played by the dynamics of molecular encounters. Both collision theory and activated complex theory are useful abstractions, although neither is a truly predictive tool. Both can reasonably explain any rate phenomenon. Nevertheless, the molecular insights presented here, incomplete though they be, are essential to sophisticated, high-tech engineering design.

References

1. C. A. Hollingsworth, *J. Chem. Phys., 20*, 921 (1952).
2. M. Manes et al., *J. Chem. Phys., 18*, 1355 (1950).
3. R. D. Levine and R. B. Bernstein, *Molecular Reaction Dynamics*, Oxford, New York, 1974.
4. R. C. Tolman, *J. Am. Chem. Soc., 42*, 2506 (1920).
5. R. V. Churchill, *Modern Operational Mathematics in Engineering*, McGraw-Hill, New York, 1972.
6. H. Bateman, *Tables of Integral Transforms*, vol. 1, McGraw-Hill, New York, 1954, p. 133.
7. H. Bateman, *Tables of Integral Transforms*, vol. 1, McGraw-Hill, New York, 1954, p. 241.
8. G. Gioumousis and D. P. Stevenson, *J. Chem. Phys., 29*, 294 (1958).
9. S. W. Benson, *Foundations of Chemical Kinetics*, McGraw-Hill, New York, 1960.
10. K. E. Russell and J. Simons, *Proc. Roy. Soc., A217*, 271 (1953).
11. D. M. Golden, *J. Phys. Chem., 83*, 108 (1979).
12. W. G. Vincenti and C. H. Kruger, *Introduction to Physical Gas Dynamics*, Wiley, New York, 1965, p. 231.
13. H. G. Bennewitz et al., *Z. Physik, 177*, 84 (1964).
14. K. G. Anlauf et al., *Discussions Fara. Soc.*, no. 44, 183 (1967).
15. J. D. Lambert, *Vibrational and Rotational Relaxation in Gases*, Oxford, London, 1977.
16. S. E. Harris et al., *Phys. Rev. Letters, 42*, 970 (1979).

Further Reading

1. S. W. Benson, *Foundations of Gaseous Kinetics*, McGraw-Hill, New York, 1960.
2. R. B. Bernstein, *Chemical Dynamics via Molecular Beam and Laser Techniques*, Oxford, New York, 1982.
3. H. S. Johnston, *Gas Phase Reaction Rate Theory*, Ronald, New York, 1966.
4. J. D. Lambert, *Vibrational and Rotational Relaxation in Gases*, Oxford, London, 1977.
5. R. D. Levine and R. B. Bernstein, *Molecular Reaction Dynamics*, Oxford, London, 1974.

Appendix

Derivation of Most Probable Distribution

Molecules may absorb energy in several modes, depending on their structural complexity. Quantum mechanics specifies a set of energy levels for each such mode, but the statistical question is how do the 10^{23} molecules distribute themselves among these levels. The distribution that will be most probable is that of maximum w subject to the constraints of a constant total number of molecules and a constant total energy. The extremum of any function subject to constraints is readily found by the technique of lagrangian undetermined multipliers, and in the present case the function is

$$\ln w = n \ln n - \Sigma n_i \ln n_i$$

and the constraints are

$$\Sigma n_i = n$$

and

$$\Sigma n_i \varepsilon_i = U$$

Straightforwardly following the lagrangian technique, we first multiply the constraint relations by a constant to be determined later and form the sum function

$$S(n_0, n_1, n_2, \ldots) = n \ln n - \Sigma n_i \ln n_i + \lambda(n - \Sigma n_i) + \mu(U - \Sigma n_i \varepsilon_i)$$

According to Lagrange, we now simply differentiate the sum function with respect to each of the n_i's and set each resulting such expression equal to zero to find thereby that set of n_i's that will maximize the sum function. The identical derivatives are

$$\frac{\partial S}{\partial n_i} = -1 - \ln n_i - \lambda - \mu\varepsilon_i = 0$$

and thus

$$n_i = e^{-(\lambda+1)}e^{-\mu\varepsilon_i}$$

But since the total molecules are constant,

$$\Sigma n_i = e^{-(\lambda+1)}\Sigma e^{-\mu\varepsilon_i} = n$$

and thus

$$n_i = \frac{ne^{-\mu\varepsilon_i}}{\Sigma e^{-\mu\varepsilon_i}}$$

This is Boltzmann's equation, and it says simply that the number of molecules possessing the energy ε_i is proportional to the exponential $e^{-\mu\varepsilon_i}$.

This is a profound result that appears throughout physical chemistry. We have but to remember that the rate of a chemical reaction is proportional to $e^{-\text{energy}/kT}$, the equilibrium conversion is proportional to $e^{-\text{energy}/kT}$, the vapor pressure of a solid or a liquid is proportional to $e^{-\text{energy}/kT}$, and so on throughout our understanding of physics and chemistry.

Appendix

B

Evaluation of the Translational and Rotational Partition Functions

The translational energy levels in the x direction are

$$\varepsilon_x = \frac{n_x^2 h^2}{8ma^2}$$

where $n_x = 1, 2, 3, \ldots$. It is notationally convenient to define what we may call a characteristic translational temperature θ_t as

$$k\theta_t = \frac{h^2}{8ma^2}$$

where θ_t is in units of kelvins. The translational partition function in the x direction may be written

$$f_{\text{trans}} = \Sigma e^{-\varepsilon_{\text{trans}}/kT} = \Sigma e^{-n_x^2\theta_t/T}$$

Although each $\varepsilon_{\text{trans}}$ is composed of a contribution in each of three directions, x, y, and z, let us, for the moment, consider only two dimensions, x and y. For each energy in the x direction, ε_x, the complementary energy ε_y may take on any of its values and give rise to terms in the sum over states of $e^{-\varepsilon_x/kT}\,\Sigma_y\, e^{-\varepsilon_y/kT}$. And with ε_x taking on any of its values, we obtain terms $\Sigma_x\, e^{-\varepsilon_x/kT}\Sigma_y\, e^{-\varepsilon_y/kT}$.

Similarly in three dimensions,

$$f_{\text{trans}} = \sum_x e^{-\varepsilon_x/kT} \sum_y e^{-\varepsilon_y/kT} \sum_z e^{-\varepsilon_z/kT}$$

or $$f_{\text{trans}} = \left(\sum_{n=1}^{\infty} e^{-n^2 \theta_t/T} \right)^3$$

This summation is just the sum of the areas of an infinite number of strips of one unit width and $e^{-n^2\theta_t/T}$ in height (see Fig. B.1).

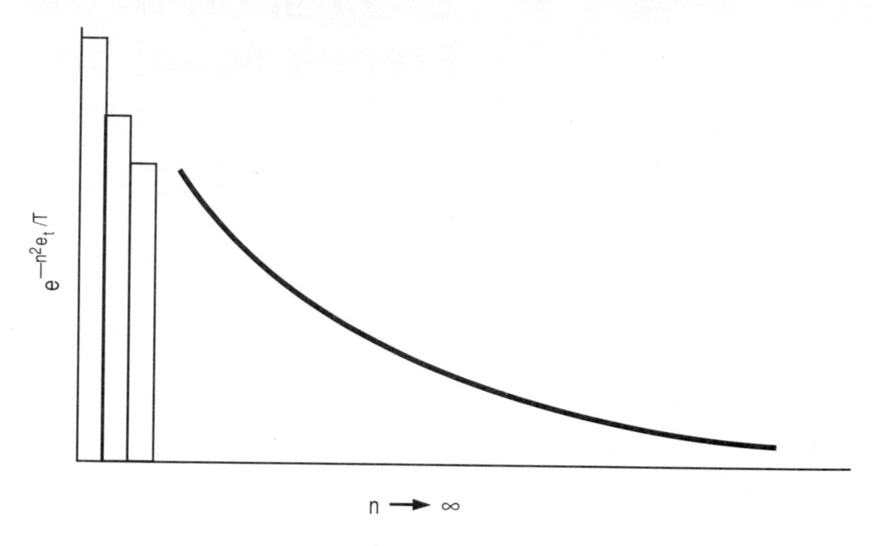

Figure B.1

We can replace the summation by an integration to find the total area if the change in height in going from one strip to the next is negligible, that is, if $\theta_t/T \ll 1$. For H atoms in a box of side 0.001 m at 10 K, $\theta_t = 2.3 \times 10^{-12}$ K and $\theta_t/T = 2.3 \times 10^{-11}$. For larger masses, higher temperatures, and larger volumes, θ_t/T can only be smaller. For small quantum numbers, the ordinate is unity, since we have almost exp (0), while for large n, the ordinate is zero. The integral is then written

$$f_{\text{trans}} = \left(\int_0^{\infty} e^{-n^2\theta_t/T}\, dn \right)^3$$

$$= \left[\frac{1}{2} \sqrt{\frac{\pi T}{\theta_t}} \right]^3$$

or $$f_{\text{trans}} = \left(\frac{2\pi mkT}{h^2} \right)^{3/2} V$$

A similar treatment to sum over the rotational levels is almost always appropriate. Solution of the Schrödinger equation for a linear rigid rotator yields eigenvalues

$$\varepsilon_{rot} = \frac{h^2}{8\pi^2 I}(J)(J+1)$$

where I is the moment of inertia, J is a quantum number 0, 1, 2,..., and each energy level has a degeneracy of $2J + 1$. As for translation, we will define a characteristic rotational temperature θ_r as

$$k\theta_r = \frac{h^2}{8\pi^2 I}$$

from which the rotational partition function may be written

$$f_{rot} = \sum_{J=0}^{\infty}(2J+1)e^{-J(J+1)\theta_r/T}$$

We have neglected one additional result from the quantum-mechanical analysis of a rigid rotator, and that is the effect of symmetry. If the rotator is asymmetric, any J is allowed, whereas if it is symmetric, then either odd J's or even J's are allowed, but not both. We shall return to this point, but first we recognize that the above summation for f_{rot} may be replaced by an integration with just the same reasoning as we applied for translation. Let $z = J(J+1)$, and since $\theta_r/T \ll 1$ for most molecules at most temperatures; the change in the value of the exponential is small for adjacent quantum numbers, and the integral

$$f_{rot} = \int_0^{\infty} e^{-z\theta_r/T}\,dz$$

becomes a reasonable approximation. Then

$$f_{rot} = \frac{T}{\theta_r} = \frac{8\pi^2 IkT}{h^2}$$

Again because the values of adjacent terms in the series for f_{rot} are so nearly the same, we can allow for any even or odd J for symmetric molecules by merely dividing the integrated result for f_{rot} by 2 for such species. We indicate this by a symbol σ called the symmetry number,

$$f_{rot} = \frac{8\pi^2 IkT}{\sigma h^2}$$

where $\sigma = 1$ for asymmetric rigid rotators and $\sigma = 2$ for symmetric species.

Appendix

C

A Guide to SI Units

The International System of Units (Le Système International d'Unités) was established as the official system of measurement in 1960 when representatives of 36 nations, including the United States, met in Paris at the eleventh general conference on weights and measures. Since that meeting, many countries and organizations have accepted SI as a modernized metric system, superior to the old cgs or MKS systems. SI is not the metric system heretofore used, and the word *metric* should be gradually phased out of existence. The basis of the SI system is a set of seven defined units of measure for seven basic quantities:

Quantity	Name of SI unit	Symbol of SI unit
Length	meter	m
Mass	kilogram	kg
Time	second	s
Temperature	kelvin	K
Amount of substance	mole	mol
Electric current	ampere	A
Luminous intensity	candela	cd

These base units have been defined with great precision, and all but the kilogram have been related to natural phenomena, which are considered to be invariant. Only the kilogram is still defined in terms of an artifact. The units are as follows:

meter The meter is the length equal to 1 650 763.73 wavelengths in vacuum of the radiation corresponding to the transition between the levels $2p_{10}$ and $5d_5$ of the krypton-86 atom.

This guide to SI units has been extensively taken from the *UOP Process Division Metric (SI) Guide*, 1st ed., 1977 and the *Metric Practice Guide*, E-380-76, American Society for Testing and Materials.

kilogram The kilogram is the unit of mass; it is equal to the mass of the international prototype of the kilogram. It is the only unit that, for historical reasons, contains a prefix in its name.

second The second is the duration of 9 192 631 770 periods of the radiation corresponding to the transition between the two hyperfine levels of the ground state of the cesium-133 atom.

ampere The ampere is that constant current which, if maintained in two straight parallel conductors of infinite length, of negligible cross section, and placed one meter apart in vacuum, would produce between these conductors a force equal to 2×10^{-7} newton per meter of length.

kelvin The kelvin, unit of thermodynamic temperature, is the fraction 1/273.16 of the thermodynamic temperature of the triple point of water.

mole The mole is the amount of substance of a system which contains as many elementary entities as there are atoms in 0.012 kilogram of carbon-12.

Note. When the mole is used, the elementary entities must be specified and these may be atoms, molecules, ions, electrons, other particles, or specified groups of such particles.

candela The candela is the luminous intensity, in the perpendicular direction, of a surface of 1/600 000 square meter of a blackbody at the temperature of freezing platinum under a pressure of 101 325 newtons per square meter.

Two purely geometrical units for plane angle measure are used in SI:

Quantity	Name of SI unit	Symbol of SI unit
Plane angle	radian	rad
Solid angle	steradian	sr

radian The unit of measure of a plane angle with its vertex at the center of a circle and subtended by an arc equal in length to the radius.

steradian The unit of measure of a solid angle with its vertex at the center of a sphere and enclosing an area of the spherical surface equal to that of a square with sides equal in length to the radius.

SI is coherent in that the product or quotient of any two unit quantities is a unit of the resulting quantity. Every derived unit is the result of some physical law. The area of a rectangle has units of m^2 since the area is defined as the product of the two sides and the meter is the fundamental unit of length. That is, in a coherent system in which foot is the unit of length, square foot is the unit of area, but acre is not. Similarly, in a coherent English system, the unit of force is the poundal, not the pound of force. Velocity is defined as distance per unit time, and its unit is then m/s. Force is defined by the first law of motion ($F = ma$) as the product of mass and acceleration, or a $kg \cdot m/s^2$, which is given the special name of a newton. Work is defined as the product of force and displacement in the direction of that force, or N · m, which is given the special name of a joule. Power is defined as work per unit time, and a force of a N · m acting for a second is a N · m/s or a watt. The SI units for force, work, and power are the same regardless of whether the process of interest is mechanical, electrical, thermal, or nuclear. Thus in SI, a force of 1 N applied through a distance of 1 m can produce energy equivalent to 1 J of heat which is the same as produced by 1 W of power in 1 s. By contrast, in English units, a force of 1 lb applied through a distance of 1 in can produce energy equivalent to 0.000 107 Btu, which is the same as produced by 1 hp in 0.000 505 h. Similarly a force of 1 N will accelerate a mass of 1 kg by 1 m/s^2. In contrast, a force of 1 kg_f accelerates a mass of 1 kg by 9.806 65 m/s^2.

Units for other than the base quantities are called derived units. They are expressed algebraically in terms of base units by means of the mathematical symbols of multiplication and division. Several derived units have been given special names and symbols that may themselves be used to express other derived units in a simpler way than in terms of the base units. Some examples of derived units are as follows:

Quantity	Derived unit	Derived unit special name	Symbol of SI derived unit
Area	m^2		
Volume	m^3		
Velocity	m/s		
Acceleration	m/s^2		
Density	kg/m^3		
Force	$kg \cdot m/s^2$	newton	N
Pressure	N/m^2	pascal	Pa
Work-energy-heat	N · m	joule	J
Power	J/s	watt	W
Quantity of electricity	A · s	coulomb	C
Electric potential	W/A	volt	V
Capacitance	C/V	farad	F
Resistance	V/A	ohm	Ω
Conductance	A/V	siemens	S
Luminous flux	cd · sr	lumen	lm
Frequency	l/s	hertz	Hz
Viscosity, dynamic	Pa · s		
Entropy	J/mol · K		
Heat capacity	J/mol · K		

A number of other units, while not officially a part of SI, are nevertheless recognized as allowable because of convenience and widespread use. Allowable units (and their symbols) include:

Distance	centimeter (cm)
Time	minute (min), hour (h), day (d), year (a)

Temperature — degree Celsius (°C) [formerly degree centigrade]. A celsius temperature (t) is related to a kelvin temperature (T) as follows:

$$T = 273.15 + t \qquad \text{(exactly)}$$

The International Practical Kelvin Temperature Scale of 1968 and the International Practical Celsius Temperature Scale of 1968 are defined by a set of interpolation equations based on the following reference temperatures:

	K	°C
Hydrogen, solid-liquid gas equilibrium	13.81	− 259.34
Hydrogen, liquid-gas equilibrium at 33,330.6 Pa (25/76 standard atmosphere)	17.042	− 256.108
Hydrogen, liquid-gas equilibrium	20.28	− 252.87
Neon, liquid-gas equilibrium	27.102	− 246.048
Oxygen, solid-liquid-gas equilibrium	54.361	− 218.789
Oxygen, liquid-gas equilibrium	90.188	− 182.962
Water, solid-liquid-gas equilibrium	273.16	0.01
Water, liquid-gas equilibrium	373.15	100.00
Zinc, solid-liquid equilibrium	692.73	419.58
Silver, solid-liquid equilibrium	1235.08	961.93
Gold, solid-liquid equilibrium	1337.58	1064.43

Except for the triple points and one equilibrium hydrogen point (17.042 K) the assigned values of temperature are for equilibrium states at a pressure

$$p_0 = 1 \text{ standard atmosphere } (101\,325 \text{ Pa})$$

In this scale the degree Celsius and the kelvin are identical, and are related by an exact difference of 273.15 degrees. It is seldom necessary to recognize the differences between the thermodynamic and practical scales, but the General Conference of Weights and Measures in 1960 established the following quantity symbols for use if desired:

Thermodynamic temperature	kelvins	T
	degrees Celsius	t
Practical temperature	kelvins	T_{int}
	degrees Celsius	t_{int}

Plane angle — degree (°), minute (′), second (″)

Volume — liter (l), a special name for a cubic decimeter (dm^3). The use of the liter is discouraged because it gives two names to the

	same quantity. In 1964 the General Conference on Weights and Measures adopted the name liter as a special name for the cubic decimeter. Prior to this decision, the liter differed slightly (previous value, 1.000 028 dm^3), and in expression of precision volume measurement this fact must be kept in mind. On most typewriters there is no difference between the lowercase ell and one, and to then avoid confusion, it is probably best to spell the word out in full. However, ml is correct for milliliter since there is no longer any possible confusion.
Mass	tonne (t), a special name for megagram (Mg) gram (g), correct name for millikilogram
Pressure	bar (bar), a special name for 1.0×10^5 Pa

Mixed units, especially those from different systems, should be avoided. For example,

Use kg/m^3 *not* kg/gal

Use 12.75 lb *not* 12 lb, 12 oz

In general, volumes should be expressed in m^3. For small volumes (samples, chemical injections) the dm^3 and cm^3 may be used. The use of the unit liter (L) (1.0 dm^3 exactly since 1964) or milliliter (ml) should be discouraged in technical work. A normal cubic meter is measured at 0.0°C and 101.325 kPa. A standard cubic meter (std m^3) is measured at 15.0°C and 101.325 kPa. A standard cubic foot (std ft^3) is measured at 60.0°F (15.6°C) and 14.696 psia.

Although the radian is the preferred unit of angular measure, the customary system is acceptable, especially for dimensions in "even" degrees, i.e., 45°, 90°, etc. Minutes and seconds, however, should be decimalized; i.e., 32° 16′ 48″ should be converted to 32.28°.

9.806 65 m/s^2 is the defined standard gravity at sea level and 45° latitude.

The unit pascal is used for either gage or absolute pressure. Since there is no agreed-on symbol for designating absolute or gage, the terms *absolute* or *gage* must be used explicitly, as in "300 kPa absolute" or "at a gage pressure of 85 kPa."

Although megagram is preferred over metric ton (t) in general, the use of metric ton (or tonne) will probably remain in denoting plant capacity, as in t/a (metric tons per annum).

The use of the term *weight* for the concept of mass is to be discouraged in technical work.

The mole as defined is equivalent to the old gram-mole. Modern usage drops the kg-mole notation and the old kg-mole becomes simply 1000 moles or a kilomole (kmol).

Prefixes

The following prefixes are used with SI units to keep numbers in a convenient range. Note that only the five largest prefixes use a capital letter for a symbol.

Factor	Prefix	Symbol
10^{18}	exa	E
10^{15}	peta	P
10^{12}	tera	T
10^{9}	giga	G
10^{6}	mega	M
10^{3}	kilo	k
10^{2}	hecto†	h†
10^{1}	deca†	da†
10^{-1}	deci†	d†
10^{-2}	centi†	c†
10^{-3}	milli	m
10^{-6}	micro	μ
10^{-9}	nano	n
10^{-12}	pico	p
10^{-15}	femto	f
10^{-18}	atto	a

†To be avoided where possible.

Guidelines for Using SI Multiples

Never use a prefix without a unit; that is, use kilogram, not kilo.

Prefixes representing multiples of 1000 are recommended. This should, in general, give a numerical value for some quantity that would fall between 0.1 and 1000. Examples:

Use kilometer (10^3 m), millimeter (10^{-3} m), and micrometer (10^{-6} m) in preference to hectometer (10^2 m) or decimeter (10^{-1} m)

Use 12.3 nA *not* 0.0123 μA

Double prefixes should not be used. Example:

Use nanometer, not millimicrometer

Prefixes should not be used in the denominator of a unit (except for the base unit kilogram). Example:

Use megagram per cubic meter, *not* kilogram per cubic decimeter and *not* gram per cubic centimeter

When prefixes are used with units raised to a power, the prefix is also raised to the same power. Examples:

Use	$mm^3 = (10^{-3}\ m)^3 = 10^{-9}\ m^3$	Correct
Not	$mm^3 = 10^{-3}(m^3)$	Incorrect
Use	$dm^3 = (10^{-1}\ m)^3 = 10^{-3}\ m^3$	Correct

Decimal multiples and submultiples of the unit kilogram are formed by attaching prefixes to the word gram: Examples:

Use 1 gram *not* 1 millikilogram

Use 10^3 kilogram = 1 megagram *not* kilokilogram

Representation of SI Units

The rules for representing SI units are quite precise and are necessary to avoid confusion. The distinction between upper- and lowercase symbols is very important. Symbols are not capitalized unless the unit is derived from a proper name. Examples:

K for Lord Kelvin, but kg for kilogram

When the full spellings are used for either units or prefixes, lowercase letters shall always be used with the one exception of degree Celsius (298 K is correct, *not* °k, and *not* °K). Unabbreviated units should form their plurals in the usual manner, but unit symbols do not change in the plural, for example, 55 kilometers or 55 km, 2.5 grams or 0.5 gram. Hertz, lux, and siemens remain unchanged in the plural, and henry becomes henries. For example, the pressure is 325 kilopascals gage or 325 kPa gage. Periods should never be used in SI symbols (55 km/h is correct but 55 km./h. is *incorrect*). In reporting numbers of five or more digits on either side of the decimal point, the digits should be placed in groups of three separated by a space instead of a comma, grouping both right and left from the decimal point. The comma to separate groups of digits is commonly used in the United States, but the comma is also used as the decimal marker in some countries, and its use to group digits could lead to confusion. Numbers with four digits on either side of the decimal point may be grouped together or separated. Examples:

3 698 741.238 68	Correct
3,698,741.23868	Incorrect (but common in the United States)
4300 or 4 300	Correct
2.4917 or 2.491 7	Correct
27 500	Correct
27500	Incorrect

The decimal point will always be symbolized by the period, and the European convention of using the comma is not to be employed. Such abbreviations as k.p.h. or kph for kilometers per hour are incorrect, rather use 55 km/h. In numbers between 1 and − 1 (but not zero), a zero should be written before the decimal point. Write 0.3 m, not .3 m, and − 0.4 kg, not −.4 kg.

When reporting derived units, one must use the center dot to indicate multiplication and slash to indicate division with only one slash used in each derived unit (for example: 1 J = 1 kg · m^2/s^2). The dot may be dispensed with when there is no risk of confusion with another unit symbol (for example, N · m or N m, but not mN). When the multiplicative dot is omitted, a space must be left between the symbols (for example, J/mol · K or J/mol K, but not J/molK). Do not use a space either before or after the product dot or the division slash. If the single slash leads to ambiguity, parentheses are acceptable, as are negative powers. Examples:

$\mathrm{m \cdot kg/(s^3 \cdot A)}$ or $\mathrm{m \cdot kg \cdot s^{-3} \cdot A^{-1}}$ *but not* $\mathrm{m \cdot kg/s^3/A}$

$\mathrm{m/s^2}$ or $\mathrm{m \cdot s^{-2}}$ *but not* m/s/s

Do not use the slash with words. Write meter per second, not meter/second or meter/s or m/second. When writing symbols or names for units having prefixes, no space is left between letters making up the symbol or the name. Write kA, kiloampere, mg, milligram. When writing a symbol after a number to which it refers, a space must be left between the number and the symbol. Write 455 kHz, 22 mg (exception 20°C).

When a quantity is used in an objective sense, it is preferable to use a hyphen between the number and the unit name except for ° and °C. Write: It is a 35-mm film, but the film width is 35 mm.

Conversion Factors

The *conversion factor* is presented in exponential form. In some cases a formula replaces the conversion factor. The symbol * denotes an exact numerical conversion. The absence of a * denotes an approximate factor to the number of significant figures shown (seven figures in general). Where the SI unit and customary unit are identical in both numerical value and unit symbol, the simple value of unity (1) is used.

Conversions should be accomplished by multiplying the quantity by the appropriate factor from the following table, followed by rounding such that the accuracy of the quantity is neither sacrificed nor exaggerated. Do not round either the quantity or the conversion factor before performing the multiplication. It is necessary to determine the intended precision of a quantity prior to converting, for the converted value must be carried to a sufficient number of digits to maintain the accuracy implied or required in the original quantity.

The inclusion of conversions for the several earlier definitions of Btu, cal, volt, etc. has been purposefully done to both alert the user to exercise care in converting to SI and to reemphasize the simplicity and value of a single coherent international system of measurements.

Classified List of Units

To convert from	To	Multiply by
	Acceleration	
foot/second^2	meter/second^2 (m/s^2)	3.048 000*E − 01
	Area	
acre	meter^2 (m^2)	4.046 856 E + 03
circular mil	meter^2 (m^2)	5.067 075 E − 10
foot^2	meter^2 (m^2)	9.290 304*E − 02
inch^2	meter^2 (m^2)	6.451 600*E − 04
yard^2	meter^2 (m^2)	8.361 274 E − 01
	Bending moment or torque	
dyne-centimeter	newton-meter (N · m)	1.000 000*E − 07
pound-force-foot	newton-meter (N · m)	1.355 818 E + 00
	Capacity (see volume)	
	Density (see mass/volume)	
	Electricity and magnetism†	
abampere	ampere (A)	1.000 000*E + 01
abcoulomb	coulomb (C)	1.000 000*E + 01
abfarad	farad (F)	1.000 000*E + 09
abhenry	henry (H)	1.000 000*E − 09
abmho	siemens (S)	1.000 000*E + 09
abohm	ohm (Ω)	1.000 000*E − 09
abvolt	volt (V)	1.000 000*E − 08
ampere, international U.S. (A_{INT-US})‡	ampere (A)	9.998 43 E − 01
ampere, U.S. legal 1948 (A_{US-48})	ampere (A)	1.000 008 E + 00
ampere-hour	coulomb (C)	3.600 000*E + 03
coulomb, international U.S. (C_{INT-US})‡	coulomb (C)	9.998 43 E − 01
coulomb, U.S. legal 1948 (C_{US-48})	coulomb (C)	1.000 008 E + 00
	Heat	
Btu (thermochemical) · in/s · ft^2 · deg F (k, thermal conductivity)	watt/meter-kelvin (W/m · K)	5.188 732 E + 02

Note: See page 423 for footnotes.

Classified List of Units (*Continued*)

To convert from	To	Multiply by
	Heat (*continued*)	
Btu (International Table) · in/ s · ft^2 · deg F (*k*, thermal conductivity)	watt/meter-kelvin (W/m · K)	5.192 204 E + 02
Btu (thermochemical) · in/ h · ft^2 · deg F (*k*, thermal conductivity)	watt/meter-kelvin (W/m · K)	1.441 314 E − 01
Btu (International Table) · in/ h · ft^2 · deg F (*k*, thermal conductivity)	watt/meter-kelvin (W/m · K)	1.442 279 E − 01
Btu (International Table)/ft^2	joule/$meter^2$ (J/m^2)	1.135 653 E + 04
Btu (thermochemical)/ft^2	joule/$meter^2$ (J/m^2)	1.134 893 E + 04
Btu (International Table)/ h · ft^2 · deg F (*C*, thermal conductance)	watt/$meter^2$-kelvin (W/ m^2 · K)	5.678 263 E + 00
Btu (thermochemical)/ h · ft^2 · deg F (*C*, thermal conductance)	watt/$meter^2$-kelvin (W/ m^2 · K)	5.674 466 E + 00
Btu (International Table)/pound-mass	joule/kilogram (J/kg)	2.326 000*E + 03
Btu (thermochemical)/pound-mass	joule/kilogram (J/kg)	2.324 444 E + 03
Btu (International Table)/ lbm · deg F (*c*, heat capacity)	joule/kilogram-kelvin (J/ kg · K)	4.186 800*E + 03
Btu (thermochemical)/lbm · deg F (*c*, heat capacity)	joule/kilogram-kelvin (J/ kg · K)	4.184 000 E + 03
Btu (International Table)/ s · ft^2 · deg F	watt/$meter^2$-kelvin (W/ m^2 · K)	2.044 175 E + 04
Btu (thermochemical)/ s · ft^2 · deg F	watt/$meter^2$-kelvin (W/ m^2 · K)	2.042 808 E + 04
cal (thermochemical)/cm^2	joule/$meter^2$ (J/m^2)	4.184 000*E + 04
cal (thermochemical)/cm^2 · s	watt/$meter^2$ (W/m^2)	4.184 000*E + 04
cal (thermochemical)/ cm · s · deg C	watt/meter-kelvin (W/m · K)	4.184 000*E + 02
cal (International Table)/g	joule/kilogram (J/kg)	4.186 800*E + 03
cal (International Table)/ g · deg C	joule/kilogram-kelvin (J/kg · K)	4.186 800*E + 03
cal (thermochemical)/g	joule/kilogram (J/kg)	4.184 000*E + 03
cal (thermochemical)/g · deg C	joule/kilogram-kelvin (J/ kg · K)	4.184 000*E + 03
	Length	
angstrom	meter (m)	1.000 000*E − 10
foot	meter (m)	3.048 000*E − 01
inch	meter (m)	2.540 000*E − 02
micron	meter (m)	1.000 000*E − 06
mil	meter (m)	2.540 000*E − 05
mile (international nautical)	meter (m)	1.852 000*E + 03
mile (U.K. nautical)	meter (m)	1.853 184*E + 03
mile (U.S. nautical)	meter (m)	1.852 000*E + 03
mile (U.S. statute)	meter (m)	1.609 344*E + 03
rod	meter (m)	5.029 200*E + 00

Classified List of Units (*Continued*)

To convert from	To	Multiply by
	Length (*Continued*)	
statute mile (U.S.)	meter (m)	1.609 344* E + 03
yard	meter (m)	9.144 000* E − 01
	Mass	
gram	kilogram (kg)	1.000 000* E − 03
hundredweight (long)	kilogram (kg)	5.080 235 E + 01
hundredweight (short)	kilogram (kg)	4.535 924 E + 01
kilogram-force-second^2/meter (mass)	kilogram (kg)	9.806 650* E + 00
kilogram-mass	kilogram (kg)	1.000 000* E + 00
ounce-mass (avoirdupois)	kilogram (kg)	2.834 952 E − 02
ounce-mass (troy or apothecary)	kilogram (kg)	3.110 348 E − 02
pound-mass (lbm avoirdupois)	kilogram (kg)	4.535 924 E − 01
pound-mass (troy or apothecary)	kilogram (kg)	3.732 417 E − 01
slug	kilogram (kg)	1.459 390 E + 01
ton (assay)	kilogram (kg)	2.916 667 E − 02
ton (long, 2240 lbm)	kilogram (kg)	1.016 047 E + 03
ton (metric)	kilogram (kg)	1.000 000* E + 03
ton (short, 2000 lbm)	kilogram (kg)	9.071 847 E + 02
tonne	kilogram (kg)	1.000 000*E + 03
	Mass/capacity (see mass/volume)	
	Mass/time (includes flow)	
pound-mass/second	kilogram/second (kg/s)	4.535 924 E − 01
pound-mass/minute	kilogram/second (kg/s)	7.559 873 E − 03
ton (short, mass)/hour	kilogram/second (kg/s)	2.519 958 E − 01
	Mass/volume (includes density and mass capacity)	
gram/centimeter^3	kilogram/meter^3 (kg/m^3)	1.000 000* E + 03
pound-mass/foot^3	kilogram/meter^3 (kg/m^3)	1.601 846 E + 01
pound-mass/inch^3	kilogram/meter^3 (kg/m^3)	2.767 990 E + 04
pound-mass/gallon (U.K. liquid)	kilogram/meter^3 (kg/m^3)	9.977 644 E + 01
pound-mass/gallon (U.S. liquid)	kilogram/meter^3 (kg/m^3)	1.198 264 E + 02
slug/foot^3	kilogram/meter^3 (kg/m^3)	5.153 788 E + 02
ton (long, mass)/yard^3	kilogram/meter^3 (kg/m^3)	1.328 939 E + 03
	Power	
Btu (International Table)/hour	watt (W)	2.930 711 E − 01
Btu (thermochemical)/second	watt (W)	1.054 350 E + 03
Btu (thermochemical)/minute	watt (W)	1.757 250 E + 01
Btu (thermochemical)/hour	watt (W)	2.928 751 E − 01
calorie (thermochemical)/second	watt (W)	4.184 000* E + 00
calorie (thermochemical)/minute	watt (W)	6.973 333 E − 02
erg/second	watt (W)	1.000 000* E − 07
foot-pound-force/hour	watt (W)	3.766 161 E − 04

Classified List of Units (*Continued*)

To convert from	To	Multiply by
	Power (*Continued*)	
foot-pound-force/minute	watt (W)	2.259 697 E − 02
foot-pound-force/second	watt (W)	1.355 818 E + 00
horsepower (550 ft · lbf/s)	watt (W)	7.456 999 E + 02
horsepower (boiler)	watt (W)	9.809 50 E + 03
horsepower (electric)	watt (W)	7.460 000*E + 02
horsepower (metric)	watt (W)	7.354 99 E + 02
horsepower (water)	watt (W)	7.460 43 E + 02
horsepower (U.K.)	watt (W)	7.457 0 E + 02
kilocalorie (thermochemical)/ minute	watt (W)	6.973 333 E + 01
kilocalorie (thermochemical)/ second	watt (W)	4.184 000*E + 03
watt, international U.S. (W_{INT-US})	watt (W)	1.000 182 E +00
watt, U.S. legal 1948 (W_{US-48})	watt (W)	1.000 017 E + 00
	Pressure or stress (force/area)	
atmosphere (normal = 760 torr)	pascal (Pa)	1.013 25 E + 05
atmosphere (technical = 1 kgF/ cm^2)	pascal (Pa)	9.806 650*E + 04
bar	pascal (Pa)	1.000 000*E + 05
centimeter of mercury (0°C)	pascal (Pa)	1.333 22 E + 03
centimeter of water (4°C)	pascal (Pa)	9.806 38 E + 01
dyne/centimeter^2	pascal (Pa)	1.000 000*E − 01
foot of water (39.2°F)	pascal (Pa)	2.988 98 E + 03
gram-force/centimeter^2	pascal (Pa)	9.806 650*E + 01
inch of mercury (32°F)	pascal (Pa)	3.386 389 E + 03
inch of mercury (60°F)	pascal (Pa)	3.376 85 E + 03
inch of water (39.2°F)	pascal (Pa)	2.490 82 E + 02
inch of water (60°F)	pascal (Pa)	2.488 4 E + 02
kilogram-force/centimeter^2	pascal (Pa)	9.806 650*E + 04
kilogram-force/meter^2	pascal (Pa)	9.806 650*E + 00
kilogram-force/millimeter^2	pascal (Pa)	9.806 650*E + 06
millimeter of mercury (0°C)	pascal (Pa)	1.333 224 E + 02
poundal/foot^2	pascal (Pa)	1.488 164 E + 00
pound-force/foot^2	pascal (Pa)	4.788 026 E + 01
pound-force/inch^2 (psi)	pascal (Pa)	6.894 757 E + 03
torr (mm Hg, 0°C)	pascal (Pa)	1.333 22 E + 02
	Speed (see velocity)	
	Stress (see pressure)	
	Temperature	
degree Celsius	kelvin (K)	$t_K = t_C + 273.15$
degree Fahrenheit	kelvin (K)	$t_K = (t_F + 459.67)/1.8$
degree Rankine	kelvin (K)	$t_K = t_R/1.8$
degree Fahrenheit	degree Celsius	$t_C = (t_F - 32)/1.8$
kelvin	degree Celsius	$t_C = t_K - 273.15$
	Time	
hour (mean solar)	second (s)	3.600 000 E + 03

Classified List of Units (*Continued*)

To convert from	To	Multiply by
	Time (*Continued*)	
hour (sidereal)	second (s)	3.590 170 E + 03
	Velocity (includes speed)	
foot/minute	meter/second (m/s)	5.080 000*E − 03
kilometer/hour	meter/second (m/s)	2.777 778 E − 01
knot (international)	meter/second (m/s)	5.144 444 E − 01
mile/hour (U.S. statute)	meter/second (m/s)	4.470 400*E − 01
	Viscosity	
centipoise	pascal-second (Pa · s)	1.000 000*E − 03
centistoke	meter^2/second (m^2/s)	1.000 000*E − 06
foot^2/second	meter^2/second (m^2/s)	9.290 304*E − 02
poise	pascal-second (Pa · s)	1.000 000*E − 01
poundal-second/foot^2	pascal-second (Pa · s)	1.488 164 E + 00
pound-mass/foot-second	pascal-second (Pa · s)	1.488 164 E + 00
pound-force-second/foot^2	pascal-second (Pa · s)	4.788 026 E + 01
slug/foot-second	pascal-second (Pa · s)	4.788 026 E + 01
	Volume (includes capacity)	
acre-foot	meter^3 (m^3)	1.233 482 E + 03
barrel (oil, 42 gal)	meter^3 (m^3)	1.589 873 E − 01
foot^3	meter^3 (m^3)	2.831 685 E − 02
gallon (Canadian liquid)	meter^3 (m^3)	4.546 090 E −03
gallon (U.K. liquid)	meter^3 (m^3)	4.546 092 E − 03
gallon (U.S. dry)	meter^3 (m^3)	4.404 884 E − 03
gallon (U.S. liquid)	meter^3 (m^3)	3.785 412 E − 03
inch^3	meter^3 (m^3)	1.638 706 E − 05
liter	meter^3 (m^3)	1.000 000*E − 03
ounce (U.K. fluid)	meter^3 (m^3)	2.841 307 E − 05
ounce (U.S. fluid)	meter^3 (m^3)	2.957 353 E − 05
pint (U.S. dry)	meter^3 (m^3)	5.506 105 E − 04
pint (U.S. liquid)	meter^3 (m^3)	4.731 765 E − 04
quart (U.S. dry)	meter^3 (m^3)	1.101 221 E − 03
quart (U.S. liquid)	meter^3 (m^3)	9.463 529 E − 04
yard^3	meter^3 (m^3)	7.645 549 E − 01
	Volume/time (includes flow)	
foot^3/minute	meter^3/second (m^3/s)	4.719 474 E − 04
gallon (U.S. liquid)/day	meter^3/second (m^3/s)	4.381 264 E − 08
	Work (see energy)	

†ESU means electrostatic cgs unit. EMU means electromagnetic cgs unit.

‡In 1948 a new international agreement was reached on absolute electrical units, which changed the value of the volt used in the United States by about 300 parts per million. Again in 1969 a new base of reference was internationally adopted making a further change of 8.4 parts per million. These changes (and also changes in ampere, joule, watt, coulomb) require careful terminology and conversion factors for exact use of old information. Terms used in this guide are:

Volt as used prior to January 1948—volt, international U.S. ($V_{INT\text{-}US}$)

Volt as used between January 1948 and January 1969—volt, US legal 1948 ($V_{US\text{-}48}$)

Volt as used since January 1969—volt (V)

Identical treatment is given the ampere, coulomb, watt, and joule. Since the henry, farad, and ohm were not changed in 1969, only conversions as listed first above are shown.

Appendix D
The Evaluation of Some Useful Integrals

Integrals of the type

$$\int_0^\infty e^{-ax^2} x^n \, dx$$

occur frequently in physical chemistry, so perhaps it would be well to develop a general solution to all such integrals.† It is also well to note that

$$\int_{-\infty}^{\infty} = 2\int_0^\infty$$

The integral is evaluated differently depending on whether n is an odd or an even integer. First, for odd values, and for $n = 1$,

$$\int_0^\infty e^{-ax^2} x \, dx$$

we make the substitution of $s = x^2$, and then $ds = 2x\,dx$. Then,

$$\int_0^\infty e^{-ax^2} x \, dx = \frac{1}{2}\int_0^\infty e^{-as} \, ds = \frac{1}{2}\left(-\frac{1}{a} e^{-as}\Big|_0^\infty \right) = \frac{1}{2a}$$

The integration for even values of n is more difficult. First, for $n = 0$, consider the related integral,

$$\int_0^\infty \int_0^\infty e^{-a(x^2 + y^2)} \, dx \, dy$$

Converting to polar coordinates and integrating, we let

$$r^2 = x^2 + y^2$$

and

$$r \, dr \, d\theta = dx \, dy$$

†Taken with only slight modification from a similar appendix in *Physical Chemistry* by E. A. Moelwyn-Hughes, Pergamon, New York, 1957.

and $\int_0^\infty \int_0^\infty e^{-a(x^2+y^2)}\,dx\,dy = \int_0^{\pi/2} \int_0^\infty e^{-ar^2}\,r\,dr\,d\theta = \frac{\pi}{2}\int_0^\infty e^{-ar^2}\,r\,dr = \frac{\pi}{4a}$

But we recognize that x and y may have any values, including, of course, $x = y$, and in that case,

$$\int_0^\infty \int_0^\infty e^{-a(x^2+y^2)}\,dx\,dy = \left[\int_0^\infty e^{-ax^2}\,dx\right]^2 = \frac{\pi}{4a}$$

or

$$\int_0^\infty e^{-ax^2}\,dx = \frac{1}{2}\sqrt{\frac{\pi}{a}}$$

Integrals for all values of n may be similarly obtained after integration by parts, and the results may be summarized as follows:

n	Integral	n	Integral
0	$\int_0^\infty e^{-ax^2}\,dx = \frac{1}{2}\sqrt{\frac{\pi}{a}}$	1	$\int_0^\infty e^{-ax^2}\,x\,dx = \frac{1}{2a}$
2	$\int_0^\infty e^{-ax^2}\,x^2\,dx = \frac{1}{4}\sqrt{\frac{\pi}{a^3}}$	3	$\int_0^\infty e^{-ax^2}\,x^3\,dx = \frac{1}{2a^2}$
4	$\int_0^\infty e^{-ax^2}\,x^4\,dx = \frac{3}{8}\sqrt{\frac{\pi}{a^5}}$	5	$\int_0^\infty e^{-ax^2}\,x^5\,dx = \frac{1}{a^3}$
⋮	⋮	⋮	⋮
Even	$\int_0^\infty e^{-ax^2}\,x^n\,dx = 1\cdot 3\cdot 5\cdots(n-1)\,\frac{(\pi a)^{1/2}}{(2a)^{(1/2)n+1}}$	Odd	$\int_0^\infty e^{-ax^3}\,x^n\,dx = \frac{[\frac{1}{2}(n-1)]!}{2a^{(1/2)(n+1)}}$

The probability integral or the Gauss error function is defined by

$$\text{erf}\,(x) = \frac{2}{\sqrt{\pi}}\int_0^x e^{-t^2}\,dt$$

Clearly, $\text{erf}\,(0) = 0$, $\text{erf}\,(\infty) = 1$, and $\text{erf}\,(-x) = -\text{erf}\,(x)$. For $x \ll 1$,

$$\text{erf}\,(x) = \frac{2}{\sqrt{\pi}}\,x\left(1 - \frac{x}{2} + \frac{x^4}{10} - \cdots\right)$$

and for $x \gg 1$

$$\text{erf}\,(x) = 1 - \frac{e^{-x^2}}{\sqrt{\pi}(x)}$$

Some typical values of erf (x) are tabulated here, and extensive such tables appear in handbooks.

x	0.2	0.4	0.6	0.8	0.9
erf (x)	0.2227	0.4284	0.6039	0.7421	0.7969
x	1.0	1.2	1.4	1.6	1.9
erf (x)	0.8427	0.9103	0.9523	0.9891	0.9928

Appendix

Problems

Chapter 2

1 Develop expressions for ln W in terms of the partition functions for Maxwell-Boltzmann and for Bose-Einstein statistics.

Chapter 3

1 Calculate the entropy of benzene as an ideal gas at 298.15 K that arises from the translational, rotational, and vibrational modes of the molecule. Sum these and compare with the calorimetric value of 269.70 J/mol · K. As a symmetric planar molecule, benzene has two moments equivalent and the third with a value exactly twice that of the other two. The doubly degenerate moment is 147.63×10^{-47} kg · m^2. The molecule has 10 nondegenerate frequencies and 10 doubly degenerate frequencies of:

Single	Double
Single: 3062 cm^{-1}	Double: 849 cm^{-1}
992	3045
1190	1485
671	1037
3063	3047
1008	1596
1520	1178
538	606
1854	1160
1145	404

2 Use Eq. (3.9) to calculate the ionization potential of atomic hydrogen and compare with the experimental value of 13.6 eV.

3 Calculate the enthalpy and entropy of BrF_3O at 400, 500, and 600 K and 1 atm. The vibration frequencies are 995, 625, 531, 345, 236, 201, 601, 394, and 330 cm^{-1}, and the rotational constants are 0.189, 0.119, and 0.086 cm^{-1}. See Christie et al.[1]

4 Calculate the C_p^0 and S^0 for tetrafluorohydrazine at 300 and 400 K. Remember that, although there has been some debate, since 1978 this species is

known to exist as one of two structural isomers, that is, in either a trans or a gauche form. These isomers have very similar but nonetheless detectably different vibration frequencies:[2]

Gauche N_2F_4	Trans N_2F_4
116, 242, 284, 300, 423, 518, 590, 737, 934, 946, 1010, 1023	131, 252, 354, 467, 494, 542, 601, 719, 873, 962, 999, 1039

The principal difference between the isomers is a center of symmetry in the trans form. The activation energy and the total energy difference between the two isomers is small, and some debate remains about whether the two isomers would be isolable at room temperature or convert one into the other. A standard data compilation gives only one set of rotational constants: 5576.21, 3189.35, and 2812.95 mHz.[3] Do you expect the entropy of the trans or gauche form to be the greater? Why? Was your expectation confirmed? *Hint*: The structure of the two isomers is best visualized by looking down the N-N line of centers:

The relative amounts of the two isomers is temperature-dependent. Which isomer would you expect to increase with increasing temperature? Why?

5 Calculate the Gibbs free energy and C_p^0 for furan at 400 K and 1 atm. The rotational constants are 9446.96, 9246.61, and 4670.88 mHz [see *J. Mol. Spect.* 9, 124 (1962)], and the vibrational frequencies are 3154, 3140, 1491, 1384, 1140, 1066, 995, 871, 863, 728, 613, 3161, 3129, 1556, 1267, 1180, 1040, 873, 838, 745, and 603 cm^{-1} as reported in the JANAF tables.

6 Very pure solid silicon for use in the electronics industry is to be made by deposition of the solid from the vapor phase. Estimate the rate at which heat is to be removed when 100 g/min of silicon gas at 5000 K condenses to a solid at room temperature (300 K). Energy data on silicon are available as follows:

ε_i, cm^{-1}	g_i
0	1
77	3
223	5
6,299	5
15,394	1
39,683	1

7 The fraction of molecules in rotational level J will show a maximum when plotted versus J. Find a general expression for this J of maximum population. Plot n_J/n versus J for HCl at 300 K.

8 Calculate the entropy and heat capacity C_p^0 of AgCl at 800 K. The molecular data are:

$^{107}Ag^{35}Cl$	3687.04 MHz, 343.6 cm^{-1}
$^{109}Ag^{35}Cl$	3670.24 MHz, 342.4 cm^{-1}
$^{107}Ag^{37}Cl$	3537.49 MHz, 336.0 cm^{-1}
$^{109}Ag^{37}Cl$	3520.14 MHz, 355.7 cm^{-1}

9 Calculate the entropy and heat capacity of CH_3OH at 300 K and 1 atm. The vibration frequencies are: 3400, 2987, 2837, 1458(4), 1370, 1110(2), 1024, and 270 cm^{-1}; the $I_AI_BI_C$ product is 1740 amu$^3 \cdot$ Å^6.

ANSWER

$$S^0 = 239.64 \text{ J/mol} \cdot \text{K} \qquad S^0(\text{exp}) = 241 \pm 2 \text{ J/mol} \cdot \text{K}$$

$$C_V^0 = 38.01 \text{ J/mol} \cdot \text{K} \qquad C_V^0(\text{exp}) = 35.58 \text{ J/mol} \cdot \text{K}$$

The discrepancy involves accounting for the hindered rotation of $-$OH as rather a low-frequency torsion at 270 cm^{-1}. If $-$OH is taken as rotating freely $C_V^0 = 34.95$ J/mol $\cdot$ K. (This problem is from Knox.[4])

Chapter 4

1 Calculate the C_p^0 for propylene at 473.15 K and S^0 at 225.35 K. A critical review of the available experimental data reveals the following selected or "best" values:[5]

Molecular weight	42.0804
Symmetry number	3
Ground electronic state	singlet
Moments of inertia	1.8133×10^{-39} g $\cdot$ cm^2
	9.0187×10^{-39} g $\cdot$ cm^2
	1.0317×10^{-38} g $\cdot$ cm^2
Reduced moment for internal rotation	3.945×10^{-40} g $\cdot$ cm^2
Potential barrier for internal rotation	1997 cal/mole
Vibrational frequencies	3091, 3022, 2991, 2973, 2932, 1653, 1459, 1414, 1378, 1298, 1178, 935, 919, 428, 2953, 1443, 1045, 990, 912, 575

ANSWER: The calculated result is $C_p^0 = 21.84$ cal/mol $\cdot$ K at 473.15 K, which compares well with an experimental value which after correction to zero pressure is $C_p^0 = 21.79$ cal/mol $\cdot$ K. This quality of agreement is repeated from 180 to 460 K.[5] The calculated entropy at 225.35 K is 59.80 cal/mol $\cdot$ K, which is in excellent agreement with an older calorimetric value of 59.77 cal/mol · K at the same temperature.[5] It is interesting to note that an early small discrepancy between the calorimetric and statistical entropies was described as arising from a random end-for-end orientation of the $CH_2=CH_2-CH_3$ molecule in the solid lattice as it froze. Better data have proven this discrepancy to be erroneous, and we conclude that solid propylene is a regular array. Some additional results from an earlier study[5] are as follows:

Ideal Gas Thermodynamic Properties for Propylene

T, K	$H^0 - H_0^0$, cal · mol^{-1}	$-\frac{(G^0 - H_0^0)}{T}$	S^0	C_p^0
		cal · K^{-1} · mol^{-1}		
298.15	3238	52.87	63.72	15.37
500	7102	59.25	73.46	22.75
800	15215	67.00	86.02	30.77

2 Calculate the entropy of butanone as an ideal gas at 298.15 K and compare with the calorimetric value of $S^0 = 80.93 \pm 0.20$ cal/mol · K. This experimental value has required an entropy excess function which demands some assumption about the equation of state of butanone vapor at 298.15 K.

Butanone has three tops, two CH_3 groups and one C_2H_5 group, and we will take all three of these rotators as moving independently of each other. The appropriate molecular data are: molecular weight = 72.1074, symmetry number = 1, electronic ground state = singlet, product of three principal moments of inertia = 6.5461×10^{-114} g^3 · cm^6, the vibrational frequencies are 2983(2), 2910(2), 2884, 1716, 1460, 1422, 1413, 1373, 1346, 1263, 1182, 1089, 997, 939, 760, 590, 413, 260, 2983(2), 2941, 1460, 1413, 1263, 1108, 952, 768, 460, and the three internal moments (in wave numbers) and their corresponding barrier heights (in cal/mol) are as follows:

CH_3	5.8201 cm^{-1}	830 cal/mol
CH_3 in C_2H_5	5.4682 cm^{-1}	2620 cal/mol
C_2H_5	1.0681 cm^{-1}	1910 cal/mol

Note that the symmetry number of butanone is not 9 as one could imagine from allowing the CH_3 of the C_2H_5 group to rotate three times for each of three equivalent rotations of the CH_3 groups on the opposite side of the carbonyl group. Rather we are here concerned with the rotation of the molecule as a whole. The internal rotations are taken account of by n of the partition function for internal rotation, Eq. (4.2). As an approximation, we will take $n = 3$ for the rotation of the ethyl group even though that group is certainly not a symmetric top.

ANSWER[6]: The statistical entropy is calculated to be 81.11 cal/mol · K at 298.15 K.

3 Calculate the enthalpy and Gibbs free energy functions for ethyl cyanide at 400 and 600 K.

Molecular weight	55.08

Electronic ground state	Singlet
Moments of inertia, $g \cdot cm^2$	31.2×10^{-40}
	173.9×10^{-40}
	194.5×10^{-40}

Note that in CH_3CH_2CN, the only rotation that can lead to different configurations is that of the CH_3 group about the connecting C—C bond.

Barrier to free rotation	5200 cal/mol
Reduced moment of inertia, $g \cdot cm^2$	4.826×10^{-40}
Vibration frequencies, cm^{-1}	3008(2), 2909, 2949, 2850, 2265, 1462(2), 1383, 1431, 1320, 1260, 1155, 1008, 840, 1078, 545, 226, 367, 783
Torsional frequency, cm^{-1}	293

ANSWER: N.E. Duncan and G.J. Janz, *J. Chem. Phys.*, *23*, 434 (1955), obtained $(H^0 - H_0^0)/T$ = 13.705 and 17.416 and $-(G^0 - H_0^0)/T$ = 59.734 and 65.992, all in cal/mol · K at 400 K and 600 K respectively.

4 A data compilation reports an internal rotational constant of 3.5678 cm^{-1}. What is I_r in $g \cdot cm^2$? What is the rotational constant in MHz?

5 Beginning with the energy levels of an unhindered internal rotor, Eq. (4.1), derive the expression for the partition function, Eq. (4.2).

6 Using data from Landolt-Börnstein, calculate the values of C_p^0 and S^0 for methanol at 500 K. Compare your result with that from a standard compilation such as the JANAF table.

7 Calculate the heat capacity C_p^0 of 1,1,1-trifluoroethane at 300 K. The several frequencies and their degeneracies are: 365(2), 541(2), 603, 830, 969(2), 1232(2), 1278, 1408, 1443(2), 2978, and 3036(2), along with a low frequency torsion at 238 cm^{-1}. The reduced moment of inertia has been reported to be 5.05×10^{-40} $g \cdot cm^2$, and the barrier to internal rotation as 3.45 kcal/mol. An experimental value at 300 K is C_V^0 = 16.83 ± 0.18 cal/mol · K (from Stull[7]).

8 Calculate the heat capacity of dimethylacetylene at 336.07 K. The several frequencies and their degeneracies are: 213(2), 317(2), 725, 1029(2), 1050(2), 1126, 1380(2), 1448(2), 1468(2), 2270, 2916, 2966(2), and 2976(3). The experimental value at this temperature is C_p^0 = 20.21 cal/mol · K. Comment on the rotation of the methyl groups (from Stull[7]).

Chapter 5

1 Develop an expression for the second virial coefficient $B(T)$ for the Sutherland potential.

$$\phi(r) = \begin{cases} \infty & \text{for } r < \sigma \\ -\,cr & \text{for } r \geq \sigma \end{cases}$$

It will probably be convenient to expand the exponential in the expression for $B(T)$ in a Taylor series. Evaluate $B(T)$ for ethylene at 260 K and compare with the Lennard-Jones results using $\sigma = 3.35$ Å and $\varepsilon/k = 222$ K. Assign the constants in the Sutherland potential as equivalent to those for Lennard-Jones.

2 Derive an expression for $(U^0 - U_0^0)/RT$ for a monatomic gas that obeys the virial equation of state.

ANSWER:

$$\frac{U^0 - U_0^0}{RT} = \frac{3}{2} - \mathrm{T}\sum_{j=1}^{\infty}\frac{1}{j}\left(\frac{dB_{j+1}}{dT}\right)\rho^j$$

where $\rho = 1/V$ and B_{j+1} is the second virial coefficient if $j = 1$, etc.

3 Calculate the second virial coefficient for the triangular potential:

$$\phi = \begin{cases} \infty & r < \sigma \\ \dfrac{\varepsilon}{\sigma(\lambda - 1)}(r - \lambda\sigma) & \sigma < r < \lambda\sigma \\ 0 & r > \lambda\sigma \end{cases}$$

Compare the predicted $B(T)$ with that for the L-J potential at some one temperature for any gas. Use the same parameters in this potential as tabulated for the Lennard-Jones potential, and for ease of computation take $\lambda = 2$. You might want to refer to Ref. 8.

ANSWER:

$$B^* = \lambda^3 + 3T^*(\lambda - 1)[2T^*(\lambda - 1)\lambda + \lambda^2 + 2(\lambda - 1)^2T^{*2}]$$
$$- 3T^*(\lambda - 1)[2T^{*2}(\lambda - 1) + 2T^*(\lambda - 1) + 1]\tau$$

Compare with the result for the square-wall potential:

$$B^* = 1 - (\lambda^3 - 1)\tau$$

In both expressions, τ is $\exp\,(T^*)^{-1}$.

4 We have given the second virial coefficient as

$$B(T) = 2\pi n\int_0^{\infty}(1 - e^{-\phi(r)/kT})r^2\,dr$$

but some texts give

$$B(T) = -\frac{2\pi n}{3kT}\int_0^{\infty} r^3\left(\frac{\partial\phi(r)}{\partial r}\right)e^{-\phi(r)/kT}\,dr$$

for the same quantity. We would hope these are equivalent. Are they?

5 Develop an expression for the second virial coefficient for the square-well potential. Use values of ε/k or σ for ethylene as determined for the Lennard-Jones potential to evaluate B for ethylene at 300 K. Take $R = 1.5$. Evaluate B

from the Lennard-Jones potential and again from a macroscopic correlation. Compare all three values.

ANSWER:

$$B(T) = b_0 [1 - \Delta(R^3 - 1)]$$

$$\Delta = e^{-\varepsilon/kT} - 1$$

6 Using the Lennard-Jones model, calculate the second virial coefficient of CO_2 at 400 K. Use six terms in the gamma function expansion and compare your result with that listed in Table 5.6. Comment on any discrepencies that may appear.

7 Show the consistency of units in Eq. (5.11).

8 Evaluate the three contributions to the intermolecular attraction for pure ammonia at 300 K and compare with the values reported in Table 5.5.

9 The parameter t^* in Fig. 5.11 is related to a so-called reduced dipole moment. Is this reduced moment dimensionless? Explain.

Chapter 6

1 Calculate the heat capacity of solid benzene at 270 K. A calorimetric measurement at the boiling point of liquid hydrogen (20 K) yields C_p = 1.84 cal/mol · K. The characteristic frequencies from vapor phase spectroscopic observation are:

993	1645(2)	1030(2)	538	1854
3062	1170(2)	1480(2)	1520	1145
608(2)	850(2)	3080(2)	1008	406(2)
3107(2)	783	1190	3063	1160(2)

The constant a in Eq. (6.19) has been estimated to be 6.45×10^{-5} $cal^{-1} \cdot mol^{-1}$. Discuss the origin of any discrepancy.

2 Develop Eq. (6.6) for the internal energy and Eq. (6.7) for the heat capacity C_V from the Debye partition function, Eq. (6.5).

Chapter 7

1 Develop expressions for the vapor pressure constants [Eqs. (7.7) and (7.9)] for both linear and nonlinear species.

2 Consider the dissociation reaction

$$\upsilon_{AB} AB = \upsilon_A A + \upsilon_B B$$

Starting with the general expression for free energy for nonlocalized particles, develop an expression for the thermodynamic equilibrium constant for this reaction. The pure components as well as the mixture may be considered to be ideal gases.

3 Show that the heat of reaction based on the elements in their usual standard states (graphite, H_2 as an ideal gas at 1 atm, etc.) is identical to the heat of reaction based upon the atoms as ideal gases at 0 K.

4 Calculate the equilibrium constant for the water gas shift reaction

$$CO + H_2O \leftrightarrows H_2 + CO_2$$

at 1295 K and compare with the experimental value of $K = 0.5$. Some perhaps useful data appear in the following table.

Property	CO_2	H_2	H_2O	CO
Heat of formation at 0 K, ΔH^0_{0f} kJ·mol^{-1}	− 393.165	0.000	− 238.935	− 113.813
Molecular weight	44.01	2.016	18.02	28.01
Moment of inertia,† kg · m^2 × 10^{47}	71.4	0.460	2.11	14.5
Vibration frequencies, ω, cm^{-1}	2350	4160	3652	2168
	1320		3756	
	668		1595	
	668			

†Note: The moment of H_2O is $I_A I_B I_C$.

Estimate the effect on the equilibrium constant of replacing H_2O with HDO. With D_2O. Can you imagine a chemistry that would result in deuterium enrichment?

5 Calculate the equilibrium constant for the dissociation of F_2 at 1115 K and compare with an experimental value of 7.6×10^{-2} atm.

6 Show the equivalence of the third terms in Eqs. (7.3) and (7.6).

7 The vapor pressure of silver is reported in the *Handbook of Chemistry and Physics* to be 10^{-5}, 10^{-4}, and 10^{-3} torr at 1040, 1121, and 1209 K respectively. Predict these vapor pressures using molecular thermodynamic arguments.

Chapter 8

1 Using theoretical predicted properties, calculate the extent of reaction of acetylene to form benzene at 500°C and 30 bar. Compare this result with that from a conventional classical thermodynamic calculation. *Note*: The required data on heats of formation, heat capacities, free energies, or whatever other data that may be needed appear in all textbooks on thermodynamics.

2 An equimolar mixture of acetylene and helium passes through a turbocompressor. The gases enter at 25°C and 1 atm and exit at 20 atm. Assuming the mixture to be an ideal gas, what is the exit temperature and what horsepower must be supplied to a unit that is to handle 1000 kg/h of this gas? The compressor has an efficiency of 0.9. The theoretically predicted frequencies of acetylene are: 488(2), 885(2), 2237, 3827, and 3770 cm^{-1}.

3 Calculate the entropy and heat capacity of benzene at 1 atm and 298 K using molecular data that have been theoretically predicted. The moments are

299.9×10^{-40} and 150×10^{-40}, and 150×10^{-40} g · cm^2. Ten frequencies are doubly degenerate: 820, 3402, 1192, 1064, 3474, 1191, 877, 590, 868, and 389 cm^{-1}, and ten are singlets: 3507, 3494, 1189, 1113, 665, 843, 1484, 553, 1689, and 1078 cm^{-1}. The moments are from Ref. 9 and the frequencies from Ref. 10.

4 Calculate the C_p^0 and S^0 at 298 K and 1 atm for pyrrole using only theoretically predicted molecular data. The moments are: 92.86×10^{-40}, 93.43×10^{-40}, and 186.2×10^{-40} g · cm^2 as calculated using MINDO/3 by Dewar and Ford.[11] The frequencies are: 3652, 3563, 3514, 1503, 1460, 1311, 1067, 1027, 715, 759, 674, 511, 3550, 3511, 1610, 1437, 1194, 1109, 1049, 708, 752, 733, 650, and 472 cm^{-1} as calculated using MINDO/3 by Dewar and Ford.[12]

5 Calculated data on N_2O appear in Ref. 13. Calculate the change in enthalpy between 600 and 800 K for this gas, and compare with the result from standard tables such as JANAF.

6 The characteristics of C_2H_2 have been calculated by the so-called INDO method, and the results appear in Ref. 13. Calculate the entropy of acetylene at 500 K and compare with that from the JANAF table.

7 Methyl acetylene isomerizes to cyclopropene. A feed stream of the substituted acetylene contains 20 percent (mol) methane and enters a reactor at 150°C. The effluent gases leave at 250°C. How much heat is added or removed from the reactor per 100 kilogram of feed? Use theoretically predicted molecular data. Compare your result with that from a conventional classical thermodynamic calculation. The needed data on these common species are available from most textbooks on thermodynamics.

Chapter 9

1 Derive the expression for the fraction of high-energy molecules, (Eq. 9.21), beginning with the distribution function, Eq. (9.18).

2 Derive any point on the Lennard-Jones curve of self-diffusivity that appears in Fig. 9.17.

3 Derive the Eucken expression for the Prandtl number, $N_{\text{Pr}} = C_p^0/(C_p^0 + 1.25R)$.

4 Using the same techniques that led to Eq. (9.18), calculate the fraction of polyatomic molecules that have total energy greater than ε while also having an energy of ε_0 or more in 2 squared terms as would be the case in an excited classical harmonic vibrator.

ANSWER:

$$\frac{dn}{n} = \frac{e^{-\varepsilon/kT}\varepsilon^{s-1}\,d\varepsilon}{(kT)^s(s-1)!}\left(\frac{\varepsilon - \varepsilon_0}{\varepsilon}\right)^{s-1}$$

5 The fraction of molecules with speeds c between 0 and ∞ may be obtained by integrating Eq. (9.5). Also find an expression for the fraction of molecules with speeds greater than c_0.

Chapter 10

1 Derive the macroscopic rate equation for the formation of O_2 according to

$$4HNO_3 \rightarrow 4NO_2 + 2H_2O + O_2$$

that is thought to occur by the mechanism

$$HNO_3 \xrightarrow{k_1} OH + NO_2$$

$$OH + NO_2 \xrightarrow{k_2} HNO_3$$

$$OH + HNO_3 \xrightarrow{k_3} H_2O + NO_3$$

$$NO_3 + NO_2 \xrightarrow{k_4} NO_2 + O_2 + NO$$

$$NO_3 + NO \xrightarrow{k_5} 2NO_2$$

Show that the initial rate is solely first-order in HNO_3. What is the rate expression when the system is flooded with NO_2?

2 Develop an expression for the rate of "close collision" between molecules that attract each other as an inverse sixth power of the separation, i.e., $-a/r^6$. The effective potential energy is

$$V_{\text{eff}} = \frac{1}{2}\mu g^2 \left(\frac{b}{r}\right)^2 - \frac{a}{r^6}$$

and close collisions are those with total energy greater than the maximum in the effective potential energy. Compare with the hard-sphere collision rate (see Johnston[14]).

ANSWER:

$$z = \frac{2^{11/6}\pi^{1/2}\Gamma(2/3)a^{1/3}(kT)^{1/6}}{\mu^{1/2}}$$

$$= \frac{8.47a^{1/3}(kT)^{1/6}}{\mu^{1/2}}$$

3 Calculate the hard-sphere collision frequency of the following ion-molecule processes and compare with the data appearing in Table 10.3: (1) $Kr^+ + H_2$, where the collision diameter is 0.31 nm, and (2) $HCl^+ + HCl$, where the collision diameter is 0.33 nm. The values from the ion-molecule model, Eq. (10.57), are typically 4 to 5 times the hard-sphere values.

4 Suppose an atom and a linear molecule form a nonlinear transition state. What is the expected temperature dependence of the preexponential factor; i.e., what is n in Eq. (10.67)?

5 Derive Eq. (10.46) beginning with maxwellian spherical molecules wherein the reaction cross section is only a function of the relative translational velocity.

6 From the viewpoint of collision theory, how would you expect the preexponential factor to depend on temperature, for a bimolecular reaction wherein the critical energy of reaction is wholly contained in relative translation?

7 Estimate the number of vibrations of the CH_3O—H stretch in methanol before 3 quanta placed into this mode have leaked out into the other vibrational modes. What is the elapsed time for this leakage to occur? Estimate the rate constant of a reaction at the O—H bond that is to take advantage of the excess energy of 3 quanta placed in the O—H stretch.

References

1. K. Christie et al., *Inorg. Chem.*, *17*, 1533 (1978).
2. D. N. Shchepkin et al., *J. Mol. Struct.*, *49*, 265 (1978).
3. "Microwave Spectral Tables—Polyatomic Molecules with Internal Rotations," Vol. III, U.S. Dept. of Commerce, NBS, 1969.
4. J. H. Knox, *Molecular Thermodynamics*, Wiley, New York, NY, 1978).
5. J. Chao and B. J. Zwolinski, *J. Phys. Chem. Ref. Data*, *4*, 251 (1975).
6. J. Chao and B. J. Zwolinski, *J. Phys. Chem. Ref. Data*, *5*, 319 (1976).
7. D. R. Stull, *Chemical Thermodynamics of Organic Compounds*, Wiley, New York, 1969.
8. M. J. Feinberg and A. G. De Rocco, *J. Chem. Phys.*, *41*, 3439 (1964).
9. M. J. S. Dewar and G. P. Ford, *J. Am. Chem. Soc.*, *99*, 7822 (1977).
10. M. J. S. Dewar and A. Komornicki, *J. Am. Chem. Soc.*, *99*, 6174 (1977).
11. M. J. S. Dewar and G. P. Ford, *J. Am. Chem. Soc.*, *99*, 7822.
12. M. J. S. Dewar and G. P. Ford, *J. Am. Chem. Soc.*, *99*, 1685 (1977).
13. B. Nelander and G. Ribbegard, *J. Mol. Struct.*, *20*, 325 (1974).
14. H. S. Johnston, *Gas Phase Reaction Rate Theory*, Ronald Press (now Wiley, New York), 1966, pp. 142*ff*.

Index

ABOUT THE AUTHOR

Henry A. McGee, Jr. received a Ph.D. in chemical engineering from the Georgia Institute of Technology in 1955. He is currently on leave with the National Science Foundation from his role as professor of chemical engineering at Virginia Polytechnic Institute and State University. Dr. McGee's primary research interests include the application of lasers in chemical processes, cryogenics, experimental and theoretical kinetics and thermodynamics, chemical and physical properties of matter at extremes of temperature, and the application of molecular engineering to problems in propulsion, polymers, hazardous waste disposal, and reaction engineering.